T0296186

*CAMBRIDGE MONOGRAPHS ON PHYSICS*



# LIQUID HELIUM

# LIQUID HELIUM

BY

K. R. ATKINS, Ph.D.

*Professor of Physics, University of Pennsylvania*
*Past Fellow of Trinity College, Cambridge*

CAMBRIDGE
AT THE UNIVERSITY PRESS
1959

**CAMBRIDGE**
**UNIVERSITY PRESS**

University Printing House, Cambridge CB2 8BS, United Kingdom

Published in the United States of America by Cambridge University Press, New York

Cambridge University Press is part of the University of Cambridge.

It furthers the University's mission by disseminating knowledge in the pursuit of education, learning and research at the highest international levels of excellence.

www.cambridge.org
Information on this title: www.cambridge.org/9781107638907



First published 1959
First paperback edition 2014

*A catalogue record for this publication is available from the British Library*

ISBN 978-1-107-63890-7 Paperback

# CONTENTS

## CHAPTER 8

## Helium Three



## CHAPTER 9

## $He^3$-$He^4$ Mixtures

# PREFACE

The liquid helium problem can be conveniently divided into two parts, a problem in statistical mechanics and a problem in hydrodynamics. The problem in statistical mechanics is essentially a many body problem, in which the interactions between the atoms cannot be ignored and the symmetry of the wave-functions plays a dominant role. An analogous situation is encountered in heavy nuclei and has received so much attention in recent years that it is very close to being solved. In the case of liquid $He^4$, Landau has proposed a scheme of elementary excitations (or spectrum of energy levels) which is able to explain the thermodynamic properties, and Feynman has derived a wave-function which gives this scheme of excitations. We are therefore far advanced in our understanding of the thermodynamic nature of the liquid, and, although it is still necessary to justify our ideas by a rigorous solution of the wave-mechanical problem of a large number of interacting helium atoms, there is every indication that such a solution will soon be forthcoming. On the other hand, the detailed behaviour of the liquid in the immediate vicinity of the $\lambda$-point is still not understood, and the indications are that some very difficult mathematical problems are involved, but that their solution will lead to a better understanding of co-operative phenomena in general.

An interesting aspect of the situation is that $He^3$ differs from $He^4$ mainly because $He^3$ atoms require antisymmetric wave-functions. Experimentally, liquid $He^3$ shows none of the unusual properties of liquid $He^4$ and therefore presents a separate problem of its own.

A thorough understanding of the statistical mechanics of liquid $He^4$ is not in itself sufficient to solve the hydrodynamical problem of superfluidity. The Onsager-Feynman concept of quantized vortex lines promises to provide the additional basic idea which is needed, but there is still much more experimental and theoretical work to be done on the details.

In these circumstances, this book has a double purpose: first, to attempt a detailed survey of the experimental results for the benefit of active research workers in the field; secondly, to present an exposition of the current state of the theory, in the belief that the ultimate solution will be along similar lines. An attempt has been made to give an exhaustive survey of all work in the field, but a certain amount of selection is inevitable and the author apologizes in advance for his personal prejudices. In discussing the theory, emphasis has been placed on physical significance rather than mathematical details. Consideration was given to the possibility of including a compilation of the numerical data presently available on the various properties of liquid helium, but it became clear that there are so many uncertainties in the exact numerical values that such a compilation might be more misleading than helpful. However, adequate references have been given to enable the reader to refer to the original sources and form his own opinion of the relative reliability of the conflicting measurements.

I should like to express my gratitude to a large number of my colleagues for many helpful discussions and for informing me of their results in advance of publication. To name them all would occupy much space and would involve the risk of an inadvertent omission. I am particularly indebted to Dr D. Shoenberg for his helpful editorial comments; to Drs H. B. Callen, H. E. Hall and W. F. Vinen for pointing out several errors and obscurities; to the National Science Foundation for a grant which enabled me to continue work in the field during the later stages of preparing this work; and especially to my wife, who gave invaluable assistance with the many tedious chores concomitant with authorship.

K.R.A.

PHILADELPHIA
*May 1958*

# CHAPTER I

# INTRODUCTION

## 1.1. The liquefaction of helium

Helium is the most difficult of all the permanent gases to liquefy. This is a consequence of the weakness of the attractive forces between helium atoms, for condensation into a liquid can occur only when the forces holding the liquid together are strong enough to overcome the disruptive influence of thermal agitation, and this is the case for helium only if the temperature is within a few degrees of the absolute zero. The critical temperature is 5·20° K. and the boiling-point is 4·21° K.

The $He^4$ atom is a particularly simple, stable and symmetrical structure. The nucleus contains two protons and two neutrons and has no resultant angular momentum or magnetic moment. The two extranuclear electrons completely fill the innermost $K$ shell and, being so close to the nucleus, are firmly bound. The ionization potential of 24·56 volts is greater than that of any other atom and transitions between atomic energy levels are not likely to be important at temperatures less than about $10^5$ ° K. The atom has no electric or magnetic dipole moment and its electric polarizability is very small. It is the last fact which explains the weakness of the interatomic forces, for van der Waals' forces are now known to be directly related to the electric polarizability of the atom, which, in its turn, depends upon the separation of the energy levels of the electrons, so we can see a direct connexion between the low boiling point of liquid helium and the high excitation energy of the helium atom.

We might therefore expect the helium atom to be the nearest approximation to the hard sphere of classical kinetic theory and liquid helium might seem the simplest liquid to investigate in order to discover the fundamental principles of the theory of the liquid state. In actual fact the behaviour of liquid helium is so remarkable that it presents an intriguing problem which is unique in physical theory. The reason for this is almost certainly connected with the low temperature of the liquid and the possibility that certain quantum effects are able to manifest themselves at such a low temperature. F. London has therefore described it as a 'quantum liquid'.

The first successful liquefaction was accomplished by Kamerlingh Onnes (1908). Helium was found to give a colourless, transparent liquid with a density about one-seventh of that of water. Solid helium cannot be made merely by reducing the temperature of the liquid, as is the case for all other substances. Liquid helium in contact with its vapour has often been reduced to temperatures as low as 0·1° K. without solidifying, and it is probable that the liquid still exists at the absolute zero. The solid can be produced only by applying a pressure of the order of 25 atmospheres. Simon (1934) has shown that this is a consequence of the large zero-point energy of the liquid and it is the first of the quantum effects that we shall have to consider.

## 1.2. The λ-transition

At 2·18° K. liquid helium undergoes a transition which may be demonstrated in the following spectacular way. Liquid at the boiling point can be cooled by reducing the pressure above its surface with a vacuum pump. During this pumping the high temperature modification, liquid helium I, boils vigorously and is clouded by a mass of small bubbles, but suddenly, as soon as the transition temperature is reached, the boiling stops abruptly and the low temperature modification, liquid helium II, appears as a transparent, quiescent liquid which refuses to boil. Liquid helium II has many unusual properties, one of which is a very high thermal conductivity (under some circumstances more than 1000 times greater than that of pure copper at room temperature) and this is the cause of the absence of boiling. If a bubble forms below the surface of a liquid, the pressure inside the bubble must be greater than the vapour pressure above the surface by the hydrostatic pressure head plus the surface tension pressure. Vapour at this increased pressure cannot be formed from the liquid unless the temperature is higher than the temperature at the surface by an amount determined by the slope of the vapour pressure curve. The thermal conductivity of liquid helium II is so large that such a temperature difference cannot be set up, and so boiling is prevented and all the evaporation takes place at the surface.

The transition from liquid helium I to liquid helium II involves no latent heat and no discontinuous change in volume, so it is not a first order transition with discontinuities in the first derivatives of the Gibbs free energy. In § 2.6 we shall discuss this in more detail and shall explain why we prefer not to describe the transition as 'of the second kind' or 'of the second order'. The term λ-transition is non-committal and arises from the resemblance between the Greek letter λ and the shape of the specific heat versus temperature curve. The transition temperature is referred to as the λ-point.

## 1.3. Superfluidity and the two-fluid theory

The most remarkable of the properties of liquid helium II is its 'superfluidity', which was discovered simultaneously by

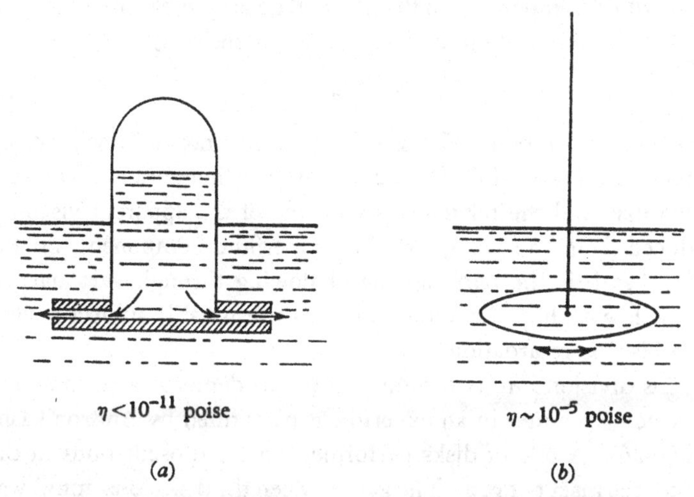


Fig. 1. The two different methods of measuring the viscosity of liquid helium II. (a) Flow through narrow channels; (b) damping of an oscillating disk.

Kapitza (1938) and by Allen and Misener (1938*a*). Fig. 1*a* illustrates the type of experiment performed by Kapitza. A very narrow gap, with a width between $10^{-5}$ and $10^{-4}$ cm., can be made by pressing together two optically polished glass surfaces. Liquid helium II flows through such a gap with a velocity of many centimetres per sec., corresponding to an effective viscosity of less than

$10^{-11}$ poise, which should be compared with $10^{-2}$ poise for water, $2\times10^{-5}$ poise for liquid helium I and $5\cdot5\times10^{-6}$ poise for helium gas at 1·64° K. However, further investigation reveals that the flow is almost independent of the pressure head and varies in a complicated way with the width of the gap, so a coefficient of viscosity is not relevant to the phenomena. Moreover, if the viscosity is estimated by observing the damping of the torsional oscillations of a disk immersed in the liquid (fig. 1 *b*), then, as long as the peripheral velocity of the disk is not too great, the behaviour is that of a normal liquid and the viscosity has a much larger value, of the order of $10^{-5}$ poise.

The solution of this dilemma is provided by the two-fluid theory originated by Tisza (1940) and developed in a slightly different form by Landau (1941). The liquid is considered to be some sort of intimate mixture of two components, a normal component and a superfluid component, so that its total density $\rho$ can be separated into a normal density $\rho_n$ and a superfluid density $\rho_s$

$$\rho=\rho_n+\rho_s. \tag{1.1}$$

The normal component is assumed to have a normal viscosity and is therefore responsible for the damping of the motion of an oscillating disk and the relatively high value of the apparent viscosity deduced from this damping. The superfluid component, on the other hand, is the part capable of flowing through very narrow channels with high velocities, and it is presumably subject to very little viscous retardation.

This division into two components was demonstrated in a very convincing manner in an experiment performed by Andronikashvili (1946). A pile of disks performed torsional oscillations in the liquid (see inset to fig. 2). The gap between the disks (0·21 mm.) was sufficiently small so that, above the $\lambda$-point, all the liquid between the disks was dragged round with them and contributed to the effective moment of inertia of the system. Below the $\lambda$-point, however, only the normal component moved with the disks, the superfluid component being subject to no frictional force which would have brought it into motion. Therefore, as the temperature was lowered below the $\lambda$-point, the period of oscillation steadily decreased and it was possible to deduce from it the fraction $\rho_n/\rho$

of liquid contributing to the moment of inertia of the system. The results are shown in fig. 2, from which it will be seen that the fraction of normal liquid, $\rho_n/\rho$, steadily decreases below the $\lambda$-point, while the fraction of superfluid, $\rho_s/\rho$, builds up from zero at the $\lambda$-point to unity at 0° K.

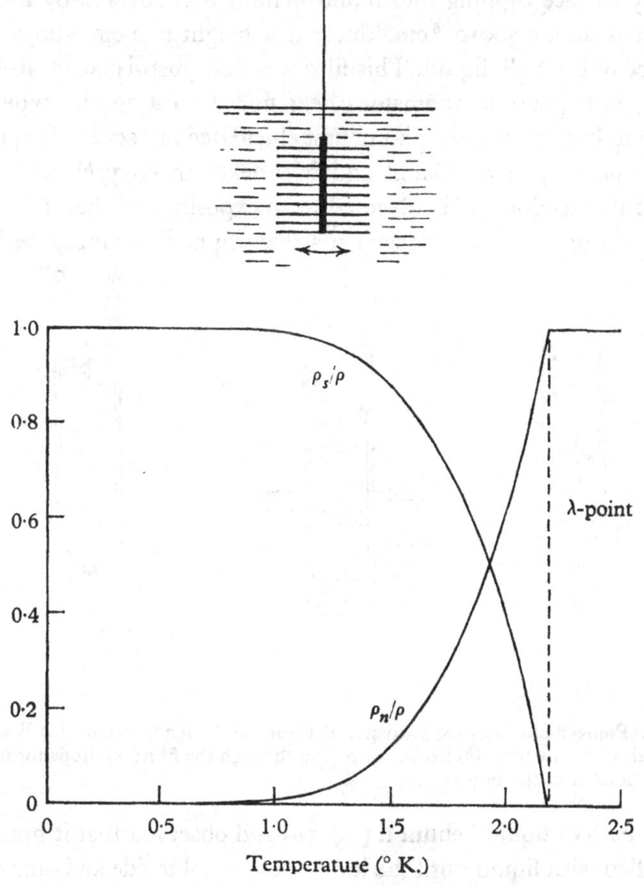


Fig. 2. Andronikashvili's experiment.

The two-fluid theory is now an essential part of the modern approach to the liquid helium problem and will be tacitly assumed throughout the rest of this monograph. It is possible, of course, that a complete theory of liquid helium will reveal that the two-fluid hypothesis is only a first approximation. However, for purposes of exposition, the concept of a mixture of superfluid and normal components is very convenient and is certainly adequate to describe the present state of theory and experiment.

## 1.4. The liquid helium film

Any surface dipping into liquid helium II is covered by a film which is about $3 \times 10^{-6}$ cm. thick at a height of 1 cm. above the surface of the bulk liquid. This film was first postulated by Rollin (1936) to explain an anomalous heat flow into a special type of cryostat, but its properties were first elucidated in a series of experiments performed by Daunt and Mendelssohn (1939*b*), and the rest of this section will be devoted to an exposition of their pioneer work. In one of their experiments they dipped an empty beaker

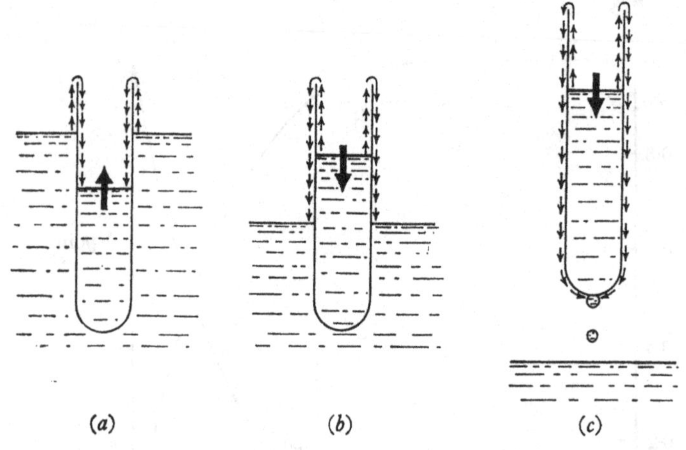


Fig. 3. Pioneer film flow experiments of Daunt and Mendelssohn. (*a*) Beaker filling through the film; (*b*) beaker emptying through the film; (*c*) drops forming on the bottom of the beaker.

into a bath of liquid helium II (fig. 3*a*) and observed that it gradually filled with liquid until the levels were equal inside and outside, presumably because superfluid flow was able to take place through the film on the wall of the beaker. When the beaker was raised, flow occurred in the opposite direction and the inside level slowly fell down to the outside level (fig. 3*b*). When the beaker was lifted completely out of the bath, it slowly emptied itself and drops of liquid could be seen forming on its bottom and dripping into the bath (fig. 3*c*).

As we shall see in chapter 4, superfluidity is most clearly manifested in flow through very narrow channels and the film presents us with an extremely narrow channel. The phenomena associated with film flow have therefore been taken to indicate the nature of 'ideal superfluidity'. The most outstanding aspect of film flow is that the rate of flow is almost independent of the pressure head or the length of the film. Fig. 4 shows the results of measurements by Daunt and Mendelssohn on the rate of emptying of a long beaker. Apart from a small region near the rim, the rate of transfer was

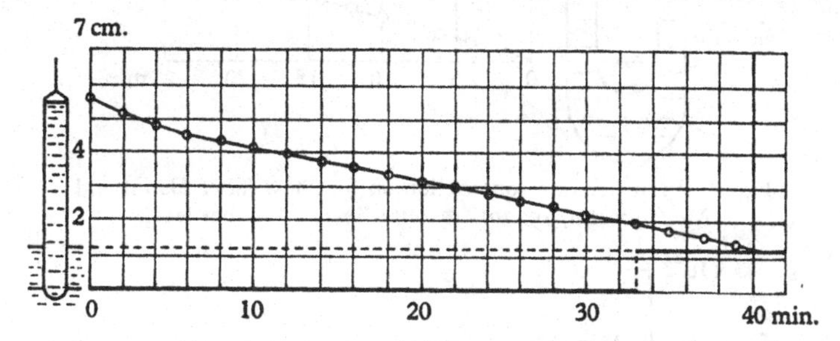


Fig. 4. Almost constant rate of film transfer from a long beaker (after Daunt and Mendelssohn, 1939*b*).

constant to within 20 % throughout the whole length of the beaker. This has led to the concept of a critical velocity $v_{s,c}$. It seems that the film is rapidly accelerated up to this velocity but that some mechanism then operates to prevent any further acceleration. Writing $d$ for the thickness of the film and assuming that the transfer is due to flow of the superfluid component with a density $\rho_s$ (provisionally assumed to be the same as $\rho_s$ in the bulk liquid), then the rate of transport of mass per unit length of the perimeter over which the film flows is

$$\rho_s v_{s,c} d \text{ g. sec.}^{-1} \text{cm.}^{-1}.$$

The critical rate of transfer, $\sigma_c$, is normally defined in terms of the volume of bulk liquid transferred to or from the inside of the vessel and is therefore

$$\sigma_c = \frac{\rho_s}{\rho} v_{s,c} d \text{ cm.}^3 \text{sec.}^{-1} \text{cm.}^{-1}. \tag{1.2}$$

This concept of a critical velocity receives support from the fact that the rate of transfer is determined by the perimeter of the

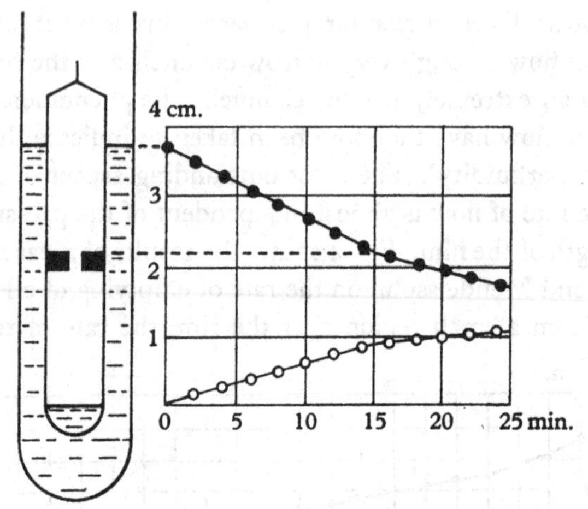


Fig. 5. Influence of a constriction on film flow (after Daunt and Mendelssohn, 1939*b*). ●, outer level; ○, inner level.

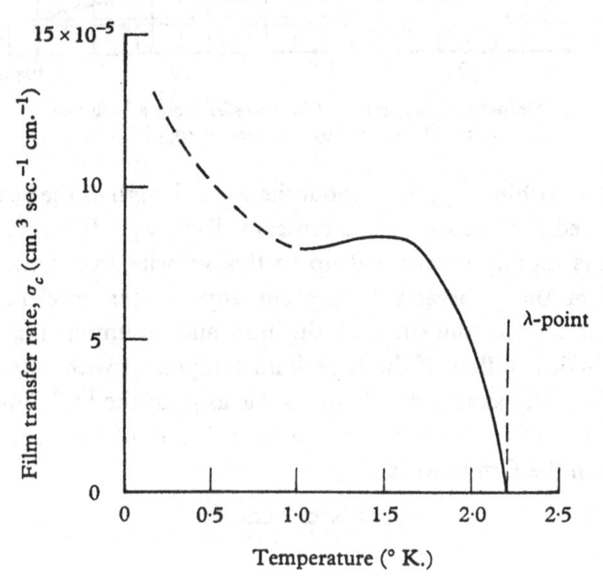


Fig. 6. Rate of transfer through the helium film as a function of temperature.

narrowest constriction higher than the upper liquid level. This was demonstrated with the apparatus of fig. 5. As long as the outer level was above the constriction the rate of transfer was determined by the inside diameter of the beaker, but as soon as the outer level fell below the constriction, the rate of transfer was reduced in exact proportion to the ratio of the perimeter of the constriction to the inside perimeter of the beaker.

The critical rate of transfer, $\sigma_c$, is shown as a function of temperature in fig. 6. Much of this temperature variation is no doubt due to the factor $\rho_s/\rho$ in equation (1.2), but $v_{s,c}$ and $d$ may vary with temperature also. Moreover, the later experiments to be described in chapter 7 indicate that the situation is quite complicated and the ideas presented in this section may have to be considered as only a first approximation.

## 1.5. Thermal effects

The flow of liquid helium II is accompanied by some unusual thermal effects, all of which may be explained by the assumption that

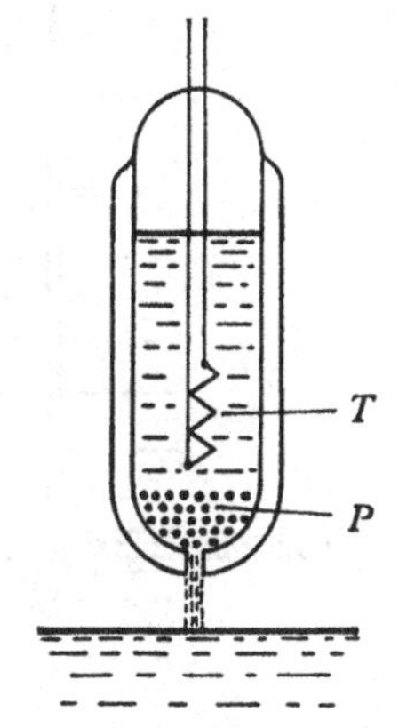


Fig. 7. Apparatus to demonstrate the mechanocaloric effect (after Daunt and Mendelssohn, 1939*a*).

the total entropy of the liquid is associated with the normal component only, and that the superfluid component is entirely devoid of entropy and behaves, in a certain sense, like a liquid at 0° K. This is directly obvious in the *mechanocaloric* effect predicted by Tisza (1938) and discovered by Daunt and Mendelssohn (1939*a*). A Dewar vessel (fig. 7) was able to empty itself through a small orifice packed with fine emery powder $P$. As the liquid flowed out of the vessel the temperature of the remaining liquid, recorded by the phosphor bronze resistance thermometer $T$, rose by about 0·1° K. The interpretation is that only the superfluid component was able to flow through the narrow channels between the grains of emery powder and so the liquid flowing out left all its entropy behind to warm up the liquid remaining inside.

The *thermomechanical* effect or 'fountain effect' was discovered by Allen and Jones (1938). The vessel of fig. 8*a* was connected to the bath via a fine capillary. When heat was supplied to the inside of the vessel the inner level rose and took up a steady position well above the bath level. Presumably the superfluid component, being

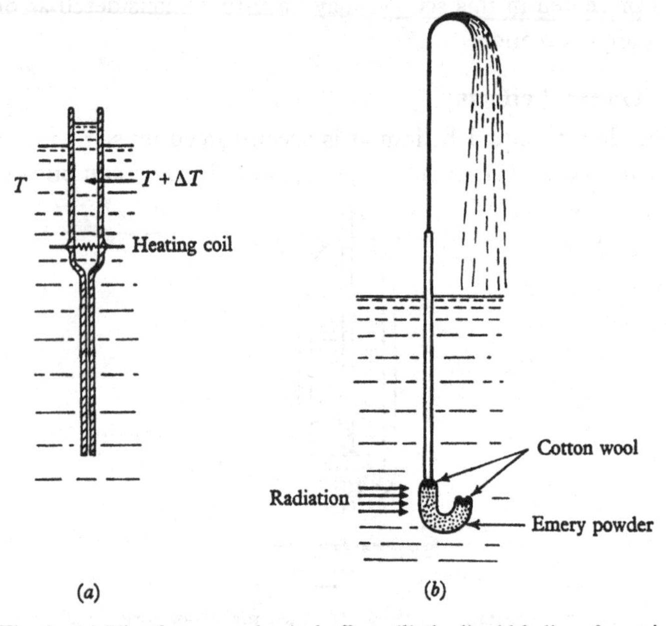


Fig. 8. (*a*) The thermomechanical effect; (*b*) the liquid helium fountain (after Allen and Jones, 1938).

cold, has a tendency to move towards regions of higher temperature and therefore flowed into the vessel until the pressure gradient forcing it out again was large enough to counterbalance the temperature gradient forcing it in. A rather pretty demonstration of this effect can be given with the apparatus of fig. 8*b*. Liquid can enter the inside of the vessel only through the packed emery powder. When a light is shone on to the apparatus the black powder absorbs heat, the inside temperature rises and the liquid rushes in with such force that it emerges from the nozzle to form a fountain which can be as much as 30 cm. high.

Once it is realized that the liquid flowing through the narrow channels has an entropy defect, the thermomechanical effect can be deduced from a thermodynamic argument. This was first done by H. London (1939) by analogy with the theory of thermoelectric effects and has been treated more rigorously by de Groot (1951) using the techniques of irreversible thermodynamics. The result is that the thermomechanical pressure difference $\Delta p$ arising from a temperature difference $\Delta T$ is given by

$$\frac{\Delta p}{\Delta T}=\frac{Q^*}{TV}, \tag{1.3}$$

where $Q^*$ is the heat transfer associated with the transfer of unit mass through the connecting channel and $V$ is the volume of unit

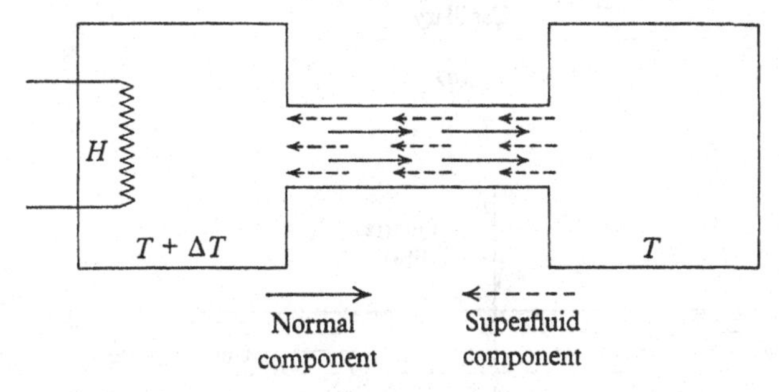


Fig. 9. Heat conduction by counterflow of the two components.

mass. In the very narrowest channels, of width $10^{-4}$ cm. or less, the flow of the viscous normal component is entirely inhibited and $Q^*$ refers to flow of the superfluid component. Under these conditions, the experiments of Kapitza (1942), which will be described in more detail in §4.3, show that

$$Q^*=TS, \tag{1.4}$$

where $S$ is the entropy of the bulk liquid. and then

$$\frac{\Delta p}{\Delta T}=\frac{S}{V}. \tag{1.5}$$

It is now possible to understand the abnormally high thermal conductivity of liquid helium II. Imagine, as in fig. 9, two vessels at different temperatures connected by a column of liquid. The cold superfluid component moves towards the vessel at the higher temperature and upon arrival there is converted into bulk liquid with absorption of heat. The hot normal component moves in the opposite direction towards the vessel at the lower temperature

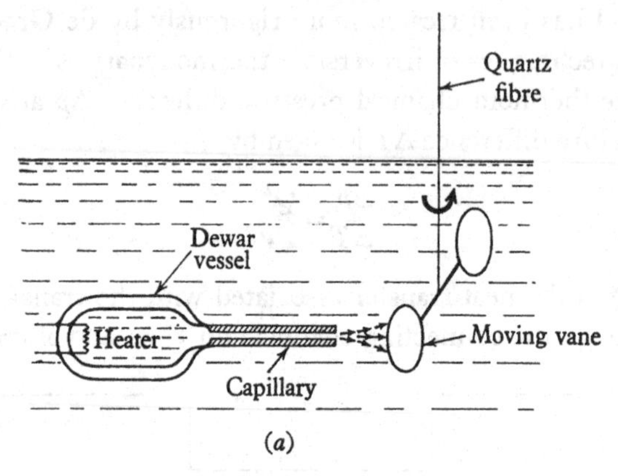


(*a*)

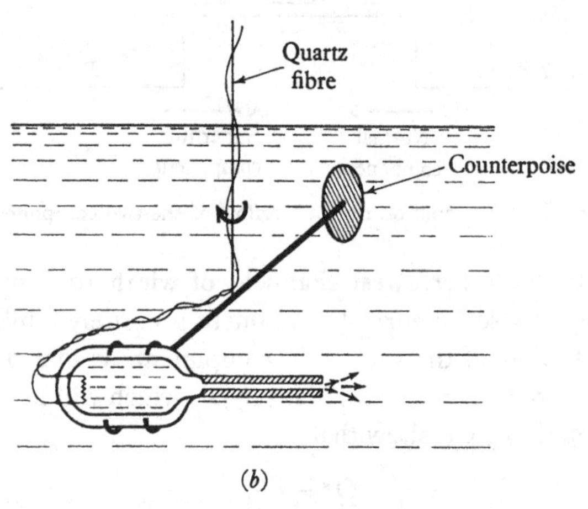


(*b*)

Fig. 10. (*a*) Force exerted on a vane by a heat current; (*b*) backward reaction of the jet on the Dewar vessel (after Kapitza, 1941).

where it gives up its excess heat. Thus, although there is no net transfer of matter, there is a very efficient mechanism for the transfer of heat by a sort of two-way convection process. In an ordinary convection process, the heat transferred by unit mass of moving liquid is $C\Delta T$, where $C$ is the specific heat, but here the heat transferred by flow of unit mass of each of the two components is the total heat content of the liquid, which is much greater.

The counterflow of the two components was made to reveal itself in some ingenious experiments performed by Kapitza (1941). In one experiment (fig. 10*a*) heat was supplied to the inside of a Dewar vessel and flowed out through a fine capillary. The normal component emerging as a jet from the end of the capillary impinged on a metal vane and exerted on it a force which could be measured by means of a quartz fibre suspension. The force was found to increase steadily with the power supplied to the inside of the vessel.

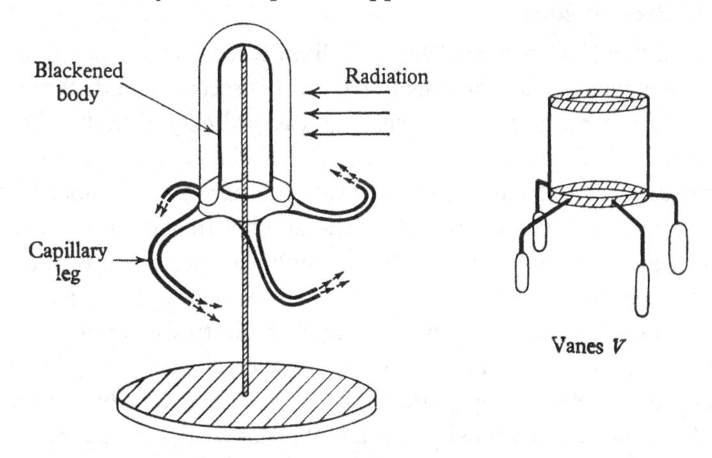


Fig. 11. Kapitza's jet-propelled spider.

When the vessel itself was attached to the quartz fibre (fig. 10*b*), the outflowing normal component exerted a backward force equal and opposite to that measured in the first experiment. Apart from revealing the flow of the normal component these experiments suggest that there is no force exerted by the superfluid component which is rushing into the capillary. Kapitza thought that the superfluid component was creeping in as a film along the wall of the capillary, but Landau (1941) favours the view that the two components move freely through one another, and points out that the superfluid component may be similar to an ideal classical liquid with zero viscosity, which exerts no force on a body round which it is flowing. The 'Kapitza Jet-propelled Spider' (fig. 11) is a novel application of these effects. The body of the spider is an inverted

Dewar vessel and is blackened inside to absorb heat when a light shines on it. The heat flows out through several capillaries which form the legs. The whole is mounted on a fine needle point and is free to rotate. When the light is switched on, the jets of normal component emerging from the legs exert a couple which starts up a rapid rotation. When the arrangement of vanes, $V$, is mounted on the spider such that there is a vane opposite the exit of each capillary, the force of reaction on the legs is equal and opposite to the force on the vanes and rotation does not take place.

## 1.6. Second sound

The two-fluid nature of liquid helium II leads to the possibility of a new type of wave propagation known as second sound. Second sound is a temperature wave, whereas ordinary sound (which will be called first sound for the sake of clarity) is, of course, a pressure wave. In a first sound wave the normal and superfluid components move in phase with one another in such a way that, although the total density of both components at a point changes, giving rise to pressure variations, there is only a second order variation in the relative concentration of the two components at the point. In a second sound wave, however, the two components move out of phase in opposite directions and there is only a second order change in the density, but a first order change in the relative concentration of the two components. Since the superfluid component is cold and the normal component hot, an increase in the concentration of the superfluid component at a point lowers the temperature there, while, half a wavelength away, an increase in the concentration of the normal component raises the temperature. The two types of wave are represented pictorially in fig. 12.

Second sound was predicted by Tisza (1940) and discovered by Peshkov (1944) using a continuous wave resonance method (fig. 13). The transmitter was a thin constantan wire wound as a flat spiral on a plastic backing. Alternating current at a frequency in the range 100 to 10,000 c./s. was fed into this wire and, because the wire had an extremely small thermal capacity, its temperature was able to follow the periodic variation in the joule heat developed in it and the temperature of the liquid in the neighbourhood was made to oscillate at a frequency of twice that of the alternating current. The resulting second sound wave travelled through the liquid and was received on a phosphor-bronze resistance thermometer wound in a similar way to the transmitter. A steady d.c. passed through the receiver wire and the temperature oscillation accompanying the

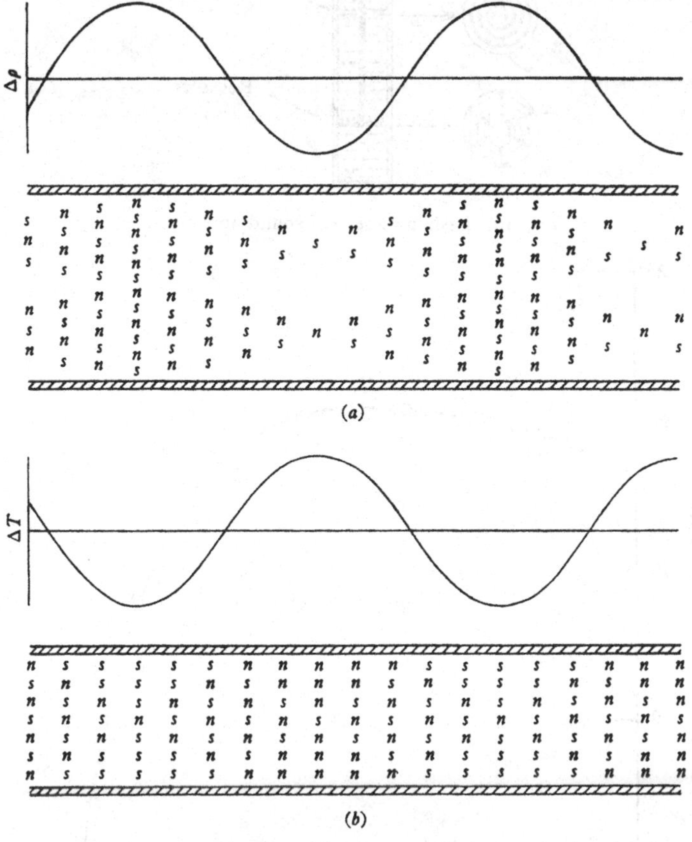


Fig. 12. Pictorial representation of (*a*) first, and (*b*) second sound. The proportions of *n* and *s* on any vertical line represent schematically the relative proportions of the normal and superfluid components. The total number of symbols on a vertical line represents the total density of the liquid. The waves are travelling in a horizontal direction.

second sound wave therefore produced a voltage oscillation which was amplified and displayed on a cathode-ray oscilloscope. Resonances were set up in the second sound tube and nodes and antinodes were observed when the receiver was moved relative to the transmitter, enabling the wavelength to be measured and the

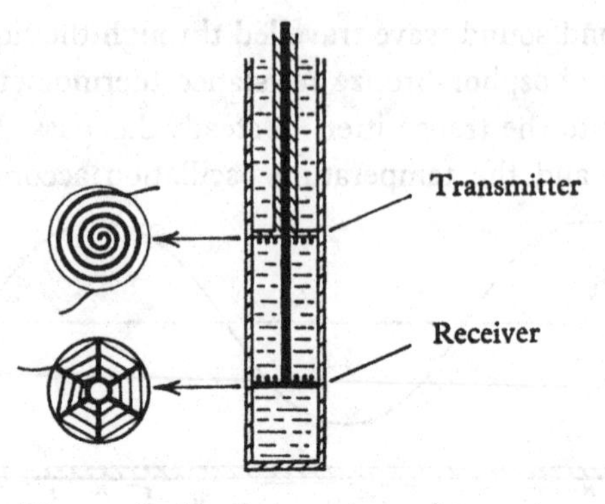


Fig. 13. Peshkov's second sound apparatus.

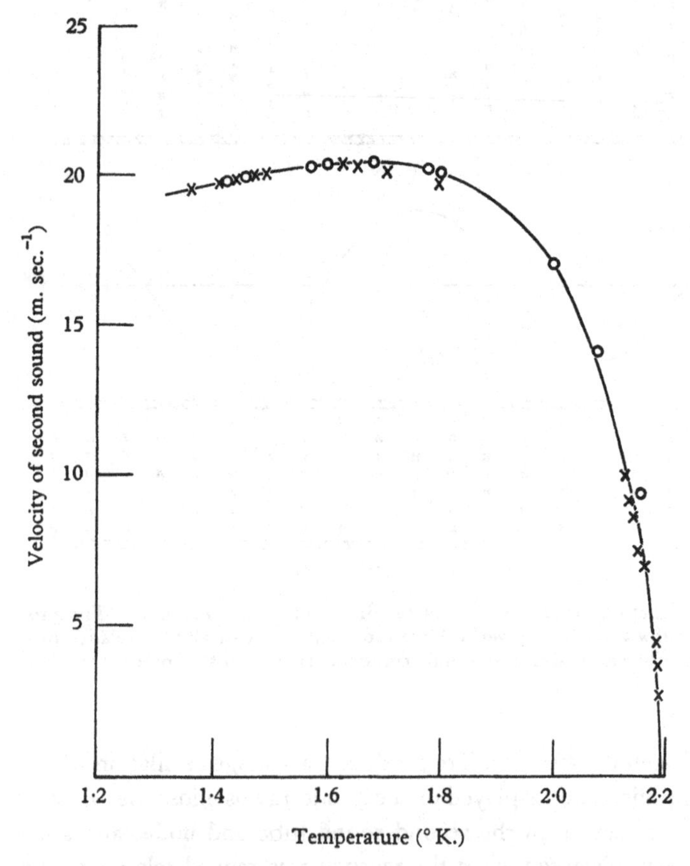


Fig. 14. The velocity of second sound, $u_2$, as a function of temperature. ×, Peshkov (1944, 1946); ○, Lane, Fairbank and Fairbank (1947).

velocity of second sound to be deduced. The velocity (fig. 14) was found to be an order of magnitude smaller than the velocity of first sound (which is about 240 m. sec.$^{-1}$), leaving no doubt that a new type of wave propagation was being investigated.

An interesting variation of Peshkov's technique was suggested by Onsager and used by Lane, Fairbank and Fairbank (1947). They used the same type of transmitter but allowed the second sound to impinge on the surface of the liquid. The temperature oscillations of the second sound wave at the liquid-vapour interface produced corresponding variations in the rate of evaporation, and the resulting puffs of vapour initiated a first sound wave which travelled through the vapour and was received on a microphone. Nodes and antinodes were obtained when the distance between the transmitter and the surface was varied, and the velocity was measured as before.

Landau (1941) derived the velocity of second sound, $u_2$, in the form

$$u_2^2 = \frac{\rho_s}{\rho_n} \frac{TS^2}{C}. \tag{1.6}$$

$S$ and $C$, the entropy and specific heat per gram of the bulk liquid, have been measured directly and so $\rho_s$ and $\rho_n$ can be deduced from $u_2$. The resulting values are in satisfactory agreement with those obtained by Andronikashvili (fig. 2). In chapter 5 we shall see that the velocity and attenuation of second sound give valuable information about the nature of liquid helium II.

## 1.7. Helium three

The $He^4$ atom contains an even number of fundamental particles and it should therefore be described by symmetric wave functions and obey Bose-Einstein Statistics. The abnormalities of liquid helium have been ascribed to the peculiarities of these statistics (London, 1938*a*, *b*; Tisza, 1940, 1947). The properties of liquid $He^3$ are therefore of considerable interest, since the $He^3$ atom contains an odd number of fundamental particles, is described by antisymmetric wave functions and obeys Fermi-Dirac statistics. In recent years pure $He^3$ has become available in usable quantities and has been investigated down to 0·1° K. without showing any sign of a $\lambda$-transition or of superfluidity. Even if this cannot be taken to provide a rigorous proof of the importance of Bose-Einstein statistics, it is certainly a fact of major importance which must be accounted for by any theory of liquid $He^4$.

Apart from this fundamental issue, $He^3$ is the only substance which is known to dissolve in liquid $He^4$. It therefore provides us with an extra variable, enabling us to check certain details of our theoretical ideas by investigating their application to $He^3$-$He^4$ solutions. This will be discussed more fully in chapter 9, where we shall see that one of the interesting aspects is that the $He^3$ forms part of the normal component and does not participate in the flow of the superfluid component.

CHAPTER 2

# EQUILIBRIUM PROPERTIES

## 2.1. The phase diagram

The phase diagram, giving the state of the substance under different conditions of pressure and temperature, is shown schematically in fig. 15. The scale has been distorted in order to

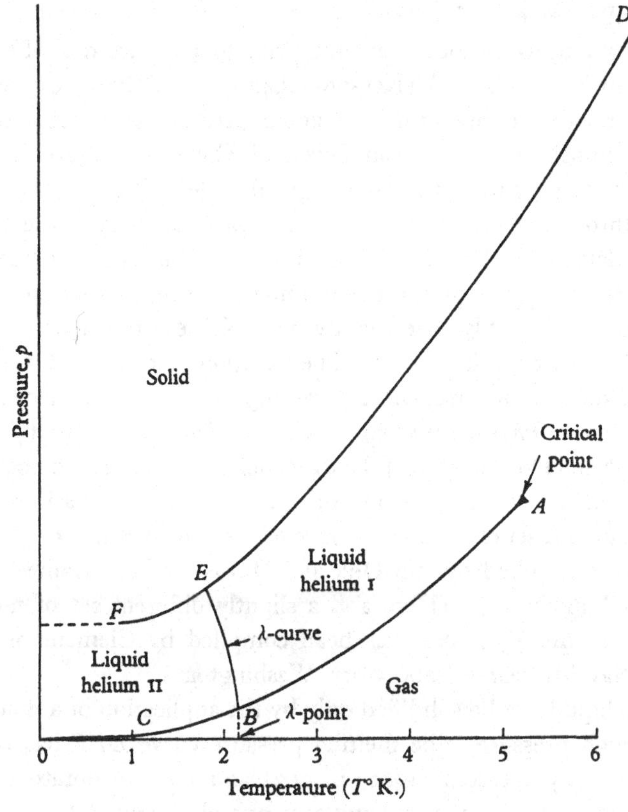


Fig. 15. The phase diagram (schematic).

give equal emphasis to all the main features. The vapour pressure curve $ABC$ starts at the critical point $A$ (5·20° K., 2·26 atmospheres) and has been followed down to the lowest temperature at which the vapour pressure is large enough to be measured. There is no sign of a triple point and it seems to be impossible to solidify the liquid by cooling it under its saturated vapour pressure. Magnetic cooling techniques have frequently been used to cool the liquid to temperatures as low as 0·1° K., and it has always been found to remain liquid and to retain its various superfluid properties. It is very probable therefore that helium remains liquid right down to the absolute zero of temperature. The existence of a liquid at 0° K. is a very peculiar fact, of great importance to the theory of the liquid state, and it requires a very careful explanation. A partial explanation will be given in §2.3.

Vapour pressure measurements prior to 1942 are described in detail in Keesom's book (Keesom, 1942, p. 186). In 1948 a set of tables relating temperature to vapour pressure was accepted by international agreement (van Dijk and Shoenberg, 1949) and is usually referred to as the '1948 agreed scale'. This scale will be used throughout this book, principally because most of the published data to which we shall have to refer are based upon this scale. More recent vapour pressure determinations suggest that the 1948 scale is considerably in error, by 0·01° K. near the $\lambda$-point and 0·02° K. near the critical point. The situation at the time of writing is described in the *Proceedings of the 1955 Paris Conference on Low Temperature Physics*, pp. 593–610; see also Keller (1956) and van Dijk and Durieux (1957). Discrepancies of the order of 0·003° K. still remain between the recent measurements of various investigators. A set of vapour pressure tables known as the $T_{L55}$ scale has been compiled by van Dijk and Durieux of the Kamerlingh Onnes Laboratory, Leiden, and a slightly different set of tables known as the $T_{55E}$ scale has been compiled by Clement of the U.S. Naval Research Laboratory, Washington.

The liquid can be solidified only by the application of a moderately high pressure. The melting pressure curve *DEF* has been measured by Keesom (1942, p. 202) and his collaborators, by Swenson (1950, 1952, 1953) and at very high pressures by Simon, Ruhemann and Edwards (1929*a*, *b*, 1930), Holland, Huggill and Jones (1951), and Mills and Grilly (1955). The feature which most concerns us here is that, near 1° K., the curve is flattening out to a pressure of 25·0 atmospheres, which is probably the melting pressure at 0° K., for there seems to be no possibility that the melting curve will meet the vapour pressure curve in a triple point.

At the saturated vapour pressure the transition from liquid helium I to liquid helium II takes place at point $B$, the $\lambda$-point.

TABLE I. *The pressure and temperature of the $\lambda$-point*

| Authors | Pressure (mm. of Hg) | Temp. (° K.) (1948 scale) |
|---|---|---|
| Schmidt and Keesom (1937) | 38·3 ± 0·2 | 2·185 |
| Long and Meyer (1951) | 38·10 ± 0·02 | 2·183 |
| Dash and Taylor (1955) | 38·00 ± 0·05 | 2·182 |

Table I gives the pressure and temperature of the $\lambda$-point as determined by various observers. The temperatures quoted are based on the 1948 scale, which may be considerably in error near the $\lambda$-point, and it is very likely that the true $\lambda$-temperature is nearer to 2·17° K. The transition temperature decreases as the pressure is increased, tracing out the $\lambda$-curve $BE$ and meeting the melting pressure curve at the point $E$ (1·764° K., 29·64 atmospheres) (Keesom and Clusius, 1931; Swenson, 1952, 1953).

## 2.2. Energy relations near 0° K.

Near 0° K. the vapour pressure decreases rapidly, the specific volume of the vapour tends towards infinity, and the gas in equilibrium with the liquid may therefore be considered ideal. There are no complications arising from nuclear spin and the energy of the gas can therefore be taken as zero at 0° K. and provides a reference point from which all other energies may be measured. The internal energy of the liquid at $p=0$, $T=0$ may then be deduced by the following argument, in which the subscripts $g$ and $l$ refer to the gas and the liquid respectively. The latent heat $L$ is

$$L = T(S_g - S_l) \tag{2.1}$$

$$= (U_g - U_l) + p(V_g - V_l). \tag{2.2}$$

At 0° K., it is obvious that $T$, $U_g$, $p$, $pV_l$ and $pV_g$ $(=RT)$ all become zero and that

$$L_0 = \operatorname*{Lt}_{T\to 0} TS_g = -U_{0,l}. \tag{2.3}$$

The internal energy of the liquid at 0° K. is therefore negative and equal in magnitude to the latent heat. Since the vapour pressure tends to zero at 0° K., the third law of thermodynamics must not be applied to the vapour, and $S_g$ actually becomes infinite at 0° K. in such a way that $TS_g$ remains finite.

$L_0$ cannot be measured directly and is best deduced from the vapour pressure curve. At sufficiently low temperatures the vapour behaves as an ideal monatomic gas and

$$\left.\begin{aligned} U_g &= \tfrac{3}{2}RT, \\ S_g &= \tfrac{5}{2}R\ln T - R\ln p + c, \\ pV_g &= RT. \end{aligned}\right\} \tag{2.4}$$

The corresponding equations for the liquid are

$$\left.\begin{aligned} U_l &= -L_0 + \int_0^T C_l\,dT, \\ S_l &= \int_0^T \frac{C_l}{T}\,dT, \end{aligned}\right\} \tag{2.5}$$

where the specific heat of the liquid $C_l = \alpha T^3$ (§ 2.5.1). Equating the Gibbs free energies of liquid and vapour, the vapour pressure curve is

$$\begin{aligned} \tfrac{5}{2}RT - \tfrac{5}{2}RT\ln T + RT\ln p - cT \\ = -L_0 + \int_0^T C_l\,dT - T\int_0^T \frac{C_l}{T}\,dT + pV_l. \end{aligned} \tag{2.6}$$

Near 0° K. the terms involving $C_l$ become negligibly small compared with the other terms and also $pV_l/RT = V_l/V_g \to 0$. The vapour equation then becomes

$$\ln p = \tfrac{5}{2}\ln T - \frac{L_0}{RT} + \left(\frac{c}{R} - \frac{5}{2}\right). \tag{2.7}$$

The vapour pressure measurements of Bleaney and Simon (1939) can be expressed in this form below 1·6° K. with $-L_0 = -14{\cdot}3$ cal. mole$^{-1}$, which is therefore the internal energy of the liquid at $T=0$, $p=0$.

Note that equations (2.4) and (2.7) give

$$S_g = \frac{L_0}{T} + \tfrac{5}{2}R, \tag{2.8}$$

whence
$$\operatorname*{Lt}_{T\to 0} TS_g = L_0 \tag{2.9}$$

in accordance with equation (2.3).

As the pressure on the liquid is increased at 0° K., the second law of thermodynamics gives

$$TdS_l = dU_l + pdV_l = 0$$

so that

$$U_l(p) - U_l(0) = -\int_0^p pdV_l. \tag{2.10}$$

The equation of state is known (Keesom and Keesom, 1933), and so the integral can be evaluated and the increase in internal energy between zero pressure and the melting pressure is found to be +1·1 cal. mole$^{-1}$. The internal energy of liquid helium on the melting curve at 0° K. is therefore −13·2 cal. mole$^{-1}$.

Turning next to the equilibrium between liquid and solid, and using the subscript $s$ to denote the solid, the Clausius-Clapeyron equation for the melting curve is

$$\left(\frac{dp}{dT}\right)_{\text{m.p.c.}} = \frac{S_l - S_s}{V_l - V_s}. \tag{2.11}$$

Experimentally $\frac{dp}{dT} \to 0$ as $T \to 0$ (fig. 15),

whence, at 0° K.

$$S_{0,l} = S_{0,s}. \tag{2.12}$$

The third law of thermodynamics is thus confirmed. At 0° K. both liquid and solid have zero entropy and are both perfectly ordered. This presents us with the additional problem of how liquid helium II can achieve perfect order and still remain a liquid.

Expressing the equality of the Gibbs free energy for the solid and the liquid along the melting curve, we have

$$T(S_l - S_s) = (U_l - U_s) + p(V_l - V_s).$$

At 0° K.

$$U_{0,l} - U_{0,s} = -p(V_{0,l} - V_{0,s}). \tag{2.13}$$

Swenson (1950) has measured the temperature variation of $V_l - V_s$ between 1·2 and 1·8° K. and has shown that, as the temperature decreases, it approaches the positive value +2·07 cm.$^3$ mole$^{-1}$. The density of the solid is therefore greater than that of the liquid at 0° K. From equation (2.13), $(U_{0,l} - U_{0,s})$ must then have the negative value −1·25 cal. mole$^{-1}$ and the internal energy of the solid at 0° K. on the melting curve must be −11·9 cal. mole$^{-1}$. We thus have the peculiar situation that the internal energy of the solid is greater than that of the liquid and this, of course, is why the liquid is the stable phase at low pressures; the solid is formed only at high pressures when its smaller volume produces a sufficient decrease of the $pV$ term in the Gibbs function $(U-TS+pV)$ to overcome the increase in $U$. The way in which this situation arises is shown in fig. 16, taken from Swenson's paper. Here $\Delta U$ and $\Delta V$ are the changes in internal energy and volume on melting, and

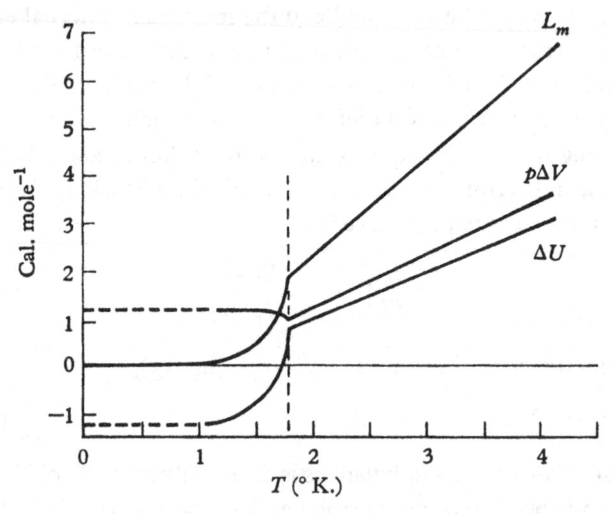


Fig. 16. Relationship between the thermodynamic properties of the liquid and the solid (after Swenson, 1950). $\Delta U = U_l - U_s$; $\Delta V = V_l - V_s$; $L_m$ = heat of melting $= \Delta U + p\,\Delta V$.

$L_m$ is the heat of melting. $\Delta U$ is seen to become negative below about 1·7° K. Well-marked anomalies at the $\lambda$-point are also apparent.

## 2.3. Theories of the structure of the liquid near 0° K.

The internal energy of *solid* helium is readily explained in terms of conventional solid state theory. X-ray analysis (Keesom and Taconis, 1938*a*) reveals that the crystalline structure of the solid is close-packed hexagonal. The forces between helium atoms are known, so that, once the crystal structure and density on the melting curve at 0° K. are known, the potential energy of the lattice can be calculated to be $-62$ cal. mole$^{-1}$. To this energy must be added the zero-point energy arising from the fact that each normal mode of vibration of the solid with a frequency $\nu$ has energy $\frac{1}{2}h\nu$ in its ground state. For a perfect Debye solid of characteristic temperature $\theta_D$ the total zero-point energy is $\frac{9}{8}R\theta_D$ cal. mole$^{-1}$. To give a resultant internal energy equal to the experimental value of $-12$ cal. mole$^{-1}$, the zero-point energy would have to be $+50$ cal. mole$^{-1}$ corresponding to a $\theta_D$ of 20° K. Specific heat measurements give a $\theta_D$ of approximately this value, but exact agreement is not to be expected because solid helium is not a perfect Debye solid (Dugdale and Simon, 1953). However, we can clearly see how the potential energy and zero-point energy almost compensate one another to give the observed value of the total internal energy.

F. London (1936, 1939, 1954) has applied somewhat similar ideas to the liquid. Assuming it to behave like a solid with a particular crystal structure, he calculated the variation of lattice potential energy with molecular volume, obtaining a curve similar to curve I of fig. 17. The estimation of the zero-point energy was a complicated matter, but the principle can be illustrated rather crudely in the following way. Imagine a point atom of mass $m$ confined within a spherical box of radius $R$ formed by its neighbours. Solving the Schrödinger equation for this situation, the ground state is found to have a zero-point energy of $h^2/8mR^2$, giving a total zero-point energy per mole of $(Nh^2/8m)(4\pi/3V)^{\frac{2}{3}}$. In actual fact the finite diameter of the atom reduces the effective volume of the box and there is also the possibility that the atom can escape through the gap between two of its neighbours, but this does not affect the qualitative form of the variation of zero-point energy with volume as shown in curve II of fig. 17. Addition of curves I and II gives the total energy as shown in curve III. The equilibrium volume $V_{l,0}$ of the liquid at $p=0$, $T=0$ corresponds to the minimum of curve III and is appreciably larger than the volume corresponding to the minimum of curve I. The zero-point energy not only reduces the absolute magnitude of the internal energy, but also 'blows out' the liquid to a large volume. In this respect the zero-point energy is equivalent to an extra repulsive force between the atoms.

The corresponding curve for the total energy of the solid is labelled IV on fig. 17. Its minimum lies above that of curve III and occurs at a smaller volume $V_{s,0}$. The diagram can be used to demonstrate how pressure produces solidification. From equation (2.10)

$$p = -\partial U/\partial V. \qquad (2.14)$$

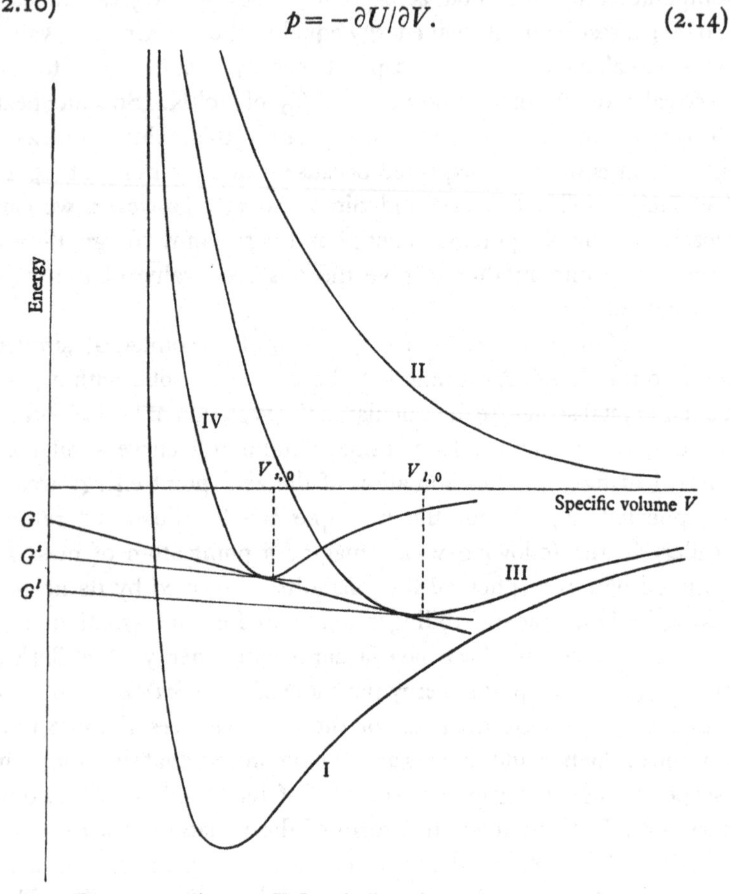


Fig. 17. Diagram to illustrate F. London's treatment of the energy of the liquid and solid at 0° K. I, potential energy of liquid; II, zero-point energy of liquid; III, total energy of liquid; IV, total energy of solid.

The pressure is therefore the slope of the tangent to the curve and the equation of this tangent is

$$U + pV = \text{constant}. \qquad (2.15)$$

The intercept on the $U$ axis therefore gives the value of the Gibbs free energy ($G = U - TS + pV = U + pV$) at the pressure in question. Starting from the minima at zero pressure and slowly increasing the pressure, the two tangents start from a horizontal position and swing round in a clockwise direction. The Gibbs free energy of the liquid is initially always less than that of the solid until we reach the common tangent when the two phases have the same pressure and the same Gibbs function and can co-exist in equilibrium. At higher pressures the Gibbs function of the liquid is greater than that of the solid; the solid is therefore the stable phase.

Within the framework of these ideas, the existence of the liquid at $0^\circ$ K. has been made plausible in the following way. Various possible spatial configurations can be considered and a $U$ versus $V$ curve plotted for each. The most likely configuration will be the one giving the lowest minimum and this will be seen to occur at an unusually large volume and to arise from a particularly loose, open structure. In this structure it is possible for an atom to escape from the cage formed by its neighbours by passing through a gap between two of them, encountering only a low potential barrier on the way. Thus the atoms are not localized, but are free to wander throughout the substance and presumably this implies fluidity. It should be noted, however, that solidification produces only a 10 % change in volume and a 3 % change in the average interatomic distance, so the fluidity cannot be entirely due to a large volume; it is perhaps very much dependent on the special nature of the spatial configuration, which must include paths of least resistance and low potential barriers.

It is very difficult to put these ideas on an exact numerical basis. London (1939) has himself pointed out a consideration of fundamental importance in connexion with the lattice potential energy. The diamond lattice is the one which gives the lowest minimum energy and which would therefore be expected to be the one realized in the actual liquid. This lattice is equivalent to a body-centred cubic in which half the sites are vacant. London showed that it is energetically favourable for an atom to leave its own site and move to a vacant site, but obviously all the atoms cannot do this or the result would be another diamond lattice. A possible solution is that the atoms continually migrate from one site to another, each site being occupied for half the time. This 50 % occupied body-centred lattice can, in fact, be shown to have a lower energy than the diamond lattice and a 50 % occupied face-centred lattice has a lower energy still. The final answer, of course, is that any solid lattice is a very inadequate approximation to the structure of the liquid, although it is possible that a very good representation might be obtained by superimposing several lattices, each with its own probability of occupation, remembering all the time that each lattice point is very much smeared out by the zero-point motion.

An exact calculation of the zero-point energy would involve a quantum mechanical treatment of a complicated many body problem involving interactions, which is beyond present theoretical resources. Pekeris (1950) has considered the problem from the point of view of the Lennard-Jones and Devonshire model of a liquid, which involves solving the Schrödinger equation for an atom in a spherically symmetrical potential well obtained by smearing the nearest neighbours uniformly over the surface of a sphere. Buckthought (1953*a*, *b*) has treated the problem from the point of view of the Hartree self-consistent field, as will be explained more fully in § 3.1.2. Although both methods are able to predict the total internal energy with moderate accuracy, they differ widely in their estimates of the lattice potential energy and the situation is therefore far from satisfactory.

## 2.4. X-ray and neutron diffraction experiments

From experiments on the scattering of X-rays or neutrons by the liquid it is possible to derive the radial distribution function $p(r)$, the significance of which is that, if we concentrate upon a particular helium atom at the origin, the average number of atoms to be found within a spherical shell of radii $r$ and $r+dr$ is $4\pi r^2 dr\, p(r)$. If $\phi(r)$ is the mutual potential energy of two helium atoms at a distance $r$ apart, it is easily seen that the potential energy per mole of the liquid is

$$\Phi = \tfrac{1}{2}N\int_0^\infty \phi(r)\, p(r)\, 4\pi r^2\, dr. \tag{2.16}$$

Since the total energy of the liquid may be deduced from the latent heat of vaporization and the equation of state of the gas, this provides us with an empirical method of obtaining both the potential and kinetic energies at all temperatures.

The broad features of the structure of the liquid were established by the early X-ray analysis experiments of Keesom and Taconis (1938*b*) and Reekie (1940, 1947), but we shall confine the present description to the more accurate work of Reekie, Hutchinson and Beaumont (1953, 1955). Their X-ray camera is illustrated in fig. 18.

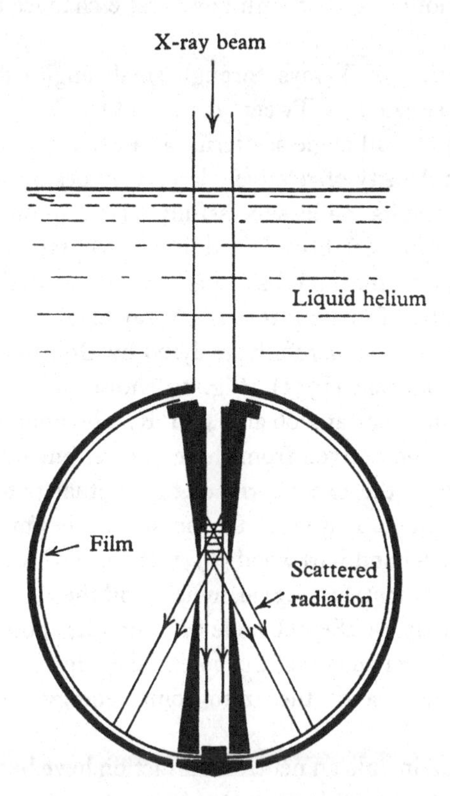


Fig. 18. The X-ray camera of Reekie, Hutchinson and Beaumont (1953, 1955).

X-radiation from a copper target was collimated into a beam 2 mm. square before traversing the cylindrical camera, which was arranged in such a way that a strip of film bent into a circle registered the angular distribution of radiation scattered from a small volume of liquid at its centre. At all temperatures down to 1·27° K. the diffraction pictures bore no resemblance to the sharp rings characteristic of a solid, but were of the broad 'liquid-ring' type. This means that, whatever the nature of the spatial correlation giving rise to short-range order, there is certainly a complete absence of the long-range order found in solids. There was no marked change in the form of the pictures upon passing through the $\lambda$-point or on cooling from the $\lambda$-point down to 1·27° K., proving that the $\lambda$-transition is not associated with any drastic changes in the spatial configuration.

The scattering of X-rays through small angles down to 1·5° has been investigated by Tweet (1954) and by Gordon, Shaw and Daunt (1954). Small angle scattering is a consequence of fluctuations in mean density of relatively large volumes of liquid and the experimental results can be successfully related to the macroscopic isothermal compressibility. In addition, accurate knowledge of the small angle scattering is essential to a correct derivation of the radial distribution function from the X-ray data.

The data have been carefully analysed by Goldstein and Reekie (1955) and Goldstein (1955). Fig. 19 shows the resulting radial distribution functions at 2·06 and 4·20° K. The potential energy of the liquid may be derived from these curves, but turns out to be very sensitive to the exact form of the mutual potential energy between two helium atoms. If one uses the potential energy equation of Slater and Kirkwood (1931) then, at 0° K., the potential energy is approximately $-43$ cal. mole$^{-1}$ and the zero-point energy is approximately $+28{\cdot}7$ cal. mole$^{-1}$. For the potential energy equation of Margenau (1939) the potential energy is approximately $-53{\cdot}9$ cal. mole$^{-1}$ and the zero-point energy approximately $+39{\cdot}6$ cal. mole$^{-1}$.

Similar experiments on neutron diffraction have been performed by Hurst and Henshaw (1955 *a*, *b*) using a monochromatic neutron beam with a wavelength of 1·04 Å. The results are in satisfactory qualitative agreement with the X-ray measurements.

The scattering of slow neutrons has the special feature that the masses of the neutron and helium atom are comparable and the scattering should therefore be influenced by the momentum distribution among the helium atoms. Goldstein, Sweeney and Goldstein (1950) have shown that a degenerate ideal Bose-Einstein gas would give a very large inelastic scattering. This effect was not present in the small angle scattering experiments of Egelstaff and

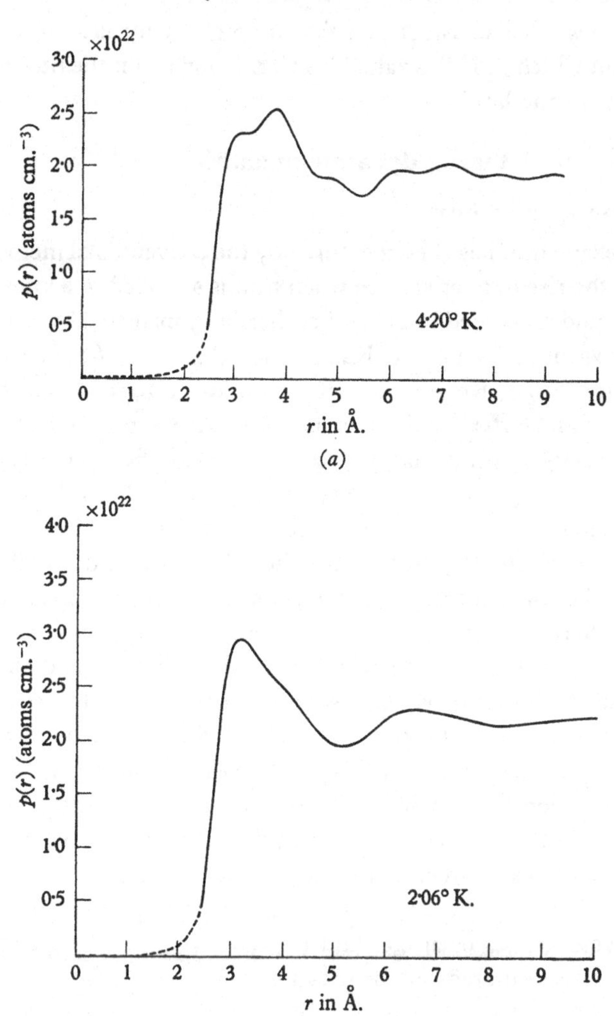


Fig. 19. The radial distribution function at (*a*) 4·20° K.; (*b*) 2·06° K. (after Goldstein and Reekie, 1955).

London (1953, 1957) with wavelengths between 4 and 6 Å., or in the measurements by Sommers, Dash and Goldstein (1955) of the total scattering cross-sections for wavelengths between 3 and 16 Å. In fact the latter experiments revealed a behaviour very similar to that observed for neutron scattering by solids.

In §3.2.1 we shall consider a particular type of neutron scattering experiment which provides valuable information about the thermal excitations in the liquid.

## 2.5. Some basic thermodynamic quantities

### 2.5.1. The specific heat

The specific heat has been measured by the conventional method of noting the rise in temperature when heat is supplied to a known mass of liquid contained in a chamber thermally insulated by a surrounding vacuum (Dana and Kamerlingh Onnes, 1926; Keesom and Clusius, 1932; Keesom and Keesom, 1932, 1935; Kramers, Wasscher and Gorter, 1952; Hercus and Wilks, 1954; Hill and Lounasmaa, 1957; Markham, Pearce, Netzel and Dillinger, 1957). A very unsatisfactory feature of the present available data is that the specific heat along the vapour pressure curve measured by Hercus and Wilks is about 10% higher than the values obtained by other workers. The main feature of the results is the $\lambda$-point anomaly (fig. 20). There is a very rapid rise up to the $\lambda$-point, an apparently discontinuous fall, a slow decrease up to about 2·5° K. and then a slow rise. A similar behaviour is observed in many other cases, such as the order-disorder transition in metallic alloys, the ferromagnetic Curie point and the specific heat anomaly of solid ammonium chloride. The detailed mechanism is different for each of these cases, but they are all co-operative phenomena involving strong interactions between the units undergoing change.

The specific heat shown in fig. 20 is not the specific heat at constant volume, $C_v$, or at constant pressure, $C_p$, but the specific heat under the saturated vapour pressure,

$$C_{\text{sat.}} = C_p - TV\alpha\left(\frac{dp}{dT}\right)_{\text{v.p.c.}}, \tag{2.17}$$

$\alpha$ being the coefficient of expansion and $(dp/dT)_{\text{v.p.c.}}$ the slope of the vapour pressure curve. Below 2·5° K. the difference between $C_{\text{sat.}}$ and $C_p$ is less than 1%, but it becomes more important at higher temperatures and amounts to about 10% at 4° K. The specific heat at constant volume, $C_v$, has been measured directly at various pressures and densities by Keesom and Clusius (1932), Keesom and Keesom (1935, 1936) and Hercus and Wilks (1954).

The results resemble those of fig. 20 except that the $\lambda$-point anomaly moves to lower temperatures at higher pressures in accordance with the negative slope of the $\lambda$-curve (fig. 15), and also $C_v$ varies more slowly with temperature than $C_{\text{sat.}}$ above 2·5° K.

To obtain values of the specific heat below 1° K. the liquid helium must be cooled by mixing it with a powdered paramagnetic salt

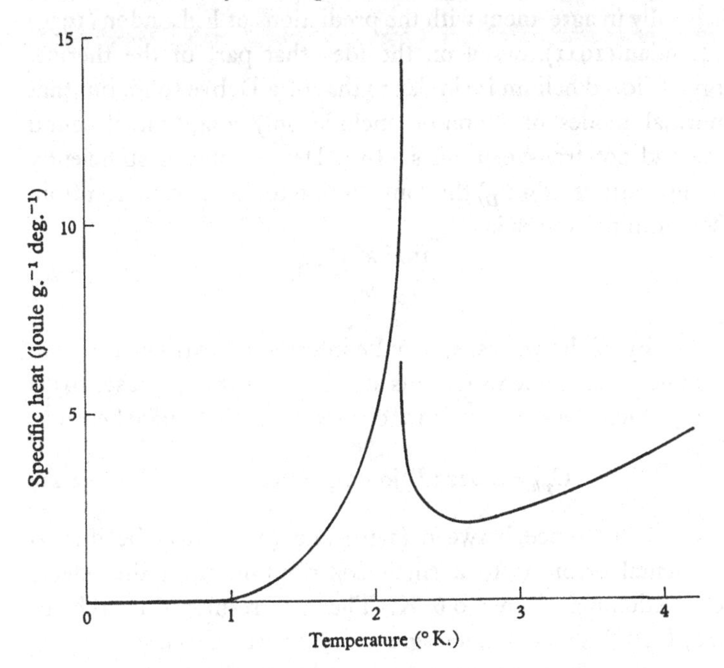


Fig. 20. The specific heat of liquid helium under its saturated vapour pressure.

and employing an adiabatic demagnetization technique. The experiment becomes increasingly more difficult at lower temperatures as the thermal capacity of the salt begins to overwhelm that of the liquid. Using iron ammonium alum, Keesom and Westmijze (1941) and later Hull, Wilkinson and Wilks (1951) found that the specific heat between 0·6° K. and 1·4° K. could be represented by the expression,

$$C = 0{\cdot}024\,T^{6{\cdot}2}\ \text{cal. g.}^{-1}\,\text{deg.}^{-1}. \qquad (2.18)$$

With a more suitable paramagnetic salt (copper potassium Tutton salt) Kramers, Wasscher and Gorter (1952) were able to extend the measurements down to 0·25° K. and they made the very important discovery that, below 0·6° K., the specific heat has the form

$$C = 0{\cdot}0235(\pm 0{\cdot}0015)\, T^3 \,\text{joule g.}^{-1}\,\text{deg.}^{-1}$$

$$= 0{\cdot}00561\, T^3 \,\text{cal. g.}^{-1}\,\text{deg.}^{-1}. \qquad (2.19)$$

This is fully in agreement with the predictions of F. London (1939) and Landau (1941), based on the idea that part of the thermal energy of liquid helium is similar to that of a Debye solid, but that its normal modes of vibration include only longitudinal sound waves, and not transverse ones. In a Debye solid at sufficiently low temperatures ($T \ll \theta_D$) the contribution to the specific heat from the longitudinal waves is

$$C_{ph} = \frac{16\pi^5}{15} \frac{k^4}{h^3} \frac{V}{c^3} T^3. \qquad (2.20)$$

The velocity of the waves, $c$, may be inferred by extrapolating the first sound measurements (Atkins and Chase, 1951*a*; Chase, 1953) to 0° K., which gives $c = 2{\cdot}39 \times 10^4$ cm. sec.$^{-1}$, so that (2.20) becomes

$$C_{ph} = 0{\cdot}0205\, T^3 \,\text{joule g.}^{-1}\,\text{deg.}^{-1}. \qquad (2.21)$$

The small difference between (2.19) and (2.21) may be due to experimental error or to a small contribution from the effects which predominate above 0·6° K. The later results of Markham, Pearce, Netzel and Dillinger agree better with (2.21).

The ability of the liquid to support longitudinal sound waves of long wavelength is, of course, beyond doubt. The important issue is the shortest wavelength which can be supported. The Debye characteristic temperature, $\theta_D$, is 20·0° K., while the $T^3$ specific heat is established only up to 0·6° K., so there is direct experimental evidence only for wavelengths down to about 30 interatomic distances. If we can assume that the Debye contribution is still present above 3° K., we arrive at the interesting result that it can explain the whole of the temperature variation of $C_v$ between 3 and 4° K., implying that the additional contribution from other effects is independent of temperature in this region (Atkins and Stasior, 1953).

### 2.5.2. The latent heat of vaporization

Dana and Kamerlingh Onnes (1926) and Berman and Poulter (1952) have determined the latent heat of vaporization directly by measuring the amount of liquid evaporated upon the addition of a known quantity of heat. The latent heat may also be calculated

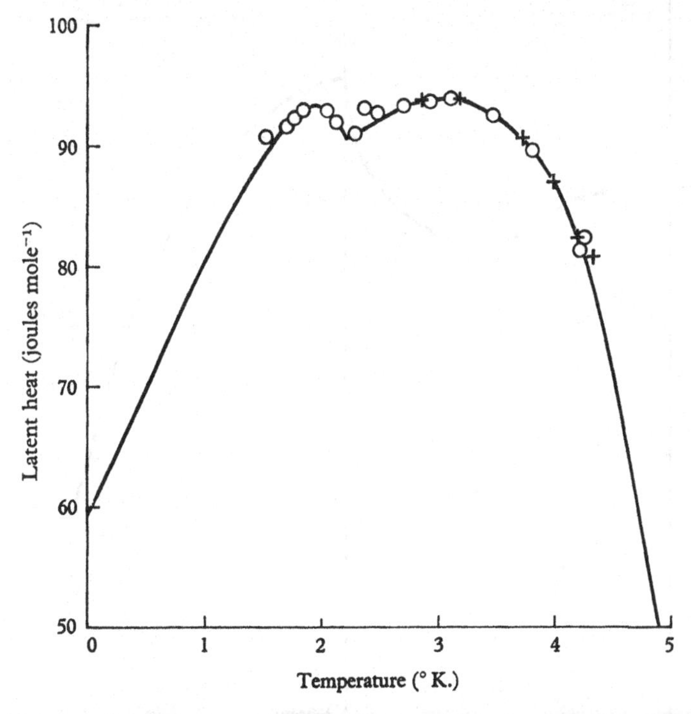


Fig. 21. The latent heat of vaporization. ○, Dana and Kamerlingh Onnes (1926); +, Berman and Poulter (1952); —, calculated from the 1955 vapour pressure scale by van Dijk and Durieux (1957).

from the vapour pressure curve with the aid of the Clausius-Clapeyron equation. Fig. 21 shows the experimental results and the calculated values derived by van Dijk and Durieux (1957) from the $T_{L55}$ vapour pressure scale. There is obviously an anomaly in the latent heat near the $\lambda$-point, but the experimental quantities are not known with sufficient accuracy to determine its exact nature.

### 2.5.3. The density, coefficient of expansion and equation of state

The density of the liquid as a function of temperature along the vapour pressure curve is shown schematically in fig. 22, which is

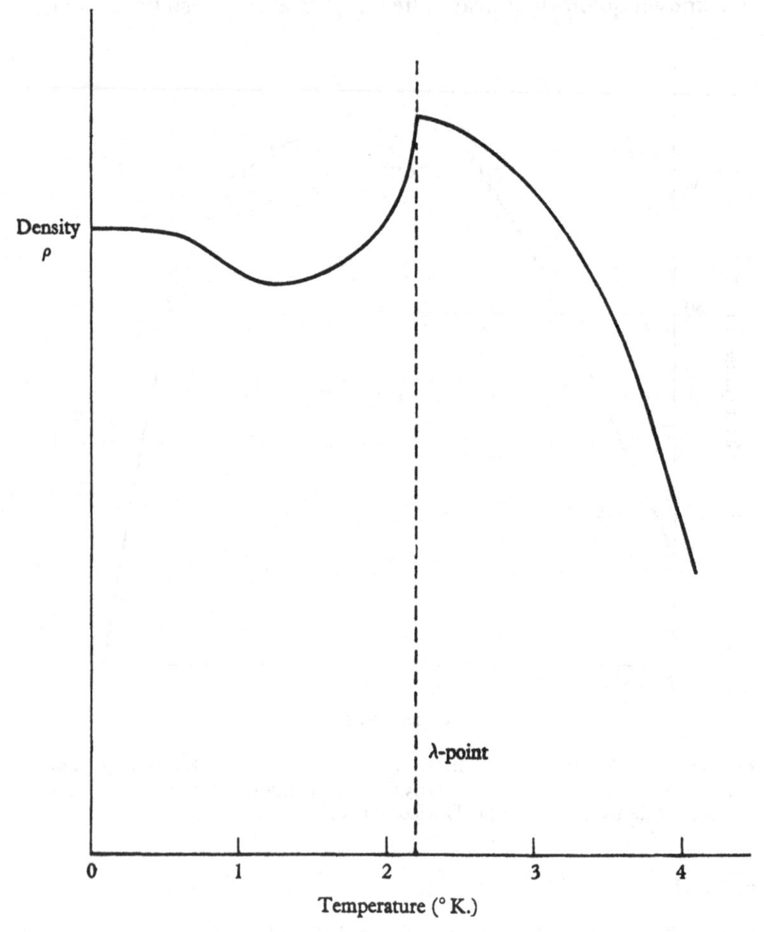


Fig. 22. The density of liquid helium as a function of temperature along the vapour pressure curve. The scale has been distorted to emphasize the minimum at 1·15° K.

based on the density measurements of Kamerlingh Onnes and Boks (1924), Mathias, Crommelin, Kamerlingh Onnes and Swallow (1925) and Kerr (1957*a*) and the coefficient of expansion measurements of Atkins and Edwards (1955). There is a sharp maximum or cusp at the $\lambda$-point and a minimum at 1·15° K.

Table II gives the most recent data obtained by Kerr (1957*a*). The $T_{55E}$ temperature scale is used. Kerr estimates that the density at the critical point is 0·0675 ± 0·0005 g. cm.$^{-3}$.

TABLE II. *The density of liquid* $He^4$ *under its own vapour pressure*

| Temp. (° K.) | Density (g. cm.$^{-3}$) | Temp. (° K.) | Density (g. cm.$^{-3}$) |
|---|---|---|---|
| 1·2 | 0·14547 | 2·2 | 0·14646 |
| 1·3 | 0·14548 | 2·3 | 0·14601 |
| 1·4 | 0·14550 | 2·4 | 0·14550 |
| 1·5 | 0·14553 | 2·6 | 0·14430 |
| 1·6 | 0·14557 | 2·8 | 0·14285 |
| 1·7 | 0·14563 | 3·0 | 0·14114 |
| 1·8 | 0·14572 | 3·2 | 0·13918 |
| 1·9 | 0·14584 | 3·4 | 0·13694 |
| 2·0 | 0·14599 | 3·6 | 0·13443 |
| 2·1 | 0·14620 | 3·8 | 0·13164 |
| 2·15 | 0·14636 | 4·0 | 0·12856 |
| $\lambda$-point | 0·14657 | 4·2 | 0·12518 |
| | | 4·4 | 0·12150 |

The density measurements show that the coefficient of expansion is positive above the $\lambda$-point, as for a normal liquid, but is negative just below the $\lambda$-point. To obtain accurate measurements of the coefficient of expansion below the $\lambda$-point, Atkins and Edwards (1955) used a dilatometer consisting of a copper chamber attached to the lower end of a glass capillary. The chamber was filled until the liquid surface stood in the glass capillary and, from the variation in the height of the meniscus with temperature, the coefficient of expansion was deduced. Just below the $\lambda$-point the coefficient had a large negative value, but decreased rapidly to zero at 1·15° K. and then went through a positive maximum near 0·9° K. The detailed behaviour is shown in figs. 24 and 31, and its theoretical significance will be discussed fully later. The positive coefficient below 1·15° K., like the $T^3$ variation of specific heat below 0·6° K., indicates that, at sufficiently low temperatures, the liquid shows a normal behaviour very similar to that of a solid.

Accurate values of the liquid and solid densities along the melting curve are available as a result of the measurements of Swenson (1950) and Mills and Grilly (1957).

The equation of state of the liquid has been determined by Keesom and Keesom (1933) over the temperature range from 1·5 to 4·2° K., the pressure range from 0 to 35 atmospheres and at densities between 0·125 and 0·183 g. cm.$^{-3}$. A vessel was filled with liquid helium to a suitable pressure and then closed off, so that the pressure $p$ could be determined as a function of temperature $T$ at constant density $\rho$. The data are presented on pp. 237–45 of Keesom (1942), which gives graphs of (1) $p$ versus $T$ at constant $\rho$ (isopycnals), (2) $\rho$ versus $T$ at constant $p$ (isobars), (3) $(\partial\rho/\partial p)_T$ versus $T$ at constant $p$, from which the isothermal compressibility $K_T=(1/\rho)(\partial\rho/\partial p)_T$ may be derived, (4) $(\partial p/\partial T)_\rho$ versus $T$ at constant $\rho$, (5) $(\partial\rho/\partial T)_p$ versus $p$ at constant $T$, from which the coefficient of expansion $\alpha_p=(1/\rho)(\partial\rho/\partial T)_p$ may be derived. Combining these data with the adiabatic compressibilities deduced from the velocity of first sound, the ratio of specific heats ($\gamma=C_p/C_v$) may be derived at various temperatures and pressures (Pellam and Squire, 1947; Findlay, Pitt, Grayson-Smith and Wilhelm, 1938; Atkins and Chase, 1951*a*; Atkins and Stasior, 1953; Atkins and Edwards, 1955).

### 2.5.4. The dielectric constant and the refractive index

These two quantities are directly related to the density by the Clausius-Mossotti formulae

$$\frac{\epsilon-1}{\epsilon+2}=\frac{4\pi}{3}\frac{\rho}{M}N\alpha_0, \tag{2.22}$$

$$\frac{n^2-1}{n^2+2}=\frac{4\pi}{3}\frac{\rho}{M}N\alpha_\lambda. \tag{2.23}$$

$\epsilon$ is the dielectric constant, $n$ the refractive index, $M$ the molecular weight, $\rho$ the density, $N$ Avogadro's number, $\alpha_0$ the electric polarizability per atom at zero frequency and $\alpha_\lambda$ the electric polarizability per atom for the wavelength $\lambda$ at which the refractive index is measured. For the free atom, $\alpha_0$ and $\alpha_\lambda$ can be determined with great accuracy from the refractive index of the gas at various wavelengths. We now consider the question of whether they have the same values in the liquid.

The dielectric constant has been measured at 500 kc./s. by observing the variation in the capacity of a condenser with liquid between its plates (Wolfke and Keesom, 1928) and at 9100 Mc./s. from the change in resonant frequency of a cavity when filled with liquid helium (Grebenkemper and Hagen, 1950). To obtain the refractive index Johns and Wilhelm (1938) measured the critical angle of reflexion, using a Wollaston cell. The results of the three investigations are collected in Table III. The values of $\alpha$ are only slightly less than for the free atom and its variation with temperature is small, all of which might be attributed to experimental errors, including errors in the density curve. However, the importance of small discrepancies and the possible value of further investigations should be emphasized. The Clausius-Mosotti equation is applicable to non-polar liquids (Fröhlich, 1949) but the polarizability $\alpha'$ is not necessarily that of the free atom, since the electronic configuration of an atom is distorted by interaction with its neighbours and this might well alter its polarizability slightly. As an extreme case, it is still an open question whether or not liquid helium contains diatomic or polyatomic molecules. The diatomic molecule is on the verge of stability because the negative potential energy arising from the attractive forces is almost exactly compensated by the positive energy of the vibrational ground state. Also, it is frequently suggested that the liquid might contain compact aggregates of atoms. Experimental values of the atomic polarizability, combined with a theoretical investigation of the effect on it of chemical or van der Waals' forces, might provide valuable information about structural questions of this nature.

Recently Edwards (1956) has made accurate determinations of the refractive index at $\lambda = 5462$ Å., using a Jamin refractometer. The method involved counting fringe shifts as the temperature was varied and therefore gave changes in $n$ with $T$, but not $n$ directly. The molar polarization was therefore assumed to be independent of temperature and equation (2.23) was differentiated with respect to $T$ to give a relationship between $dn/dT$ and the coefficient of expansion

$$\alpha = \frac{6n}{(n^2-1)(n^2+2)}\frac{dn}{dT}. \tag{2.24}$$

The absolute value of $n$ at one particular temperature was chosen to give agreement with previously measured values of $\alpha$ below the $\lambda$-point. The measurements were then used to derive the values of the coefficient of expansion above the λ-point which are shown in fig. 24. The molar polarization was found to be 0·1245 ± 0·0002 (private communication).

Still more recently Maxwell, Chase and Millet (private communication) have measured the dielectric constant at 100 kc./s., with particular emphasis on the vicinity of the λ-point. Their results suggest that the coefficient of expansion approaches zero just above the λ-point and may even become negative there. Their values of the molar polarization vary slightly with temperature between 0·1220 and 0·1225.

TABLE III. *The molar polarization of liquid helium*

| From the dielectric constant at 500 kc./s. | | From the dielectric constant at 9100 Mc./s. | | From the refractive index of the liquid at $\lambda = 5462$ Å | |
|---|---|---|---|---|---|
| Temp. (° K.) | $N\alpha_0'$ | Temp. (° K.) | $N\alpha_0'$ | Temp. (° K.) | $N\alpha_\lambda'$ |
| 1·97 | 0·1194 | 1·62 | 0·1228 | — | — |
| 2·16 | 0·1195 | 1·97 | 0·1228 | — | — |
| 2·19 | 0·1198 | 2·19 | 0·1231 | 2·18 | 0·117 |
| λ-point | | | | | |
| 2·25 | 0·1195 | 2·25 | 0·1231 | 2·26 | 0·117 |
| 2·64 | 0·1198 | 2·64 | 0·1231 | — | — |
| 3·04 | 0·1199 | 3·04 | 0·1233 | — | — |
| 3·58 | 0·1200 | — | — | — | — |
| 4·21 | 0·1206 | 4·21 | 0·1236 | 4·22 | 0·104 |

From measurements of the refractive index of the gas at various frequencies
$N\alpha_0 = 0{\cdot}1234$ extrapolated to zero frequency;
$N\alpha_\lambda = 0{\cdot}1246$ at $\lambda = 5462$ Å.

From Edwards' recent measurements of the refractive index of the liquid
$N\alpha_\lambda' = 0{\cdot}1245$ at $\lambda = 5462$ Å.

### 2.5.5. The surface tension

Allen and Misener (1938*b*) used the capillary rise method to obtain the surface tension curve shown in fig. 23. At the λ-point there is no discontinuity but there may be a small anomaly. As required by the third law of thermodynamics, the surface tension becomes independent of temperature as 0° K. is approached. The curve extrapolates to zero at the critical temperature.

Atkins (1953*b*) has discussed the possible influence on the surface tension of surface modes of vibration. Since changes in density may produce large changes in the contribution to the surface energy from the mutual potential energy of the surface atoms, the discussion was restricted to temperatures below the λ-point, where the density does not change by more than 0·7 %. Following a suggestion by Frenkel (1940*b*, 1946), the surface modes were taken to be similar to surface tension waves, and these waves were treated in a manner which is familiar in connexion with the Debye theory of solids. The free energy associated with the modes was estimated and shown to give a negative contribution to the surface tension

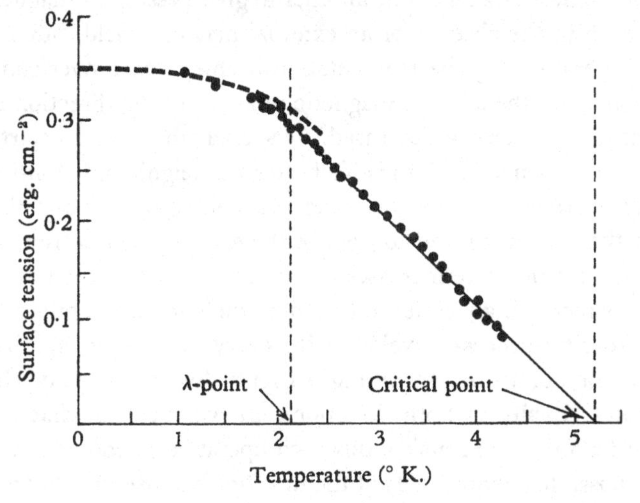


Fig. 23. The surface tension of liquid helium. ●, measurements of Allen and Misener (1938*b*); - - - -, theory of Atkins (1953*b*).

varying approximately as $T^{\frac{7}{3}}$. In this way a major part of the variation with temperature below the λ-point was explained. The zero-point energy of the modes was found to represent an appreciable fraction of the total surface energy at 0° K.

## 2.6. The nature of the λ-transition

### 2.6.1. The Ehrenfest relations

The type of specific heat anomaly shown in fig. 20 also occurs during the transition from ferromagnetism to paramagnetism and during the order-disorder change in certain alloys. These are both examples of co-operative phenomena and the liquid helium λ-transition probably also arises from a situation incorporating the main features of a co-operative phenomenon. A co-operative phenomenon results from co-operative interactions between the units undergoing change, but there is more to it than this, since the behaviour of a solid, for example, involves strong interatomic forces but can be treated quite simply in terms of normal modes which interact only weakly with one another. The main features of a co-operative phenomenon can be illustrated by considering in detail a semi-classical treatment of ferromagnetism. At 0° K. the electron spins are aligned parallel to one another to give a resultant magnetization even in the absence of an external magnetic field, but at a finite temperature the thermal agitation is able to turn over some of the spins and the average magnetic moment in the direction of magnetization is thereby decreased. As soon as this process starts, an electron chosen at random is likely to have neighbours pointing against the direction of magnetization as well as with it and this reduces the energy needed to reverse the spin of the electron, so that, as the temperature increases and more spins are turned over, it becomes increasingly easier to turn over the remaining spins and the disordering process develops with every increasing rapidity. At the Curie point the disordering is eventually complete and the spins point equally in both directions. It will be seen that the essential feature which makes this a co-operative phenomenon is that the possible energy levels of the unit (in this case the electron spin) are influenced by the behaviour of the neighbouring units. In the liquid helium case, no mechanism has been postulated in precise detail, so we must confine ourselves to some general remarks concerning the experimental facts which have to be explained and the features which might be expected to be present by analogy with other $\lambda$-transitions.

Ehrenfest (1933) put forward certain thermodynamic considerations based on the concept of a second order transition. A first order transition, such as a change of state, involves a latent heat and a change of volume, which means that, although the Gibbs free energy $G$ is unchanged by the transition, there is a discontinuous change in each of its first order derivatives, the entropy $S=-(\partial G/\partial T)_p$ and the volume $V=(\partial G/\partial p)_T$. In a second order transition the first order derivatives are continuous, but there is a discontinuous change in the second order derivatives, the specific heat at constant pressure $C_p = -T(\partial^2 G/\partial T^2)$, the coefficient of expansion $\alpha = (1/V)(\partial^2 G/\partial T \partial p)$, and the compressibility $K_T = -(1/V)(\partial^2 G/\partial p^2)$. Obviously it is possible to define a transition of the $n$th order, in which the first $n-1$ derivatives of $G$ are continuous but the $n$th order derivatives are discontinuous.

Let us assume that the $\lambda$-transition in liquid helium is of the second order and use the fact that the entropy is continuous on crossing the $\lambda$-curve. At any point $A$ on the $\lambda$-curve

$$S_{A,\mathrm{I}} = S_{A,\mathrm{II}},$$

the subscripts referring to liquid helium I and liquid helium II. At a neighbouring point $B$

$$S_{B,\mathrm{I}} = S_{B,\mathrm{II}}$$

or

$$S_{A,\mathrm{I}} + \delta S_{\mathrm{I}} = S_{A,\mathrm{II}} + \delta S_{\mathrm{II}},$$

whence

$$\delta S_{\mathrm{I}} = \delta S_{\mathrm{II}},$$

that is

$$\left(\frac{\partial S_{\mathrm{I}}}{\partial T}\right)_p \Delta T + \left(\frac{\partial S_{\mathrm{I}}}{\partial p}\right)_T \Delta p = \left(\frac{\partial S_{\mathrm{II}}}{\partial T}\right)_p \Delta T + \left(\frac{\partial S_{\mathrm{II}}}{\partial p}\right)_T \Delta p,$$

$\Delta T$ and $\Delta p$ being restricted to movement along the $\lambda$-curve. Therefore

$$\frac{\Delta p}{\Delta T} = -\frac{(\partial S_{\mathrm{II}}/\partial T)_p - (\partial S_{\mathrm{I}}/\partial T)_p}{(\partial S_{\mathrm{II}}/\partial p)_T - (\partial S_{\mathrm{I}}/\partial p)_T}$$

which may be rewritten by using

$$C_p = T(\partial S/\partial T)_p \quad \text{and} \quad \alpha = (1/V)(\partial V/\partial T)_p = -(1/V)(\partial S/\partial p)_T,$$

so that

$$\left(\frac{dp}{dT}\right)_\lambda = \frac{C_{p,\mathrm{II}} - C_{p,\mathrm{I}}}{TV(\alpha_{\mathrm{II}} - \alpha_{\mathrm{I}})}. \tag{2.25}$$

This relates the slope of the $\lambda$-curve to the discontinuities in the specific heat and the coefficient of expansion. Using the fact that the volume $V$ is also continuous across the $\lambda$-curve, a second Ehrenfest relation may be derived:

$$\left(\frac{dp}{dT}\right)_\lambda = \frac{\alpha_{\mathrm{II}} - \alpha_{\mathrm{I}}}{K_{T,\mathrm{II}} - K_{T,\mathrm{I}}}, \tag{2.26}$$

where $K_T$ is the isothermal compressibility.

Keesom (1933, 1942) has shown that the experimental data for liquid helium are consistent with equation (2.25), but he fully appreciated that the specific heat is varying so rapidly in the vicinity of the λ-point that it is impossible to make a justifiable extrapolation in order to determine the magnitude of its discontinuity. He therefore proved (1936) that equations (2.25) and

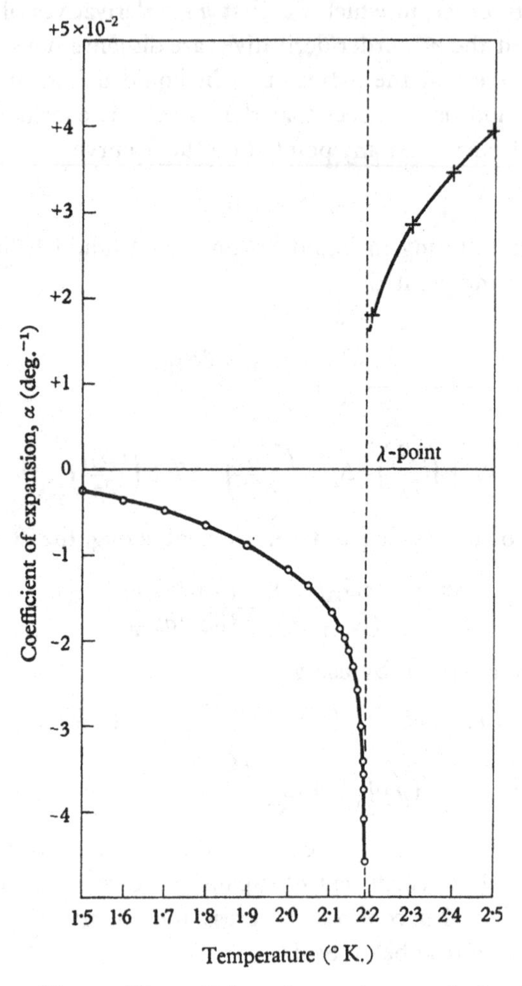


Fig. 24. The coefficient of expansion near the λ-point. o, Atkins and Edwards (1955); +, Edwards (1956).

(2.26) are still approximately true if $C_p$, $\alpha$ and $K_T$ are measured at two temperatures on opposite sides of the λ-point separated by a small temperature interval. However, the coefficient of expansion measurements in fig. 24 suggest another possibility, first envisaged by Tisza (1951), that the second order derivatives of the Gibbs free energy tend to infinity at the transition temperature. If this is so, Keesom's arguments are no longer rigorous and the situation can no longer be elucidated without knowing the exact functional form of the various quantities near the $\lambda$-point.

Equation (2.26) cannot be tested directly because there are no detailed measurements of the compressibility $K_T$, but we do have data on the velocity of first sound, which is related to the compressibility by $u_1^2 = \gamma/\rho K_T$. A discontinuity of about 5 % is to be

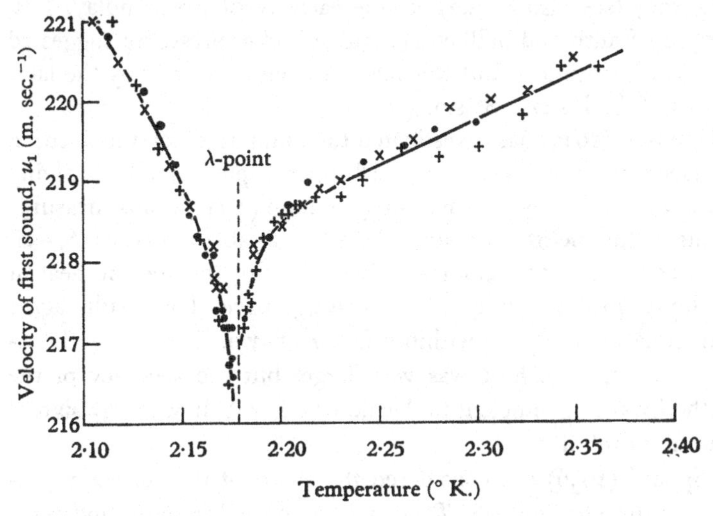


Fig. 25. The velocity of first sound in the vicinity of the $\lambda$-point (after Atkins and Chase, 1951*a*).

expected in $K_T$ and this might lead to a $2\frac{1}{2}$ % discontinuity in $u_1$, although it is impossible to be quite certain of this without knowing how variations in $\gamma$ ($= C_p/C_v$) affect the situation. Measurements along the vapour pressure curve have provided no evidence for a discontinuity (Findlay, Pitt, Grayson-Smith and Wilhelm, 1938; Pellam and Squire, 1947; Atkins and Chase, 1951*a*). The detailed investigation of Atkins and Chase at a frequency of 14 Mc./s. gave the curve of fig. 25, which again reveals the fundamental difficulty that $u_1$ is decreasing so rapidly near the $\lambda$-point that it is really impossible to make any estimate of the discontinuity. In fact the curve is quite consistent with the possibility that the compressibility is tending to infinity and the velocity of first sound to zero.

The situation is further complicated by the presence of a large attenuation near the $\lambda$-point which makes the signal disappear completely. Chase (1953) used a lower frequency (2 Mc./s.) and was able to follow the signal all the way through the $\lambda$-transition and to demonstrate qualitatively that there was no detectable discontinuity and that the velocity did not become zero. This result is not entirely conclusive, because there are undoubtedly relaxation effects giving rise to attenuation and dispersion, and a conclusive experiment would have to be performed at vanishingly small frequency (see also §5.2.2). Some early results of Findlay, Pitt, Grayson-Smith and Wilhelm (1939) at higher pressures suggested a $2\frac{1}{2}\%$ discontinuity, but this has not been borne out by the later results of Atkins and Stasior (1953).

Swenson (1953) has investigated the point at which the $\lambda$-curve intersects the melting curve (the 'upper triple point'). Applying the Clausius-Clapeyron equation (equation (2.11)) to his measurements of the melting pressure curve he was able to deduce $S_l - S_s$ and hence study the anomalous behaviour of the specific heat of the liquid in the vicinity of the melting curve. The results agree with equation (2.25) to within a factor of two. Just below the $\lambda$-curve the specific heat was very large, but the accuracy of the method was not sufficient to decide whether it showed any sign of tending to infinity.

Pippard (1956) has considered the shape of the surface representing the entropy $S(p, T)$ as a function of pressure and temperature, and has suggested that in the vicinity of a $\lambda$-curve it takes the form

$$S(p, T) = S_0 + aT + f\left[\left(\frac{dp}{dT}\right)_\lambda T - p\right]. \tag{2.27}$$

One can then readily show that the equations

$$C_p = VT_\lambda \left(\frac{dp}{dT}\right)_\lambda \alpha + \text{constant}, \tag{2.28}$$

$$\alpha = \left(\frac{dp}{dT}\right)_\lambda K_T + \text{constant} \tag{2.29}$$

apply immediately below and immediately above the $\lambda$-point with the same values of the constants. Ehrenfest's equations (2.25) and (2.26) are special cases of these more general relations. These equations are in very good agreement with the measurements which have been made on the $\lambda$-transition in solid ammonium chloride. They can be made to agree with the measurements on $\alpha$ and $K_T$ (from $u_1$) for liquid helium if it is assumed that the $\lambda$-point for one set of these measurements was in error by $0{\cdot}001^\circ$ K., which is not unlikely.

We therefore arrive at two tentative conclusions. First, although the Ehrenfest relations may not give an exact description of the situation, there is probably some sense in which they represent an approximation to the detailed behaviour. Secondly, we should give serious consideration to the possibility that the second order derivatives of the Gibbs function tend to infinity at the $\lambda$-point. This is certainly consistent with the specific heat data of fig. 20, the coefficient of expansion data of fig. 24 and the first sound data of fig. 25. Moreover, Onsager (1944) has considered a two-dimensional model of ferromagnetism and finds that its specific heat tends to infinity at the Curie point as $\ln(T_c - T)$. It is interesting to note that the coefficient of expansion of liquid helium II near the $\lambda$-point may be represented by a similar expression (Atkins and Edwards, 1955)

$$\alpha = 0{\cdot}0008 + 0{\cdot}0148 \log_{10}(T_\lambda - T). \qquad (2.30)$$

Buckingham, Fairbank and Kellers (private communication) have recently made very careful measurements of the specific heat near the $\lambda$-point. The helium was permanently sealed into a large copper block made in such a way that the liquid was everywhere within 0·003 in. of the copper, and temperature uniformity was therefore ensured. Heat was supplied at a steady rate and the temperature of the liquid was followed as a function of time by means of a carbon resistor also embedded in the copper. The temperature resolution was estimated to be about $10^{-6}\,^\circ$ K. and it was possible to make measurements within $10^{-5}\,^\circ$ K. of the $\lambda$-point. Both above and below the $\lambda$-point the results could be approximately fitted by the curve

$$C_v = \Delta - 3{\cdot}1 \log_{10} |T - T_\lambda| \text{ joule g.}^{-1} \text{ deg.}^{-1}, \qquad (2.31)$$

with $\Delta \simeq 3{\cdot}1$ below the $\lambda$-point and $-1{\cdot}7$ above the $\lambda$-point. Equations (2.30) and (2.31) are obviously consistent with equation (2.28). The discontinuity of 4·8 joule g.$^{-1}$ deg.$^{-1}$ in $\Delta$ may perhaps be related to a similar discontinuity in $\alpha$ by the Ehrenfest equation.

### 2.6.2. Fluctuations near the λ-point

The shape of the specific heat curve immediately above the λ-point implies that, whatever the process responsible for the λ-transition, its effects linger for a few tenths of a degree above the λ-point, and Keesom and Keesom (1935) have suggested that this might be a consequence of statistical fluctuations. They envisage an 'ideal' specific heat curve rising very rapidly below the λ-point, but having no anomaly above the λ-point. If, now, the liquid just above the λ-point can be divided into small domains interacting only weakly with one another, temperature fluctuations in these domains will carry some of them below the λ-point, so that the liquid may be visualized as a matrix of liquid helium I containing a few inclusions of liquid helium II. The large specific heat of the inclusions is assumed to increase the average specific heat of the mixture, which is therefore anomalously high just above the λ-point and decreases rapidly as the λ-point is left behind and possible fluctuations result in fewer and fewer inclusions. Just below the λ-point the liquid is to be imagined as a mixture of inclusions of helium I in helium II with an average specific heat less than the 'ideal' value, the overall result being to smooth out the specific heat curve into a sharp maximum rather than a finite discontinuity or an infinite singularity. These ideas obviously demand a more rigorous analysis and, in particular, the concept of independent domains of arbitrarily chosen size requires critical examination, but the suggestion may be a very fruitful one and we shall see in § 5.2.2 how Pippard (1951) has developed it to give a very plausible explanation of the behaviour of first sound near the λ-point.

Further light is thrown on the matter by considering the analogous situation in ferromagnetism, for here also there is an excess specific heat above the Curie temperature and its explanation is fairly well understood. Below the Curie point the electron spins are aligned in the direction of magnetization over large regions, i.e. there is long-range order. Above the Curie point this long-range order has disappeared, but there is still short-range order, and it is variations in this short-range order which give rise to the excess specific heat. Just above the Curie point, then, we must imagine that for a large specimen the spins point equally in both directions on the average, but, if we consider small regions containing a few spins, then in some of these regions there is a predominance of spins pointing in one direction, in some other regions a predominance of spins pointing in the opposite direction and in still other regions the spins point equally in both directions. It is obvious that this situation has a formal similarity to that envisaged by the fluctuation theory. It is also possible to define a parameter $\delta$, the 'average range of short-range order', such that all regions of linear dimensions comparable with or less than $\delta$ have an appreciable probability of being ordered, and $\delta$ will be large close to the transition temperature but will decrease rapidly with increasing temperature, being the analogue of the size of the inclusions in the fluctuation theory. The difficulty in carrying over these ideas to the liquid helium case is, of course, that we have a very incomplete understanding of the ordering process responsible for the $\lambda$-transition. Bowers and Mendelssohn (1949, 1950*a*) have suggested that the short-range order occurs in momentum space. We note, however, that this might be accompanied by short-range order in configuration space. We could, if we wished, consider ferromagnetism as an ordering in angular momentum space.

Other properties of liquid helium also display an anomaly just above the $\lambda$-point, for instance, the coefficient of expansion (fig. 24) and the velocity of first sound (fig. 41). Bowers and Mendelssohn (1949, 1950*a*) have measured the viscosity of liquid helium I by a capillary flow method and have found a sharp dip in anticipation of the $\lambda$-point. That this was an anomaly in the true viscosity and not an onset of superfluidity was proved by the fact that the viscosity had the same value in very narrow channels, whereas superfluidity is usually more pronounced in narrow channels. These viscosity measurements have been extended to higher pressures by Tjerkstra (1952). Fig. 26 was constructed by him from his own results and those of Bowers and Mendelssohn and shows the viscosity as a function of temperature at various constant liquid densities. The dip just above the $\lambda$-point is pronounced in all cases. It is interesting to notice that at low densities there is a gas-like behaviour, with the viscosity increasing with increasing temperature, but at higher densities there is a transition to a

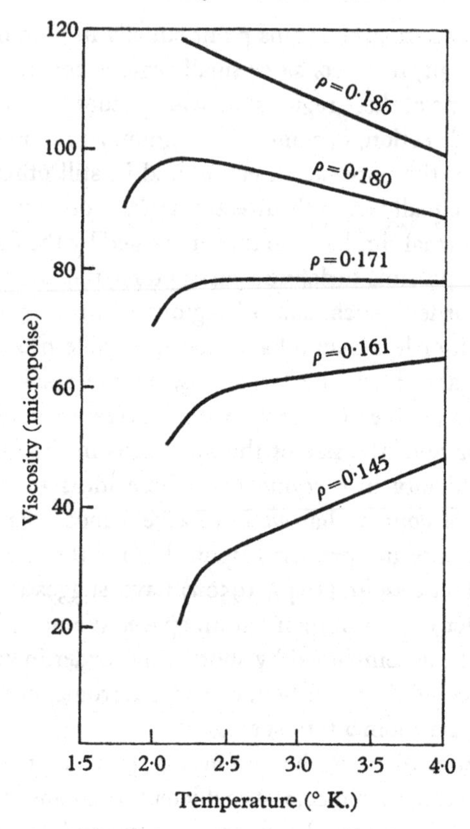


Fig. 26. The viscosity of liquid helium I at various constant densities (after Tjerkstra, 1952).

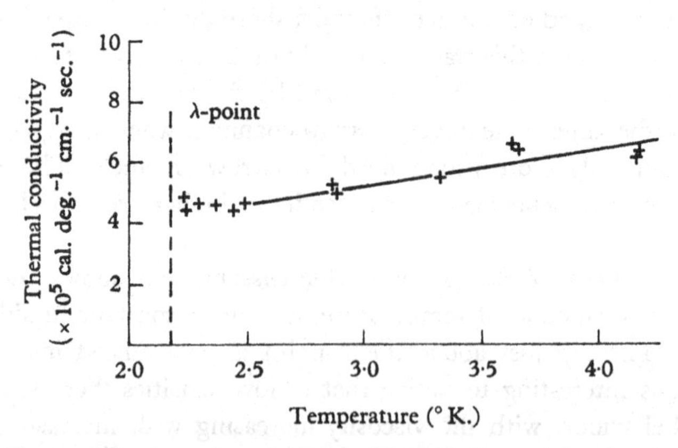


Fig. 27. The thermal conductivity of liquid helium I (after Grenier, 1951).

liquid-like behaviour and the viscosity decreases with increasing temperature.

Bowers and Mendelssohn (1951, 1952*a*) and Grenier (1951) have searched for a similar dip in the thermal conductivity just above the $\lambda$-point, but fig. 27, which is taken from the paper by Grenier, indicates that either the dip is not present or it is too small to be measured.

Density fluctuations in the vicinity of a critical point are known to give rise to the striking phenomenon of critical opalescence. Lawson and Meyer (1954) searched unsuccessfully for an anomalously large scattering of light near the $\lambda$-point of liquid helium. Their results were in complete accord with Rayleigh's classical theory of scattering by small density fluctuations. In fact their experiment supplements the results obtained from X-ray and neutron diffraction in the limit of very long wavelengths.

# CHAPTER 3

# THEORIES

## 3.1. Bose-Einstein condensation

### 3.1.1. Bose-Einstein condensation in an ideal gas

Although there is no completely satisfactory theory of the liquid state, the theories of the gaseous and solid states are well established on a firm foundation. It is therefore tempting to treat a liquid either as a very imperfect gas in which the intermolecular forces have become important or as a broken down solid in which the binding forces are too weak to localize the atoms near lattice points. The theory of liquid helium to be discussed in the present section is based on the phenomenon of Bose-Einstein condensation in an ideal gas. Landau's theory, which will be discussed later in this chapter, approaches the problem from the opposite point of view of the theory of the solid state.

The justification for treating liquid helium like a gas is that its atoms are far apart compared with most liquids and solids and each atom moves in a shallow potential field which does not change appreciably until the atom approaches within one atomic diameter of a neighbouring atom. The viscosity of a gas increases with increasing temperature and liquid helium I shows a similar behaviour at pressures near the vapour pressure (fig. 26), whereas the viscosity of a normal liquid decreases with increasing temperature, because, as one layer of atoms slides over an adjacent layer, the atoms have to surmount potential barriers of height $A$ and their ability to do so increases with increasing temperature in accordance with the factor $e^{-A/kT}$. One might infer that the potential barriers are not very important in liquid helium I and that its atoms are able to move with comparative freedom to give a gas-like viscosity

$$\eta = \tfrac{1}{3}\rho v l, \tag{3.1}$$

where $v$ is an appropriate mean velocity of the atoms and $l$ is their mean free path. With $\eta \sim 3 \times 10^{-5}$ poise and $v \sim 2 \times 10^{4}$ cm./sec., $l$ is $3 \times 10^{-8}$ cm., which is the order of magnitude of the average interatomic distance in the liquid.

The $He^4$ atom contains an even number of fundamental particles (2 protons, 2 neutrons and 2 electrons) and must therefore obey Bose-Einstein statistics. F. London (1938*a*, *b*, 1939) pointed out that a high order transition very similar to that in liquid helium occurs in an ideal Bose-Einstein gas. In Bose-Einstein statistics the mean occupation number $\bar{n}_i$ of the *i*th state with energy $\epsilon_i$ is

$$\bar{n}_i = \frac{1}{\zeta e^{\epsilon_i/kT} - 1}. \tag{3.2}$$

The parameter $\zeta$ is determined by summing to obtain the total number of atoms $N$,

$$N = \sum_i \frac{1}{\zeta e^{\epsilon_i/kT} - 1}. \tag{3.3}$$

For an ideal gas contained within a volume $V$, the density of states is

$$g(\epsilon) = \frac{2\pi V}{h^3}(2m)^{\frac{3}{2}}\epsilon^{\frac{1}{2}}, \tag{3.4}$$

and it can be shown that a sufficient approximation to (3.3) is

$$N = \frac{1}{\zeta - 1} + \int_0^\infty \frac{2\pi V}{h^3}(2m)^{\frac{3}{2}} \frac{\epsilon^{\frac{1}{2}}\,d\epsilon}{\zeta e^{\epsilon/kT} - 1}. \tag{3.5}$$

The first term on the right-hand side represents the mean occupation number of the ground state and the summation over all higher states has been replaced by a continuous integral. The ground state must be introduced separately because its contribution will be found to be especially important, and if the approximation of equation (3.4) were used throughout, the ground state would be assigned zero statistical weight ($g = 0$ when $\epsilon = 0$).

The transition from (3.3) to (3.5) is not easy to justify with complete mathematical rigor and there has been much discussion of this point. Fraser (1951) has presented what appears to be a rigorous proof and also gives a list of references to earlier work.

As negative occupation numbers are not permissible, $\zeta \geqslant 1$, and the integral in equation (3.5) therefore has a maximum possible value equal to

$$N_n = \int_0^\infty \frac{2\pi V}{h^3}(2m)^{\frac{3}{2}} \frac{\epsilon^{\frac{1}{2}}\,d\epsilon}{e^{\epsilon/kT} - 1}$$

$$= \frac{2{\cdot}612\,V(2\pi mkT)^{\frac{3}{2}}}{h^3}. \tag{3.6}$$

If $N > N_n$, no more than $N_n$ atoms can be in the excited states and the remaining $N - N_n$ atoms must be in the ground state. There is therefore a transition temperature $T_c$ given by

$$N = \frac{2{\cdot}612 V (2\pi m k T_c)^{\frac{3}{2}}}{h^3}$$

or

$$T_c = \frac{h^2}{2\pi m k}\left(\frac{N}{2{\cdot}612 V}\right)^{\frac{2}{3}}. \quad (3{\cdot}7)$$

If $T > T_c$, $N_n$ can be equal to $N$ and the atoms are distributed amongst the excited states, with a negligible number in the ground state.

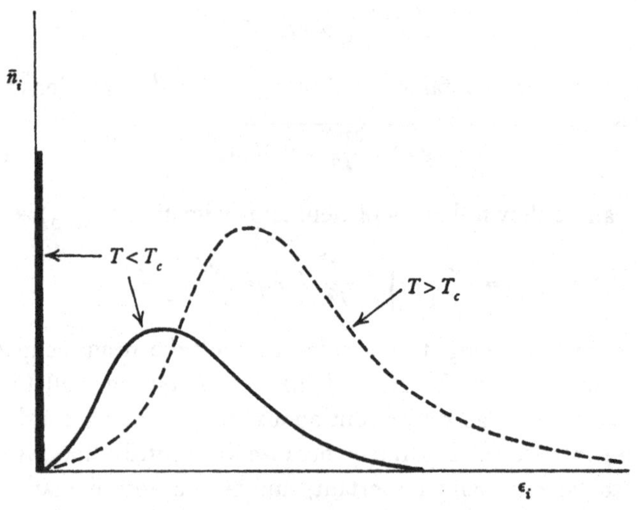


Fig. 28. The distribution function for the atoms of an ideal Bose-Einstein gas (schematic).

This is the situation represented by the dashed curve in fig. 28. If $T < T_c$, the number of atoms in the excited states is obtained from (3·6):

$$N_n = N\left(\frac{T}{T_c}\right)^{\frac{3}{2}}. \quad (3{\cdot}8)$$

The remaining atoms

$$N_s = \frac{1}{\zeta - 1} = N\left[1 - \left(\frac{T}{T_c}\right)^{\frac{3}{2}}\right] \quad (3{\cdot}9)$$

are in the ground state. This is the situation represented by the full curve of fig. 28. A finite number of atoms therefore 'condenses' out into the ground state, the fraction so condensed varying from 0 at $T_c$ to 1 at 0° K. The obvious next step is to associate the atoms in the ground state with the superfluid component and the atoms in the excited states with the normal component. The zero entropy of the superfluid component is thereby immediately explained.

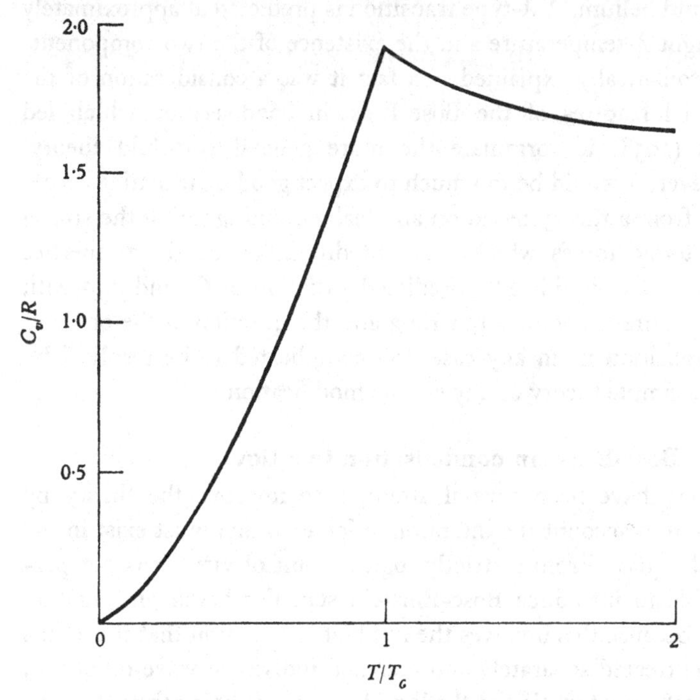


Fig. 29. The specific heat at constant volume of an ideal Bose-Einstein gas (after F. London, 1954).

The transition temperature $T_c$ depends upon the density through the factor $(N/V)^{\frac{2}{3}}$. Using the measured density of liquid helium, $T_c = 3{\cdot}13$° K., which is not very far from the observed $\lambda$-point of 2·18° K. It should be noted, however, that equation (3.7) implies that $T_c$ increases with increasing density, whereas the $\lambda$-curve actually has a negative slope (fig. 15).

Once the occupation numbers have been determined, the thermodynamic functions may be easily derived. Fig. 29 shows the specific heat at constant volume as a function of temperature. Below $T_c$ the specific heat per mole is $C_v = 1{\cdot}926R(T/T_c)^{\frac{3}{2}}$, and when $T \gg T_c$,

$C_v = \frac{3}{2}R$, as in classical theory. At $T_c$ there is a third order transition, but the shape of the theoretical specific heat curve is quite different from the experimental curve of fig. 20.

The phenomena predicted for an ideal Bose-Einstein gas are therefore in qualitative agreement with the phenomena observed in liquid helium. A $\lambda$-type transition is predicted at approximately the right $\lambda$-temperature and the existence of the two components is automatically explained. In fact it was a consideration of the essential features of the Bose-Einstein condensation which led Tisza (1938) to formulate the more general two-fluid theory. However, it would be too much to expect good quantitative agreement from a theory based on an ideal gas and ignoring the strong interatomic forces which undoubtedly influence the properties of the actual liquid. The predicted variation of $C_v$ and $\rho_s/\rho$ with temperature is completely wrong and the question of the order of the transition is, in any case, too complicated to be resolved by such a simple theory or any simple modification of it.

### 3.1.2. Bose-Einstein condensation in a liquid

There have been several attempts to improve the theory by taking into account the interatomic forces which must exist in the actual liquid. From a strictly logical point of view, it is not permissible to introduce Bose-Einstein statistics based on equation (3.2), because this involves the implicit assumption that the atoms can be treated separately and assigned individual wave functions, the total wave function of the liquid as a whole being then built up from the one atom wave functions. This is certainly not permissible as soon as interatomic forces are introduced. However, one might hope to show that, in some sense, the formalism of the previous section is an adequate approximation, with perhaps a different effective mass and a different form for the spectrum of states.

If, for example, one chooses the more general spectrum

$$\epsilon(p) = Ap^{1/r} \tag{3.10}$$

(F. London, 1939, 1943; Dingle, 1952*a*) better agreement with the observed specific heat data can be obtained and a discontinuity in specific heat at the $\lambda$-point obtained. $A$ and $r$ are adjustable parameters and the procedure tends to become merely curve-fitting, with no physical justification. Buckthought (1953*a*, *b*) has represented the $He^4$ atom by a Bloch wave moving through the quasi-crystalline field obtained by smoothing out the interatomic forces due to all the other atoms, rather like an electron in a metal, except that one must use Bose-Einstein rather than Fermi-Dirac statistics. This gives rather better values for the thermodynamic functions and, in particular, predicts a negative coefficient of expansion for He II and a negative slope for the $\lambda$-curve. The theory still suffers from the major defect that it gives a specific heat varying as $T^{\frac{3}{2}}$ near absolute zero, whereas the observed specific heat is much smaller and varies as $T^3$.

More success is achieved by the introduction of an energy gap $\Delta$ and an effective mass $\mu$:

$$\epsilon = \Delta + p^2/2\mu. \tag{3.11}$$

One then obtains for the thermodynamic quantities below the $\lambda$-point

$$N_n = \frac{V(2\pi\mu kT)^{\frac{3}{2}}}{h^3} e^{-\Delta/kT}, \tag{3.12}$$

$$S = \frac{kV(2\pi\mu kT)^{\frac{3}{2}}}{h^3} e^{-\Delta/kT}\left(\frac{5}{2} + \frac{\Delta}{kT}\right), \tag{3.13}$$

$$C_v = \frac{kV(2\pi\mu kT)^{\frac{3}{2}}}{h^3} e^{-\Delta/kT}\left[\frac{15}{4} + \frac{3\Delta}{kT} + \left(\frac{\Delta}{kT}\right)^2\right]. \tag{3.14}$$

The exponential factors make it much easier to fit these formulae to the experimental results and very good agreement is obtained if $\Delta/k = 8{\cdot}6^\circ$ K. and $\mu = 8{\cdot}8 m_{He}$, where $m_{He}$ is the mass of the helium atom. There is a discontinuity in specific heat at the $\lambda$-point similar to the experimental one shown in fig. 20, and above the $\lambda$-point:

$$C_{v,\mathrm{I}} = \frac{3}{2}\frac{k}{\mu} \text{ per g.} \tag{3.15}$$

There is some experimental evidence that $C_v$ *is* constant above 3° K., when the 'tail' just above the $\lambda$-point has died out (Atkins and Stasior, 1953), but equation (3.15) gives a value about four times too small if the same value of $\mu$ is used above and below the $\lambda$-point.

Mikura (1955) has shown that the values of the negative slope of the $\lambda$-curve, the coefficient of expansion and the variation of the various thermodynamic quantities with pressure can all be approximately fitted if one assumes that the variation of $\Delta$ and $\mu$ with pressure takes the form

$$\left.\begin{aligned} \Delta &\propto \left(\frac{\rho}{\mu}\right)^{0\cdot 4}, \\ \frac{\mu}{m_{He}} &= 8\cdot 8 + 0\cdot 144(p - 0\cdot 05), \end{aligned}\right\} \tag{3.16}$$

with $p$ in atmospheres.

Toda (1951) has made the interesting suggestion that the energy gap arises from the formation of small molecular clusters, or crystallites, containing 8 or 9 atoms. $\Delta$ is then the excess energy needed to produce such a cluster and $\mu$ is the mass of the cluster. At 0° K. no clusters exist, but as the temperature is raised the clusters gradually form until, at the $\lambda$-point, the liquid consists entirely of clusters. As the clusters must be more dense than the surrounding monatomic liquid, the negative coefficient of expansion is immediately explained and this type of explanation agrees with the experimental result that the increase in density is approximately proportional to the density of the normal component (the number of clusters) (Atkins and Edwards, 1955). Bijl (1940), Bijl, de Boer and Michels (1941) and Matsubara (1951) have presented quantum mechanical arguments which suggest that such a clustering might occur in a Bose-Einstein system.

Another possible pictorial view of the situation can be based on the treatment of liquid helium as a loose, broken-down solid. In such a pseudo-solid an atom might be forced into something like an interstitial position by the addition of an activation energy $\Delta$. Because of the looseness of the structure, the rate of diffusion of the interstitial atom would be high and it might be a good approximation to treat it like a freely moving gas atom with an effective mass determined by the necessity to push aside the other atoms during its progress.

## 3.2. Landau's theory

### 3.2.1. Phonons and rotons

There is a school of thought which believes that a liquid is best considered as a type of solid in which the molecules are able to stray freely from the lattice sites. This approach has been presented forcefully by Frenkel (1946) in his book *The Kinetic Theory of Liquids*. Landau (1941, 1947) approaches the liquid helium problem from this point of view, at least as far as the elementary excitations are concerned. In the theory of Bose-Einstein condensation emphasis was placed on the wave-functions of the individual atoms. In the theory of solids it is not possible to give the wave function for an individual atom, but one discusses instead the normal modes of motion of the solid as a whole. The modes are analogous to standing sound waves (Debye waves); in fact the modes of wavelength large compared with the interatomic separation are identical with the ultrasonic waves of the same wavelength which could be excited experimentally in the solid. A mode of frequency $\nu$ has an energy

$$\epsilon_n = (n + \tfrac{1}{2})\, h\nu, \tag{3.17}$$

where $n$ is an integer. Alternatively, one may say that $n$ *phonons*, each of energy $h\nu$, have been excited in the liquid, for the same reasons that one may say, in the case of black body radiation in a constant temperature enclosure, that an electromagnetic mode with energy $(n+\frac{1}{2})\,h\nu$ may be replaced by $n$ *photons*. The phonon is a 'particle of sound'. It may also be described as an 'elementary thermal excitation'. Its momentum is

$$p = h\nu/c, \tag{3.18}$$

where $c$ is the velocity of the sound, and therefore its energy $\epsilon$ is related to its momentum $p$ by

$$\epsilon = cp. \tag{3.19}$$

A liquid can support longitudinal, but not transverse, sound waves, and longitudinal phonons of long wavelength therefore certainly exist in liquid helium. The important question is—what other modes of motion may also exist? Hypersonic transverse phonons having wavelengths comparable with the interatomic separation may possibly exist, but this is a controversial matter. One interesting possibility is that in a liquid, as opposed to a solid, stirring of the atoms is possible and may give rotational modes. Landau (1947) showed that the properties of liquid helium II can be most readily explained if the modes have the form shown in fig. 30, which relates the energy $\epsilon$ of an elementary excitation to its momentum $p$. In the region of small $p$ the only excitations are the longitudinal phonons and in this region the curve is a straight

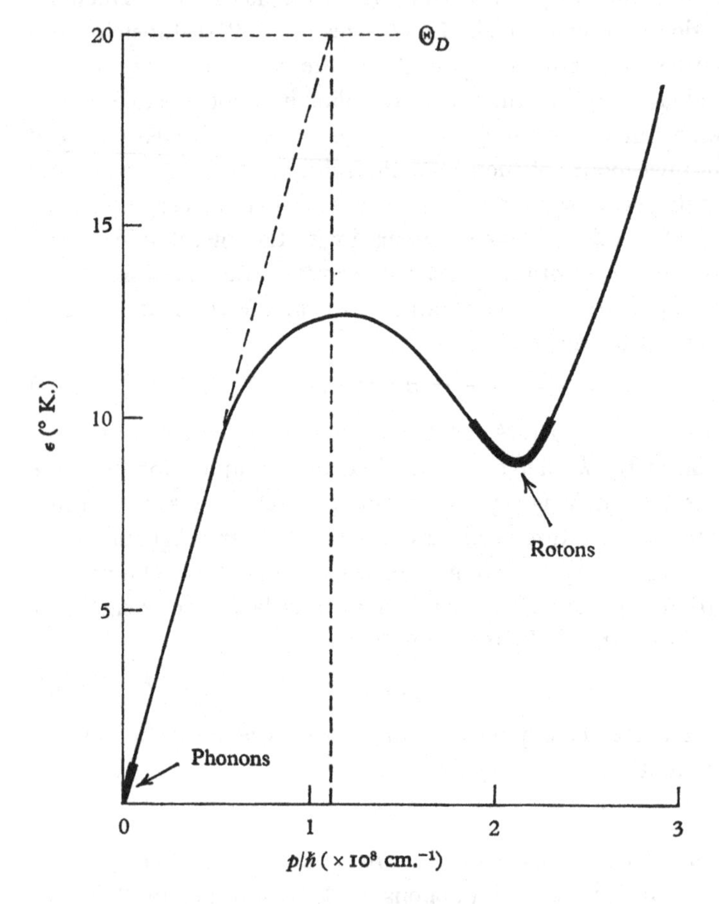


Fig. 30. The elementary excitations of the Landau theory. The dashed curve represents the phonon spectrum for a simple Debye theory. The energy $\epsilon$ has been expressed in °K. by dividing by Boltzmann's constant $k$.

line, in accordance with equation (3.19). At higher values of $p$ there is a minimum in the curve, near which

$$\epsilon = \Delta + \frac{(p - p_0)^2}{2\mu}. \tag{3.20}$$

The excitations in this region were named rotons by Landau, because he suspected that they corresponded to rotational modes of motion. The thickened portions of the curve indicate which excitations are strongly excited at 1° K. The rotons are more plentiful than the high energy phonons of the same energy because the density of states is proportional to $dp/d\epsilon$, which is large near the minimum. The vertical broken line shows the value of $p$ corresponding to the cut-off wavelength in the simple form of the Debye theory applicable to a continuum with no dispersion of sound but a finite number of normal modes. The wavelengths of the roton excitations are smaller than this Debye cut-off wavelength and of the order of $3 \times 10^{-8}$ cm. With these excitations it is therefore possible to construct a wave packet of small dimensions, which is consistent with those theories that visualize the roton as a highly localized entity comparable in size with the interatomic separation.

In his original theory Landau (1941) assumed

$$\epsilon = \Delta + \frac{p^2}{2\mu}. \tag{3.21}$$

In a later paper (1947) he modified this to the form of equation (3.20) without theoretical justification in order to obtain better numerical agreement with the experimental results. The quantities $\Delta$, $p_0$ and $\mu$ have to be determined empirically to give the best fit to the measured values of $S$ and $\rho_n/\rho$. The values quoted below are those given by Khalatnikov (1952*b*, *c*), but there is some uncertainty in the experimental data from which they are derived, particularly in the entropy:

$$\frac{\Delta}{k} = 8{\cdot}9 \pm 0{\cdot}2^\circ\ \text{K.}, \tag{3.22}$$

$$\left.\begin{aligned} p_0 &= (2{\cdot}1 \pm 0{\cdot}05) \times 10^{-19}\ \text{deg. cm. sec.}^{-1}, \\ \frac{p_0}{\hbar} &= 1{\cdot}99\ \text{Å.}^{-1}, \end{aligned}\right\} \tag{3.23}$$

$$\begin{aligned} \mu &= (1{\cdot}72 \pm 0{\cdot}68) \times 10^{-24}\ \text{g.} \\ &= 0{\cdot}26 m_{\text{He}}. \end{aligned} \tag{3.24}$$

Direct evidence for the existence of rotons has been obtained from a neutron scattering experiment suggested by Cohen and

Feynman (1957) and performed by Palevsky, Otnes, Larsson, Pauli and Stedman (1957). When neutrons with wavelengths in the vicinity of 4 Å. are incident on liquid helium II, the scattering can be shown to be mainly due to an inelastic process in which each scattered neutron creates a single roton. At a fixed incident neutron wavelength and fixed angle of scattering, the principles of conservation of energy and momentum imply that the created roton has definite values of $\epsilon$ and $p$ and that all the scattered neutrons experience the same shift in wavelength. This shift in wavelength was measured and from it both $\epsilon$ and $p$ were deduced. By varying the incident wavelength, $\epsilon$ was determined as a function of $p$ and the excitation curve was traced out in the vicinity of its minimum. The results provide striking confirmation of the excitation curve suggested by Landau, with values of $\Delta$, $p_0$ and $\mu$ in approximate agreement with (3.22), (3.23) and (3.24) (Palevsky, private communication).

If the roton travels undisturbed for an average time $\tau$ before experiencing some sort of collision, its energy must be uncertain to about $\hbar/\tau$, and there is a corresponding broadening of the scattered neutron wavelengths. At 1·2° K. the broadening was too small to be measured, proving that the roton mean free path is not less than $2 \times 10^{-6}$ cm. at this temperature, which is consistent with the value quoted in §4.5.2. Nearer the $\lambda$-point the time $\tau$ between roton-roton collisions becomes much smaller and observable broadening should be observed.

### 3.2.2. The thermodynamic functions

Once the nature of the elementary excitations has been determined the derivation of the thermodynamic functions is a straightforward problem in statistical mechanics. The free energy is

$$F = kT \int_0^\infty \frac{4\pi p^2 dp\, V}{h^3} \ln(1 - e^{-\epsilon/kT}).$$

The exponential factor favours excitations of low energy and the dominant contributions to the integral come from the phonon and roton regions indicated by the thickening of the curve in fig. 30. Since the rotons have higher energy than the phonons, the exponential factor suppresses them at temperatures well below 1° K., but at temperatures above 1° K. they become more important than the phonons because they occupy a larger region of phase space. The entropy and specific heat per gram due to the phonons are

$$S_{ph} = \frac{16\pi^5 k^4 T^3}{45 h^3 c^3 \rho}, \tag{3.25}$$

$$C_{ph} = \frac{16\pi^5 k^4 T^3}{15 h^3 c^3 \rho}. \tag{3.26}$$

The corresponding expressions for the rotons are

$$S_r = \frac{2\mu^{\frac{1}{2}} k^{\frac{1}{2}} p_0^2 \Delta}{(2\pi)^{\frac{3}{2}} \rho T^{\frac{1}{2}} \hbar^3} \left(1 + \frac{3kT}{2\Delta}\right) e^{-\Delta/kT}, \tag{3.27}$$

$$C_r = \frac{2\mu^{\frac{1}{2}} p_0^2 \Delta^2}{(2\pi)^{\frac{3}{2}} \rho k^{\frac{1}{2}} T^{\frac{3}{2}} \hbar^3} \left[1 + \frac{kT}{\Delta} + \frac{3}{4}\left(\frac{kT}{\Delta}\right)^2\right] e^{-\Delta/kT}. \tag{3.28}$$

Because of the existence of the energy gap $\Delta$ and the resulting factor $e^{-\Delta/kT}$, very few rotons are excited at the lowest temperatures. In § 2.5.1 we saw that the specific heat measurements of Kramers, Wasscher and Gorter (1952) are in good agreement with equation (3.26) below 0·6° K., proving that the only important elementary excitations of small energy are the longitudinal phonons. Between 0·6 and 1° K. the rotons begin to take over and at 1° K. $S_r$ is approximately equal to $S_{ph}$. Above 1° K. the rotons dominate the situation, $S_r$ being about ten times greater than $S_{ph}$ at 1·6° K. It is an important feature of the behaviour of liquid helium that the phonons are the only excitations which need be considered near 0° K., whereas above 1° K., there is another type of excitation (the roton?) which seems to produce the $\lambda$-transition.

Above 1·6° K. the entropy can no longer be represented by adding the values of $S_{ph}$ and $S_r$ given by equations (3.25) and (3.27). If we realize that the rotons excited at a temperature $T$ lie in the energy range from $\Delta$ to about $\Delta + kT$, it is obvious that, when $T \sim 2°$ K., we ought to consider more carefully the exact shape of the curve given in fig. 30. The definition of a roton contained in equation (3.20) implies that the region near the *minimum* can be approximated by a parabola and this is permissible only at sufficiently low temperatures. Moreover, it is possible that, when $T \sim 2°$ K., we ought not to ignore the region near the *maximum*, where $dp/d\epsilon$ is large and there is consequently a high density of states. However, remedying these defects probably would not give a $\lambda$-type transition, and Landau has suggested that the major difficulty arises from interactions of the rotons with one another. Near the $\lambda$-point there would be so many rotons that they would inevitably overlap, and the suggestion is that their mutual interactions would lead to a modification of the $\epsilon$-$p$ curve and $\Delta$, $p_0$ and $\mu$ would become temperature dependent. This is presumably analogous to the effect of anharmonicities on strongly excited normal modes of a solid.

This inability to give a satisfactory description of the $\lambda$-transition is a serious shortcoming of the Landau theory. The introduction of an energy gap into the theory of Bose-Einstein condensation (§3.1.2) results in an apparent formal analogy with the Landau theory, but the important difference there is that one is discussing particle-like states and in addition to the excited states above the energy gap there is also a *ground state*. It is the condensation of the 'particles' into this ground state which gives rise to the $\lambda$-transition, and no such process can occur in the Landau theory.

Nevertheless, Landau's theory has been very successful in explaining many properties of liquid helium below 1·6° K. One such property is the coefficient of expansion (Atkins and Edwards, 1955) which illustrates rather neatly the transition from a roton region above 1° K. to a phonon region below 1° K. The experimental details were given in §2.5.3 and we might anticipate that the positive coefficient of expansion below 1·15° K. is mainly due to the phonons, whereas the negative coefficient of expansion above 1·15° K. is mainly due to the rotons.

From simple thermodynamics the coefficient of expansion at constant pressure can be expressed in the form

$$\begin{aligned}\alpha_p &= \frac{1}{V}\left(\frac{\partial V}{\partial T}\right)_p \\ &= -\frac{1}{V}\left(\frac{\partial S}{\partial p}\right)_T \\ &= -\frac{1}{V}\left(\frac{\partial S_{ph}}{\partial p}\right)_T - \frac{1}{V}\left(\frac{\partial S_r}{\partial p}\right)_T.\end{aligned} \tag{3.29}$$

Differentiating equation (3.25) with respect to the pressure, one obtains for the phonon contribution

$$\alpha_{ph} = \frac{16\pi^5 k^4 T^3}{15h^3c^3}\left(\frac{1}{c}\frac{\partial c}{\partial p}+\tfrac{1}{3}K_T\right) \tag{3.30}$$

$$= +(1.08 \pm 0.04)\times 10^{-3}T^3\,\text{deg.}^{-1}. \tag{3.31}$$

The measurements of $c$ and $(1/c)(\partial c/\partial p)$ are discussed in § 5.2.1. $\alpha_{ph}$ is positive and is represented by the dashed curve in fig. 31.

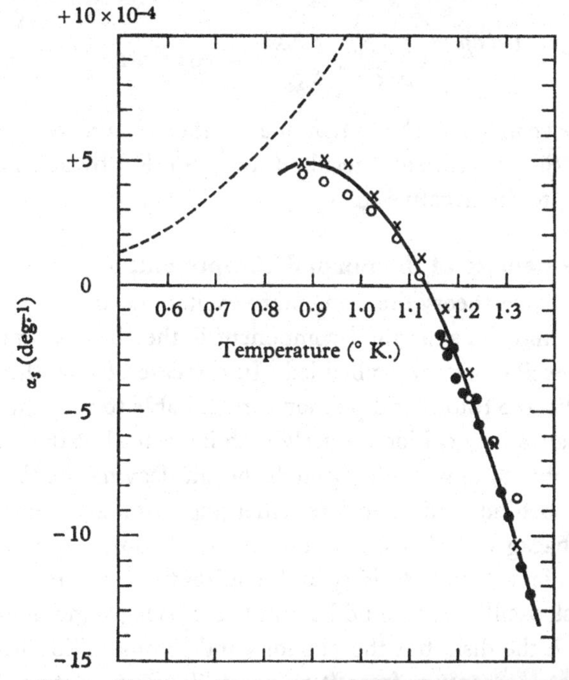


Fig. 31. The coefficient of expansion near 1° K. ○ × ●, experimental results; ----, phonon contribution; ——, Landau's theory; (after Atkins and Edwards, 1955).

Differentiating equation (3.27), one obtains for the roton contribution

$$\alpha_r = -\frac{2k^{\frac{1}{2}}\mu^{\frac{1}{2}}p_0^2\Delta}{(2\pi)^{\frac{3}{2}}\rho T^{\frac{1}{2}}\hbar^3}\left(1+\frac{3kT}{2\Delta}\right)e^{-\Delta/kT}$$

$$\times\left[\frac{1}{2}\frac{\rho}{\mu}\frac{\partial\mu}{\partial\rho}+2\frac{\rho}{p_0}\frac{\partial p_0}{\partial\rho}+\frac{\rho}{\Delta}\frac{\partial\Delta}{\partial\rho}-1-\frac{\rho}{\Delta}\frac{\partial\Delta}{\partial\rho}\frac{\Delta}{kT}\left\{1+\frac{\frac{3}{2}\left(\frac{kT}{\Delta}\right)^2}{1+\frac{3kT}{2\Delta}}\right\}\right]. \tag{3.32}$$

Using known values of $\Delta$, $p_0$ and $\mu$, the values of $\frac{\rho}{\Delta}\frac{\partial\Delta}{\partial\rho}, \frac{\rho}{p_0}\frac{\partial p_0}{\partial\rho}, \frac{\rho}{\mu}\frac{\partial\mu}{\partial\rho}$ must be chosen to fit the experimental results. The important factor is

$$\frac{\rho}{\Delta}\frac{\partial\Delta}{\partial\rho} = -0{\cdot}57 \pm 0{\cdot}06. \tag{3.33}$$

It is the decrease of the energy gap with increasing density which makes $\alpha_r$ negative. In addition one must have

$$\frac{1}{2}\frac{\rho}{\mu}\frac{\partial\mu}{\partial\rho} + 2\frac{\rho}{p_0}\frac{\partial p_0}{\partial\rho} + \frac{\rho}{\Delta}\frac{\partial\Delta}{\partial\rho} = -0{\cdot}95 \pm 0{\cdot}2. \tag{3.34}$$

The full curve in fig. 31 shows that one can then obtain good agreement with the experimental results (up to 1·6° K.) by adding the positive $\alpha_{ph}$ to the negative $\alpha_r$.

### 3.2.3. The density of the normal component, $\rho_n$

In the Landau theory the elementary excitations are the normal component and the superfluid component is the 'background' in which the excitations are embedded. In the case of flow through a narrow slit, the rotons and phonons are not able to pass through the slit because they collide diffusely with its walls, but the 'background' is able to flow freely through the slit, leaving the thermal excitations behind and therefore emerging with zero entropy. That the 'background' is able to do this is, of course, the fundamental problem of superfluidity and requires further explanation. When a disk oscillates in liquid helium II the background does not interact with the disk, but the phonons and rotons collide with it and abstract momentum from it to damp down its motion. The elementary excitations are therefore responsible for the normal viscosity, $\eta_n$. In the Andronikashvili experiment (fig. 2) the background does not move with the pile of disks, but the rotons and phonons are dragged round with the disks and have an inertial effect which gives rise to the density of the normal component, $\rho_n$.

To calculate the effective density of the phonons, let the temperature be so low that the rotons can be ignored and imagine the liquid helium to be contained within an infinitely long tube. The axis of the tube is taken as the $z$-direction and the walls of the tube are then given a velocity $v_z$ in this direction. The superfluid component remains at rest, but the phonons acquire the same drift velocity $v_z$ as the walls. An observer in a frame of reference fixed with respect to the superfluid component who wishes to apply statistical mechanics to the phonons must recognize the existence of two constants of the motion, the energy $E$ and the linear momentum $\mathbf{P}$ per cm.$^3$. If we consider a distribution in which there are $n_s$ phonons in states which have energy $\epsilon_s$ and momentum $p_{s,z}$ in the $z$-direction, the restrictive equations are

$$\sum_s n_s \epsilon_s = E, \tag{3·35}$$

$$\sum_s n_s p_{s,z} = P_z. \tag{3·36}$$

The number of phonons per cm.$^3$ is not necessarily constant. If there are $g_s$ modes with energy $\epsilon_s$ and momentum $p_{s,z}$, the probability of this distribution is

$$W = \prod_s \frac{(n_s + g_s - 1)!}{(g_s - 1)!\, n_s!}. \tag{3·37}$$

The actual distribution is found by maximizing $W$ subject to the restrictive equations, which may be done by the method of Lagrangian multipliers and yields the result

$$n_s = \frac{g_s}{\exp\,[\beta\epsilon_s - \gamma_z p_z] - 1}. \tag{3·38}$$

The multiplier $\beta$ is well known to be equal to $1/kT$. To obtain $\gamma_z$, the average velocity of the phonon gas is

$$v_z = \frac{\displaystyle\iiint \frac{c}{h^3} \cos\theta \frac{dp_x\, dp_y\, dp_z}{\exp\,[(\epsilon - \omega_z p_z)/kT] - 1}}{\displaystyle\iiint \frac{1}{h^3} \frac{dp_x\, dp_y\, dp_z}{\exp\,[(\epsilon - \omega_z p_z)/kT] - 1}}, \tag{3·39}$$

where $\omega_z = kT\gamma_z$ and $\theta$ is the angle between the $z$-axis and the direction of motion of the phonon ($\cos\theta = p_z/p$). Performing the integration one finds

$$\omega_z = v_z. \tag{3·40}$$

The distribution function is therefore

$$n_s = \frac{g_s}{\exp\,[(\epsilon_s - p_{s,z} v_z)/kT] - 1}. \tag{3·41}$$

The energy density is

$$E = \iiint \frac{\epsilon}{h^3} \frac{dp_x dp_y dp_z}{\exp[(\epsilon - p_z v_z)/kT] - 1}$$

and this can be shown to reduce to

$$E = \frac{1 + \frac{1}{3}(v_z/c)^2}{[1 - (v_z/c)^2]^3} \iiint \frac{\epsilon}{h^3} \frac{dp_x dp_y dp_z}{e^{\epsilon/kT} - 1}$$

$$= \frac{1 + \frac{1}{3}(v_z/c)^2}{[1 - (v_z/c)^2]^3} E_0. \qquad (3.42)$$

$E_0$ is the energy the phonon gas would have at rest ($v_z = 0$). The momentum density is

$$P_z = \iiint \frac{p_z}{h^3} \frac{dp_x dp_y dp_z}{\exp[(\epsilon - p_z v_z)/kT] - 1}$$

$$= \frac{4}{3} \frac{E_0}{c^2} \frac{v_z}{[1 - (v_z/c)^2]^3}$$

$$= \frac{4}{3} \frac{E}{c^2} \frac{v_z}{1 + \frac{1}{3}(v_z/c)^2}. \qquad (3.43)$$

But

$$P_z = \rho_{ph} v_z, \qquad (3.44)$$

where $\rho_{ph}$ is the effective mass density of the phonons. Hence

$$\rho_{ph} = \frac{4}{3} \frac{E}{c^2} \frac{1}{1 + \frac{1}{3}(v_z/c)^2}, \qquad (3.45)$$

$$\simeq \frac{4}{3} \frac{E}{c^2} \qquad (3.46)$$

as $v_z$ is usually small compared with $c$. Not all authors agree about the second order corrections in $(v_z/c)^2$; see, for example, Khalatnikov (1952*a*). More explicitly, we may substitute for $E$ and write

$$\rho_{ph} = \frac{16\pi^5 k^4 T^4}{45 h^3 c^5}. \qquad (3.47)$$

A similar calculation gives the effective density of the rotons as

$$\rho_r = \frac{2\mu^{\frac{1}{2}} p_0^4}{3(2\pi)^{\frac{3}{2}} (kT)^{\frac{1}{2}} \hbar^3} e^{-\Delta/kT}. \qquad (3.48)$$

The density of the normal component is then

$$\rho_n = \rho_{ph} + \rho_r. \qquad (3.49)$$

$\rho_r$ is unimportant up to 0·6° K., but then takes over rapidly and at 1° K. is about 100 times larger than $\rho_{ph}$. It is interesting to notice that the number of rotons per cm.³ is

$$N_r = \frac{2\mu^{\frac{1}{2}}(kT)^{\frac{1}{2}}p_0^2}{(2\pi)^{\frac{3}{2}}\hbar^3}e^{-\Delta/kT} \qquad (3.50)$$

so that the effective mass per roton is not $\mu$ but

$$\frac{\rho_r}{N_r} = \frac{p_0^2}{3kT}. \qquad (3.51)$$

The same type of argument may be applied to rotational motion. If the liquid helium is contained within a cylindrical vessel rotating with an angular velocity $\mathbf{\Omega}$, then the superfluid component remains at rest, but the phonons and rotons rotate with the vessel. The distribution function for the phonons can be shown to be

$$n_s = \frac{g_s}{\exp[(\epsilon_s - \mathbf{m}_s.\mathbf{\Omega})/kT] - 1}, \qquad (3.52)$$

where $\mathbf{m}_s$ is the angular momentum of a single phonon. The total angular momentum of the liquid can then be calculated and shown to be determined by an effective liquid density $\rho_n = \rho_{ph} + \rho_r$, where $\rho_{ph}$ and $\rho_r$ have the same values as before.

The Landau theory is not able to deal adequately with the $\lambda$-transition, but an approximate value for the $\lambda$-temperature may be obtained by noticing that, at the $\lambda$-point,

$$\rho_n(T_\lambda) = \rho, \qquad (3.53)$$

or, since $\rho_{ph}$ is negligible above 1° K.,

$$\frac{2\mu^{\frac{1}{2}}p_0^4}{3(2\pi)^{\frac{3}{2}}(kT_\lambda)^{\frac{1}{2}}\hbar^3}e^{-\Delta/kT_\lambda} = \rho, \qquad (3.54)$$

whence $T_\lambda = 2{\cdot}55°$ K.; this is of course only approximate, since the formulae used are not valid above 1·6° K., because of roton interactions. Atkins (1955*a*) has extended this argument to obtain an approximate value of the slope of the $\lambda$-curve by differentiating equation (3.54) with respect to $\rho$:

$$\frac{\rho}{T_\lambda}\frac{\partial T_\lambda}{\partial \rho}\left(\frac{\Delta}{kT_\lambda} - \frac{1}{2}\right) = \frac{\Delta}{kT_\lambda}\frac{\rho}{\Delta}\frac{\partial \Delta}{\partial \rho} - \frac{1}{2}\frac{\rho}{\mu}\frac{\partial \mu}{\partial \rho} - \frac{4\rho}{p_0}\frac{\partial p_0}{\partial \rho} + 1. \qquad (3.55)$$

From the coefficient of expansion and the variation of the velocity of second sound with pressure, we can obtain

$$\frac{\rho}{\Delta}\frac{\partial \Delta}{\partial \rho} = 0{\cdot}57 \pm 0{\cdot}06, \tag{3.56}$$

$$\frac{\rho}{p_0}\frac{\partial p_0}{\partial \rho} = +0{\cdot}26, \tag{3.57}$$

$$\frac{\rho}{\mu}\frac{\partial \mu}{\partial \rho} = -1{\cdot}8. \tag{3.58}$$

The errors are large, but difficult to estimate. Equation (3.55) then yields $(\rho/T_\lambda)(\partial T_\lambda/\partial \rho) = -0{\cdot}42$, as compared with the experimental value of $-0{\cdot}37$. Both the negative slope of the $\lambda$-curve and the negative coefficient of expansion seem to be caused primarily by the negative value of $(\rho/\Delta)(\partial \Delta/\partial \rho)$.

## 3.3. The nature of the elementary excitations

### 3.3.1. Quantum hydrodynamics

Rotons were originally derived by Landau (1941) from a procedure which he called the quantization of hydrodynamics. The fundamental idea is that the thermal motion of a liquid may be represented by specifying the density $\rho$ and the mass current $\mathbf{j}$ at all points in the liquid. Introducing the three-dimensional delta-function $\delta$, the density at the point $\mathbf{R}$ is

$$\rho(\mathbf{R}) = \sum_\alpha m_\alpha \delta(\mathbf{r}_\alpha - \mathbf{R}), \tag{3.59}$$

where $m_\alpha$ and $\mathbf{r}_\alpha$ are the mass and radius vector of the $\alpha$th particle. Similarly the mass current $\mathbf{j}(\mathbf{R})$ is

$$\mathbf{j}(\mathbf{R}) = \tfrac{1}{2}\sum_\alpha [\mathbf{p}_\alpha \delta(\mathbf{r}_\alpha - \mathbf{R}) + \delta(\mathbf{r}_\alpha - \mathbf{R})\,\mathbf{p}_\alpha], \tag{3.60}$$

where $\mathbf{p}_\alpha = m_\alpha \mathbf{v}_\alpha$ is the momentum of the $\alpha$th particle. Note that $\rho(\mathbf{R})$ is the local density, which is not necessarily the same as the average density if, for example, there is a Debye wave excited in the liquid. Similarly $\mathbf{j}(\mathbf{R})$ may have a local value even though there is no macroscopic flow, as, for example, if a set of small microscopic vortices is excited in the liquid. To apply the formalism of quantum theory, $\rho$, $\mathbf{j}$ and $\mathbf{v}$ must be considered as operators. One obvious step is to write $\mathbf{p}_\alpha = (h/2\pi i)\nabla_\alpha$. The Hamiltonian of the total assembly is

$$\mathscr{H} = \int [\tfrac{1}{2}\mathbf{v}.\rho\mathbf{v} + \rho U(\rho)]\, dV, \qquad (3.61)$$

$\frac{1}{2}\mathbf{v}.\rho\mathbf{v}$ representing the kinetic energy and $U(\rho)$ the internal energy per unit mass. If $f$ is any operator

$$\frac{ih}{2\pi}\frac{\partial f}{\partial t} = f\mathscr{H} - \mathscr{H}f. \qquad (3.62)$$

Various commutation relations may be derived between these operators. One of the most important of these is that the components of curl **v** do not commute with one another

$$(\text{curl}\, v)_x(\text{curl}\, v)_y - (\text{curl}\, v)_y(\text{curl}\, v)_x \neq 0. \qquad (3.63)$$

Landau compares this with the well-known result for the components of angular momentum

$$M_x M_y - M_y M_x = \frac{ih}{2\pi} M_z \qquad (3.64)$$

and suggests, by analogy, that vortex motion, like angular momentum, is quantized. Also, from equation (3.62)

$$\frac{ih}{2\pi}\frac{\partial}{\partial t}(\text{curl}\, \mathbf{v}) = (\text{curl}\, \mathbf{v}).\mathscr{H} - \mathscr{H}.(\text{curl}\, \mathbf{v}). \qquad (3.65)$$

Therefore, if curl $\mathbf{v} = 0$, it remains zero—if the motion is initially irrotational throughout the whole liquid, it will always be irrotational. This is the analogue of Lagrange's theorem in classical hydrodynamics. It follows that the liquid possesses stationary states for which curl $\mathbf{v} = 0$ everywhere and also, because curl **v** is quantized, that it is not possible to pass continuously from these irrotational states to rotational states by introducing arbitrarily small values of curl **v**. There must be an energy gap $\Delta$ between the ground state of the irrotational motions and the ground state of the rotational motions. To apply these ideas to liquid helium, it is necessary to assume that the ground state of the rotational motions lies higher than the ground state of the irrotational motions, but the ideas just presented do not tell us either the sign or magnitude of $\Delta$. We do, however, form the picture that the excitation of phonons gives irrotational states while the excitation of rotons give rotational states, and the roton may therefore be imagined as a small vortex.

The rigorous formulation and development of the above ideas leads to difficult mathematical complications. Not even the properties of phonons are obvious, since the hydrodynamical equations are non-linear, whereas the electromagnetic equations are linear, and so the analogue between phonons and photons is more heuristic than logical. In fact Lord Rayleigh (1905) has shown that the pressure of a sound wave cannot be related to the energy flow in as simple a way as for light. This problem has been solved by Kronig and Thellung (1952) and independently by London and Rosen (see F. London (1954), *Superfluids*, vol. II, p. 111). Any arbitrary state of the sound field may be Fourier analysed into a linear superposition of plane waves

$$\rho=\rho_0+\sum_k [A_k e^{i(\omega_k t-\mathbf{k}.\mathbf{R})}+A_k^* e^{-i(\omega_k t-\mathbf{k}.\mathbf{R})}], \tag{3.66}$$

$$\mathbf{v}=\mathbf{v}_0+\sum_k [\mathbf{B}_k e^{i(\omega_k t-\mathbf{k}.\mathbf{R})}+\mathbf{B}_k^* e^{-i(\omega_k t-\mathbf{k}.\mathbf{R})}]. \tag{3.67}$$

If the hydrodynamical equations were linear the $A_k$'s and $\mathbf{B}_k$'s would be constants and the plane waves would be independent of one another. The effect of the non-linearity is that when two sound waves pass over one another they interact and scatter one another, the scattered waves then having different values of $\mathbf{k}$. This can be taken into account by making the $A_k$'s and $\mathbf{B}_k$'s functions of time and, fortunately, the non-linearities are small enough so that they can be considered as *slowly varying* functions of time. The analysis into plane waves is therefore an adequate approximation and quantization leads to a verification of the simple relationship $\epsilon=cp$. The effect of the non-linearities in varying the $A_k$'s and $\mathbf{B}_k$'s is equivalent to changing the number of phonons associated with each frequency $\omega_k$ and is therefore connected with those processes in which phonons collide with one another and are scattered, created or annihilated. Such processes are important in the theory of the viscosity of the phonon gas and the theory of the absorption of first and second sound (§§4.5.2 and 5.2.3).

Thellung (1953) and Ziman (1953*a*) have extended the analysis to rotational motions. They show that Landau's commutation relations can indeed be justified. Also, the Hamiltonian splits up into several parts, one part giving rise to phonons, a second part giving rise to rotons and the remaining parts being related to the various types of interaction between these elementary excitations.

A fundamental defect of the whole approach is that it is based essentially on a liquid continuum and ignores the atomic nature of the actual liquid. Ziman has pointed out that the energy gap $\Delta$ for the formation of rotons would be infinite in a continuous quantum liquid. Similar difficulties arise in the Debye theory of solids when the solid is treated as a continuum, and these difficulties are partly removed by the introduction of a cut-off wavelength comparable with the interatomic separation. Ziman introduces a cut-off wave vector $k_0$ in a similar way and obtains a finite $\Delta$. Allcock and Kuper (1955) have carefully assessed the situation and have shown that, when the Hamiltonian is correctly diagonalized, a dominant role is played by the interaction between rotons and phonons. This interaction is a consequence of the effect on the roton of the fluctuations in density accompanying the phonon. For an incompressible fluid $\Delta$ would be zero. When a small compressibility is included, $\Delta$ varies as $k_0^9/c$. The extreme sensitivity to $k_0$ emphasizes that the liquid continuum approach is likely to be a poor approximation and an atomistic theory is called for.

### 3.3.2. Feynman's theories

The only completely satisfactory approach to the problem is obviously to solve the Schrödinger equation for all the $N$ atoms in the liquid, taking their interactions into account. This equation is

$$H\psi = -\frac{\hbar^2}{2m}\sum_i \nabla_i^2\psi + \sum_{ij} V(\mathbf{r}_{ij})\,\psi = E\psi, \tag{3.68}$$

where $V(\mathbf{r}_{ij})$ is the mutual potential energy of the $i$th and $j$th atoms when they are separated by a distance $\mathbf{r}_{ij}$. An exact solution would be prohibitively difficult, but Feynman has shown how physical intuition can be used to make plausible guesses at the nature of the solutions. He has published three papers in which his views have been successively developed and, to some extent, modified. He has given an excellent summary of his final position in an article in *Progress in Low Temperature Physics*, vol. 1, edited by C. J. Gorter (1955). Here the three papers will be discussed in chronological order, since the evolution of the ideas is instructive.

In his first paper (Feynman, 1953*a*), he attempts to show that the Bose-Einstein condensation occurs even when the interatomic forces are taken into account. Using the space-time approach to quantum mechanics (Feynman, 1948), he transforms the partition function $Z$ into an integral over trajectories

$$Z=\frac{1}{N!}\sum_{P}\int d^{N}\mathbf{z}_i\int_{trP}\exp\left\{-\int_0^{\beta}\left[\frac{m}{2\hbar^2}\sum_i\left(\frac{dx_i}{du}\right)^2+\sum_{ij}V(\mathbf{x}_i-\mathbf{x}_j)\right]du\right\}D^{N}\mathbf{x}_i(u), \quad (3.69)$$

where $u=it/\hbar, \quad \beta=1/kT.$

The integral $\int_{trP}$ is to be taken over all the trajectories $\mathbf{x}_i(u)$ of all the particles starting from an initial co-ordinate $\mathbf{z}_i=\mathbf{x}_i(0)$ and ending on a final co-ordinate $\mathbf{x}_i(\beta)=P\mathbf{z}_i$ which was the initial co-ordinate of any one of the particles. This feature, corresponding to a permutation of the positions of the particles, arises from the fact that the helium atoms are indistinguishable and have symmetrical wave functions. The sum $\sum_P$ is to be taken over all possible permutations and the integral $\int d^N\mathbf{z}_i$ over all initial configurations $\mathbf{z}_i$.

Equation (3.69) is exact but intractable. It is therefore necessary to introduce some physical argument which will select the important trajectories and then to concentrate on these. If two helium atoms approach too closely (within about 2·6 Å. of one another) strong repulsive forces come into play, $V(\mathbf{x}_i-\mathbf{x}_j)$ becomes large and the exponential term becomes small. It therefore seems a plausible approximation to treat the atoms like hard spheres of diameter 2·6 Å., and to ignore all initial configurations and trajectories for which the spheres overlap. The idea is that, since we are dealing with a loose liquid structure, the motion of an atom is not opposed by potential barriers because the other atoms can move out of the way to make room for it. This readjustment of the positions of the other atoms merely increases the effective mass of the moving helium atom, in rather the same way that a hard sphere of mass $m$ moving through an ideal classical liquid acquires an effective mass $>m$ because of the kinetic energy imparted to the liquid flowing round it. This brings us to something very like a gas of hard spheres having effective masses $m'$, and the mathematical situation turns out to be very similar to that which gives rise to the Bose-Einstein condensation of an ideal gas. A third order transition occurs and the rise in specific heat just above the $\lambda$-point is qualitatively explained. However, the theory predicts a specific heat varying as $T^{\frac{3}{2}}$ near 0° K., badly at variance with the observed $T^3$ variation.

The mathematical problems connected with the evaluation of Feynman's partition function have been discussed by Kikuchi (1954), ter Haar (1954) and Chester (1954, 1955). There seems to be general agreement that, to the next order of approximation, the $\lambda$-transition is a second order transition with a discontinuity in the specific heat.

The difficulty of the low temperature specific heat led Feynman in his second paper to a consideration of the low energy excitations which exist in the liquid near 0° K. (Feynman, 1953*b*). Any solution $\psi(\mathbf{r}_1, \mathbf{r}_2, \ldots, \mathbf{r}_N)$ of the Schrödinger equation is a function of the positions $\mathbf{r}_1, \ldots, \mathbf{r}_N$ of the $N$ atoms, and for any fixed configuration it is a real number which is related to the probability that the configuration occurs. Consider first the ground state wave function $\phi$. By analogy with known solutions for simpler systems, we expect $\phi$ to have no nodes and to be everywhere real and positive. Interchange of two helium atoms leaves $\phi$ unaltered, so it must be symmetrical in the co-ordinates $\mathbf{r}_i$. Consider how $\phi$ varies when one atom is moved inside its 'cell' and all the other atoms are fixed. If the atom is forced to one side of the cell so that it touches a neighbouring atom, the strong repulsive forces operate to make $\phi$ small. As the atom is moved inwards $\phi$ increases, reaches a maximum near the centre of the cell, and then falls to a small value again on the opposite side of the cell. Thus $\phi$ is largest when the atoms are well spaced and this is the most probable type of configuration. The kinetic energy of the atom is related to the curvature of the curve just described and this is the origin of the zero-point energy. Over a small region the density may fluctuate from the mean, the probability of this occurring falling off exponentially with the square of the density change. This corresponds to the zero-point energy of the Debye modes, since the density fluctuations may be Fourier analysed into compressional waves.

Turning now to excited states with energy very close to the ground state, we must obviously include amongst these compressional waves of long wavelength, i.e. phonons. It can be shown that the wave function for such a compressional wave is

$$\psi_{\text{phonon}} = \phi \sum_i e^{i\mathbf{k}.\mathbf{r}_i} \tag{3.70}$$

summed over all the atoms. The physical significance of this form becomes obvious if we notice that, when the density is uniform, $e^{i\mathbf{k}.\mathbf{r}_i}$ is as often positive as negative and $\psi = 0$, but, when the density varies with a wavelength $\lambda = 2\pi/k$, there are more atoms where $e^{i\mathbf{k}.\mathbf{r}_i}$ is positive than where it is negative and $\psi$ has its maximum value. The presence of $\phi$ as a multiplying factor ensures that the excited state, like the ground state, has a well-spaced arrangement of atoms and that, in fact, the excited state is similar to the ground state with a small perturbation (the phonon). A long wavelength phonon involves the motion of many atoms, each of which moves only a short distance. The neglect of such motions led to the absence of the $T^3$ specific heat term in Feynman's first paper. It is still necessary to eliminate the $T^{\frac{3}{2}}$ specific heat term by showing that the phonons are the only low energy excitations.

The wave function $\psi$ for an excited state must be orthogonal to the ground state wave function $\phi$. As $\phi$ is always positive, $\psi$ must be positive for half the configurations and negative for the other half. $\psi$ must also be orthogonal to all the phonon wave functions and must therefore alternate from positive to negative for configuration changes which do not alter the density and therefore correspond simply to stirring the atoms. Consider a configuration for which $\psi$ has its maximum positive value and label the sites of the atoms $\alpha$ sites. For a configuration which gives $\psi$ its largest negative value and which is obtained from the previous configuration by a pure stirring without change of density, let the sites be called $\beta$ sites. Because the atoms are indistinguishable, it does not matter which atom is on which site, it only matters that the site is occupied. It therefore seems plausible that any $\alpha$ configuration can be transformed into any $\beta$ configuration by moving each atom through a short distance to a neighbouring $\beta$ site, without having to move any atom through a distance much larger than the average interatomic separation. The wave function can therefore be transformed from maximum positive to minimum negative by making only small changes in the co-ordinates, and since the kinetic energy is related to the curvature of the wave function, the energy of the state cannot be small. The phonons are therefore the only low lying states. They are exceptions to the above argument because they involve the motion of a large number of atoms, each through a distance small compared with the interatomic separation, with an accompanying density variation.

In his third paper (Feynman, 1954) this argument is extended to give quantitative information about the spectrum of excitations. To obtain a smooth variation of $\psi$ between the $\alpha$ and $\beta$ configurations, let us try

$$\psi = \phi \sum_i f(\mathbf{r}_i), \tag{3.71}$$

where $f(\mathbf{r}_i)$ is $+1$ if the $i$th atom is on an $\alpha$ site and $-1$ if it is on a $\beta$ site. The energy of the state is

$$\epsilon = \frac{\int \psi^* H \psi \, d^N \mathbf{r}_i}{\int \psi^* \psi d^N \mathbf{r}_i}. \tag{3.72}$$

Applying the variational principle, the function $f(\mathbf{r})$ must be chosen so that $\epsilon$ is a minimum. This leads to $f(\mathbf{r}) = e^{i\mathbf{k}.\mathbf{r}_i}$ and so

$$\psi = \phi \sum_i e^{i\mathbf{k}.\mathbf{r}_i}. \tag{3.73}$$

This is exactly the equation (3.70) used to describe phonons of long wavelength, but excitations of large **k** and short wavelength (comparable with the average interatomic separation) are now included in the scheme. The momentum of the excitation is

$$\mathbf{p} = \hbar \mathbf{k}. \tag{3.74}$$

The energy may be transformed into

$$\epsilon = \frac{\hbar^2 k^2}{2m \int p(\mathbf{r}_{ij}) \, e^{i\mathbf{k}.\mathbf{r}_{ij}} \, d^3 \mathbf{r}_{ij}}, \tag{3.75}$$

$$= \frac{\hbar^2 k^2}{2m S(k)}, \tag{3.76}$$

where $p(\mathbf{r}_{ij})$ is the radial distribution function giving the probability per unit volume that the atom $j$ will be found at a distance $r_{ij}$ from the atom $i$. $S(k)$, the Fourier transform of $p(r_{ij})$, is the liquid structure factor, which is obtained directly from the experiments on the scattering of X-rays or neutrons by the liquid (see § 2.4).

The radial distribution function or, as it is sometimes called, the pair correlation function, has already been given in fig. 19. The liquid structure factor $S(k)$ is shown in fig. 32*a*. The first maxi-

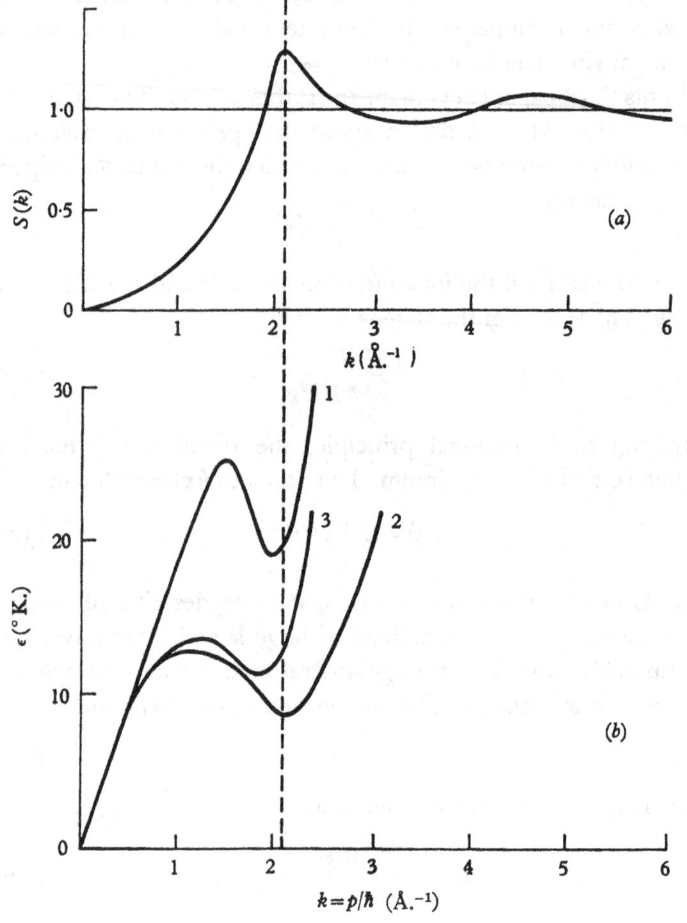


Fig. 32. The elementary excitations in Feynman's theory. (*a*) The liquid structure factor $S(k)$. (*b*) The energy $\epsilon$ of an excitation plotted against its wave-vector $k=p/\hbar$: 1, calculated from the original form of Feynman's theory (equation (3.76)); 2, empirical form required in Landau's theory to fit the experimental data; 3, calculated from the improved wave function of Feynman and Cohen (1956) (equation (3.81)). The energy $\epsilon$ has been expressed in °K. by dividing by Boltzmann's constant $k$. (The reader should distinguish carefully between the use of $k$ as Boltzmann's constant, or as the magnitude of a wave vector.)

mum near $k=2$ Å.$^{-1}$ corresponds to a wavelength of 3·1 Å., which is the distance apart of first nearest neighbours in the liquid as given by the first maximum in the radial distribution curve. Curve 1 of fig. 32*b* is the result of inserting this structure factor into equation (3.76) to obtain the energy of an elementary excitation as a function of its wave vector $k$ $(=p/\hbar)$, while curve 2 has the form previously postulated by Landau to fit the known properties of the liquid. The two curves have similar shapes, but the numerical agreement is poor. The minimum in the curve, which gives rise to the rotons, corresponds to the first maximum in the $S(k)$ curve and therefore to first nearest neighbours in the $p(r)$ curve.

It is interesting to notice that, when the wavelength of the excitation is long and $k$ is small, equation (3.73) then represents the phonons, which are thereby automatically included in the scheme. Near the origin the $\epsilon - k$ curve must therefore become linear with a slope $\hbar c$. The reason for this is that the value of $S(k)$ for small values of $k$ is determined by long wavelength components in the Fourier analysis of $p(r)$, and these are related to density fluctuations over regions of the liquid containing many atoms and are nothing more than Debye modes. In terms of this analysis it is easy to prove that $S(k)=\hbar k/2mc$, whence $\epsilon=\hbar ck$.

The value of $\Delta$ corresponding to the minimum of curve 1 in fig. 32*b* is 19·1° K., which agrees poorly with the experimental value of 8·9° K. The explanation is probably that the wave function of equation (3.73) has too simple a form, and Feynman and Cohen (1956) have therefore attempted to find a better wave function using the following arguments. To represent the roton as a localized excitation using the wave functions of equation (3.73) we can form a wave packet

$$\psi=\phi\sum_i h(\mathbf{r}_i)\,e^{i\mathbf{k}.\mathbf{r}_i} \tag{3.77}$$

choosing the modulating function $h(\mathbf{r}_i)$ to be, for example, a Gaussian function falling off rapidly after a distance larger than $2\pi/k$. This wave function can be shown to give no appreciable change in the local liquid density, but to give a local mass current at the point $\mathbf{a}$ equal to

$$\mathbf{j}=\rho\frac{\hbar\mathbf{k}}{m}\,|\,h(\mathbf{a})\,|^2 \tag{3.78}$$

which always has the same direction as $\mathbf{k}$, but falls off in the same way as $h(a)$. Such a situation is incompatible with the equation of continuity

$$\operatorname{div}\mathbf{j}+\frac{\partial\rho}{\partial t}=0 \qquad (3.79)$$

and cannot represent a stationary state or, in the hydrodynamical analogy, a steady state of flow. Feynman and Cohen therefore looked for wave functions which give a return flow of the atoms and satisfy the continuity equation. Such a wave function is

$$\psi=\phi\sum_i e^{i\mathbf{k}.\mathbf{r}_i}\exp\left\{i\sum_{j\neq i} g(\mathbf{r}_j-\mathbf{r}_i)\right\}. \qquad (3.80)$$

The function $g$ should be chosen to give a minimum energy, according to the variational principle, but this proved to be too difficult, so they worked with an approximate wave function which corresponds to a dipole pattern of backflow

$$\psi=\phi\sum_i e^{i\mathbf{k}.\mathbf{r}_i}\left[1+i\sum_{j\neq i}\frac{A\mathbf{k}.(\mathbf{r}_j-\mathbf{r}_i)}{|\mathbf{r}_j-\mathbf{r}_i|^3}\right]. \qquad (3.81)$$

Applying the variational principle, the energy is a minimum when $A=-3.5$ Å.$^3$ and the resulting relationship between $\epsilon$ and $k$ is shown by curve 3 of fig. 32*b*. The new values of Landau's parameters are $\Delta=11.5^\circ$ K., $p_0/\hbar=1.85$ Å.$^{-1}$ and the agreement with experiment is much better than in the earlier form of Feynman's theory. Moreover, the modified wave function still goes over to the phonon wave function for small values of $k$.

The new wave function suggests that the roton is the quantum mechanical analogue of a microscopic vortex ring. The motion of the helium atoms is to be pictured as similar to the flow of air around a smoke-ring. Atoms pass through the centre of the ring and then complete a closed path by flowing in the opposite direction outside the ring. The diameter of the ring has the same order of magnitude as the average interatomic spacing.

### 3.3.3. Quantum hydrodynamics and the excitation curve

Pitaevskii (1956) has shown that equation (3.76) for the excitation curve may be derived by the methods of Landau's quantum hydrodynamics. He first expresses the Hamiltonian of equation (3.61) in terms of the number density of helium atoms, $n$,

$$\mathscr{H}=\tfrac{1}{2}m\int \mathbf{v}n\mathbf{v}\,d\tau+\mathscr{H}'(n). \qquad (3.82)$$

$\mathscr{H}'(n)$ is the Hamiltonian for the liquid at rest and $\mathbf{v}$ is the 'hydrodynamic velocity'. If $\bar{n}$ is the average number density and we expand the Hamiltonian in the vicinity of $\bar{n}$ by putting $n=\bar{n}+\delta n$, the criterion of the equilibrium density is that the first derivative of $\mathscr{H}$ with respect to $n$ should vanish. Expanding as far as second order terms,

$$\mathscr{H}=\mathscr{H}'(\bar{n})+\tfrac{1}{2}\bar{n}m\int \mathbf{v}^2\,d\tau+\int\phi(\mathbf{r},\mathbf{r}')\,\delta n\,\delta n'\,d\tau\,d\tau', \qquad (3.83)$$

$\phi$ is the second functional derivative of $\mathscr{H}'$ with respect to $n$. Expressing $\delta n$ in terms of its Fourier components

$$\delta n=\sum_{\mathbf{k}} n_{\mathbf{k}}e^{i\mathbf{k}.\mathbf{r}}, \qquad (3.84)$$

and using the hydrodynamical equation of continuity

$$\frac{\partial n}{\partial t}+\bar{n}\,\mathrm{div}\,\mathbf{v}=0, \qquad (3.85)$$

equation (3.83) can be transformed to

$$\mathscr{H}=\mathscr{H}'(\bar{n})+\sum_{\mathbf{k}}\left(\frac{m}{2\bar{n}k^2}\left|\frac{\partial n_{\mathbf{k}}}{\partial t}\right|^2+\tfrac{1}{2}\phi_{\mathbf{k}}\,|\,n_{\mathbf{k}}\,|^2\,V\right). \qquad (3.86)$$

The expression inside the bracket has the same form as the Hamiltonian of an oscillator with angular frequency $\omega$, where

$$\omega^2=(k^2\phi_{\mathbf{k}}\,\bar{n}/m)\,V. \qquad (3.87)$$

The average value of the potential energy of this oscillator in its ground state is

$$\tfrac{1}{2}\phi_{\mathbf{k}}\,\overline{|\,n_{\mathbf{k}}\,|^2}\,V=\tfrac{1}{4}\hbar\omega. \qquad (3.88)$$

From these last two equations, the energy of the excitation corresponding to this oscillator is

$$\epsilon=\hbar\omega=\frac{\hbar^2k^2}{2m}\,\frac{\bar{n}}{\overline{|\,n_{\mathbf{k}}\,|^2}}. \qquad (3.89)$$

But $\overline{|\,n_{\mathbf{k}}\,|^2}/\bar{n}$ is already known to be equal to the structure factor $S(k)$, so that equation (3.89) is equivalent to

$$\epsilon=\frac{\hbar^2k^2}{2mS(k)} \qquad (3.90)$$

which is the expression obtained in the earlier form of Feynman's theory.

## 3.4. The quantum mechanical many body problem

Feynman's justification of the Landau excitation curve uses empirical knowledge of the liquid structure factor $S(k)$. The ultimate theoretical aim is to solve the Schrödinger equation and determine the wave functions and energies of the ground state and all the excited states. The radial distribution function and liquid structure factor could then be deduced theoretically. Recently considerable interest has been shown in this type of problem, the quantum mechanical many body problem involving a dense aggregate of strongly interacting particles under circumstances where quantum mechanical effects are important. Liquid $He^4$ is the outstanding example of interacting Bosons, but there are three important cases of interacting Fermions, the nucleus, electrons in a metal and liquid $He^3$.

We shall discuss the progress that has been made in solving the Schrödinger equation for a series of models which approximate more and more closely to the actual nature of liquid $He^4$. All these treatments assume two body forces and therefore ignore the possibility that the force between two helium atoms may be modified by the presence of other neighbouring atoms. Moreover, in all cases the interatomic forces have been assumed to be entirely repulsive, and since the attractive part of the force is responsible for the condensation into a liquid, the theories are obliged to realize the actual density of the liquid artificially by confining the atoms within a box under pressure. Under these circumstances it is clear that the resulting ground state wave function and energy can be at best only crude approximations: so we shall not discuss the details of the mathematical procedures or the results, but will consider a few qualitative features of interest.

The Bose-Einstein gas with weak repulsive forces has been considered by Bijl (1940), Bogoliubov (1947), Bogoliubov and Zubarev (1955) and Zubarev (1955). They are able to calculate the ground state wave function, the ground state energy and the radial distribution function. In the zeroth approximation, the wave function describing an excitation is identical with that postulated by Feynman (equation (3.73)) and the energy of this excitation is given by equation (3.76). In the ground state a finite fraction of the atoms have zero momentum and the remainder are distributed amongst non-zero momentum states. For an ideal Bose-Einstein gas in its ground state all the atoms have zero momentum. Penrose and Onsager (1956) have discussed this point in detail and conclude that, whatever the exact form of the interatomic forces, there is always a finite fraction of atoms with identical momenta (perhaps zero but not necessarily so) at temperatures below the $\lambda$-point. This is the analogue of Bose-Einstein condensation in an ideal gas. However, a rough calculation shows that the fraction of atoms with identical momenta at 0° K. is only about 8% at the density of liquid helium, so it is not clear what significance must be attached to this feature.

Huang and Yang (1957), Huang, Yang and Luttinger (1957) and Lee, Huang and Yang (1957) have considered the particular case of hard spheres, the interatomic force being zero when the two spheres are separated and infinite when they overlap. The condition of weak forces must now be replaced by the requirement that the atomic diameter $a$ shall be small compared with the average distance between atoms $b$. (In liquid $He^4$ the atomic diameter is 2·6 Å. and the average interatomic separation is 3·6 Å., so this condition is far from being satisfied). The ground state energy per atom is

$$E_0 = \frac{2\pi\hbar^2 a}{mb^3}\left[1 + \frac{128}{15\pi^{\frac{1}{2}}}\left(\frac{a}{b}\right)^{\frac{3}{2}}\right]. \qquad (3\cdot91)$$

At 0° K. the fraction of atoms with zero momentum is

$$\frac{\langle n_0 \rangle}{n} = 1 - \frac{8}{3\pi^{\frac{1}{2}}}\left(\frac{a}{b}\right)^{\frac{3}{2}} \qquad (3\cdot92)$$

which would give the value 0·08 if it could be extrapolated to the values of $a$ and $b$ relevant to liquid helium. The relationship between the energy and momentum of an excitation is of the form

$$\epsilon = \frac{p^2}{2m}\left(1 + \frac{16\pi\hbar^2 a}{b^3}\frac{1}{p^2}\right)^{\frac{1}{2}}. \qquad (3\cdot93)$$

This is quite unlike the Landau excitation curve, but it has the following very interesting feature. For large values of $p$, $\epsilon \simeq \Delta + p^2/2m$ and the excitations are free particles, but with an energy gap $\Delta = 4\pi\hbar^2 a/mb^3$ above the ground state. However, for small $p$

$$\epsilon \simeq \left(\frac{4\pi\hbar^2 a}{m^2 b^3}\right)^{\frac{1}{2}} p. \tag{3·94}$$

Moreover the wave function has the form given by equation (3·73) and it is clear that the excitations are similar in character to phonons. It is readily shown that the factor of proportionality between $\epsilon$ and $p$ is the velocity of sound in the gas. At 0° K. the pressure of the gas is $p = -\left(\frac{\partial U}{\partial V}\right)_S = -\frac{\partial E_0}{\partial v}$, where $v = b^3$ is the volume per atom. Ignoring terms of order $a/b$,

$$p = \frac{2\pi\hbar^2 a}{mv^2}. \tag{3·95}$$

The velocity of sound at 0° K. is $c = \left(\frac{\partial p}{\partial \rho}\right)_S^{\frac{1}{2}}$, where $\rho = m/v$. Whence

$$c = \left(\frac{4\pi\hbar^2 a}{m^2 b^3}\right)^{\frac{1}{2}}. \tag{3·96}$$

The condition for the excitation to be a phonon is

$$p^2 \ll 16\pi\hbar^2 a/b^3,$$

or the phonon wavelength

$$\lambda_{ph} \gg (\pi b^3/4a)^{\frac{1}{2}}.$$

Since sound of long wavelength can be propagated in a gas, it is not unreasonable that phonons of small momentum should exist in the gas, but the classical condition that the sound should not be appreciably attenuated is that the wavelength should be large compared with the mean free path of the atoms, which has the order of magnitude of $b^3/a^2$.

Brueckner and Sawada (1957) have extended the analysis to higher densities and stronger interactions by employing the mathematical techniques developed by Brueckner to deal with the problem of the nucleus. The Brueckner method starts with a 'model wave function', which in the helium case consists of plane waves and therefore corresponds to the ideal Bose-Einstein gas. The actual wave function is obtained from the model wave function by means of an exceedingly complicated 'model operator' which takes into account the effect of the strong interactions. This model operator includes all the important processes which occur when the atoms interact with one another. For example, a helium atom excited out of the ground state while travelling through the liquid may interact with an unexcited atom so that both are excited, and the two excited atoms may interact with one another once or several times and interact frequently with the unexcited atoms before eventually one of them falls back into an unexcited state giving up its energy to the other, which then proceeds on its way. The number of possible processes is very large.

To obtain numerical results, the method was applied to a collection of hard spheres. Consider an atom moving through the liquid with momentum **p**. If we consider only those direct and exchange interactions with the unexcited particles which do not produce excited pairs, the zeroth order energy is

$$\epsilon^{(0)} = \frac{p^2}{2m} + \frac{4\pi a\hbar^2}{mb^3}. \tag{3.97}$$

If we now introduce the possibility that the first particle can excite two unexcited particles to states of momentum **p** and $-$**p**, we find that the average number of excited particles at any instant is $\frac{\hbar}{p}\left(\frac{4\pi a}{b^3}\right)^{\frac{1}{2}}$; but the fluctuation around the average value is $\frac{\sqrt{(3)}}{2}\frac{\hbar}{p}\left(\frac{4\pi a}{b^3}\right)^{\frac{1}{2}}$. For a state of low momentum the average number of excited particles is therefore large and the fluctuations around the mean are also large. Such a state bears little relationship to a single excited particle: it is actually a phonon. However, for values of $p$ near the roton momentum $p_0$ the number of excited particles is of the order of unity, which conforms with the idea that a roton is a small localized entity involving very few atoms. The relationship between the energy and momentum of an excitation is

$$\epsilon = \frac{p^2}{2m}\left[1 + \frac{2\lambda^2 \sin(pa/\hbar)}{(pa/\hbar)^3}\right]^{\frac{1}{2}}. \tag{3.98}$$

$\lambda$ is a monotonic increasing function of $(a/b)$ and has a value of about 6 for liquid helium. For large $p$, $\epsilon = p^2/2m$, but for small $p$, $\epsilon = cp$ as for phonons. For small values of $\lambda$ (low densities) the curve is similar to that given by equation (3.93), with $\frac{\partial \epsilon}{\partial p}$ always positive and increasing steadily with $p$. However, with $a = 2{\cdot}2$ Å. and $b = 3{\cdot}6$ Å., corresponding approximately to the case of liquid helium, the $\epsilon$-$p$ curve has a maximum and a minimum and bears a very strong resemblance to the Landau curve of fig. 30, with $\Delta \sim 7°$ K. and $p_0 \sim 1{\cdot}75$ Å.$^{-1}$. Of course the initial slope is not given by the observed value of the velocity of first sound, since attractive forces have been ignored. For the same reason, the value deduced for the ground state energy ($+50$ cal. mole$^{-1}$) cannot be compared with the experimental value. It is not yet clear whether the introduction of attractive forces would seriously modify the roton region of the spectrum.

CHAPTER 4

# THE NEW THERMOHYDRODYNAMICS

## 4.1. The need for a new hydrodynamics

The more important features of the unusual flow properties of liquid helium II were introduced in chapter 1. From what has already been said there it should be obvious that the new phenomena cannot be explained merely by the introduction of a zero viscosity, or by any simple modification of classical hydrodynamics. In fact the experimental situation is so complicated that one might doubt whether it is possible to make any successful approach along the lines developed by Euler, Navier and Stokes to describe the flow of normal liquids. Fortunately, as long as the velocities are less than a certain critical velocity, the ideas underlying the two-fluid theory (§ 1.3) introduce a great simplification and it seems to be possible to write down two hydrodynamical equations, one for the superfluid component and the other for the normal component. These equations will be introduced later in this chapter and it will be seen that they are formally similar to the equations of classical hydrodynamics, except that they include extra terms to take into account the thermal effects such as the mechanocaloric effect and the thermomechanical effect (§ 1.5).

By themselves the hydrodynamical equations provide only a partial description of the situation. They must be supplemented by the boundary conditions and other important concepts such as the idea of a critical velocity. Moreover, when the critical velocity is exceeded the situation becomes much more complicated and the hydrodynamical equations must then either be modified or perhaps even abandoned completely. It is therefore very important to emphasize the basic experimental facts underlying the theoretical treatment and so, before formulating the equations, we shall consider in detail the nature of superfluidity and also the nature of the thermal effects which have such an important influence on the flow.

## 4.2. Superfluidity

### 4.2.1. Isothermal flow through narrow channels

It is not easy to obtain satisfactory quantitative data on isothermal flow through narrow channels. It is technically difficult to produce very narrow channels (down to $10^{-5}$ cm. wide) with a uniform cross-section and their average cross-section cannot be measured accurately. Moreover, only the superfluid component flows strongly through narrow channels and the heat defect of this component gives rise to temperature differences, which in turn react back on the flow, so that it is very difficult to ensure that the flow is really isothermal. The most complete of the early measurements are those of Allen and Misener (1939). Their research was conducted before the thermal effects were properly understood, and so there is some doubt about the accuracy of their numerical values. It seems probable, however, that this does not invalidate their qualitative conclusions, which will serve as a convenient starting point for the present discussion.

They made their narrowest channels by inserting a bundle of wires into a metal tube, which was then drawn down until the wires were so closely packed that the cross-section of each wire was hexagonal and the spaces between the wires were approximately rectangular. The cross-section for flow was determined by observing the flow of He gas through the channels at room temperature, and the mean width of the channels was obtained by dividing this cross-section by half the total perimeter of the wires. Glass capillaries were used to study flow through wider channels. The method was the straightforward one of attaching the glass capillary or wire-filled tube to the lower end of a glass reservoir and observing the rate of emptying or filling of the reservoir when it was partially immersed in a bath of liquid helium II. A repeat experiment with the reservoir closed at the bottom showed that the film flow over the rim of the reservoir represented only a small correction.

For the wider capillaries, at temperatures very near the $\lambda$-point, Poiseuilles' law was approximately obeyed, presumably because the flow of the normal component predominated. Otherwise the dependence of the rate of flow on the pressure head, the width of the channel and the length of the channel was very complicated.

Qualitatively, the results can be explained by superimposing the viscous flow of the normal component on the more complicated flow of the superfluid component. The nature of this superfluid flow can be seen from the observations at 1·2° K., where $\rho_n/\rho$ is only about 3 % and the flow of the normal component can be ignored. Fig. 33 shows the mean velocity of flow as a function of the pressure head for channels of various widths at 1·2° K. When

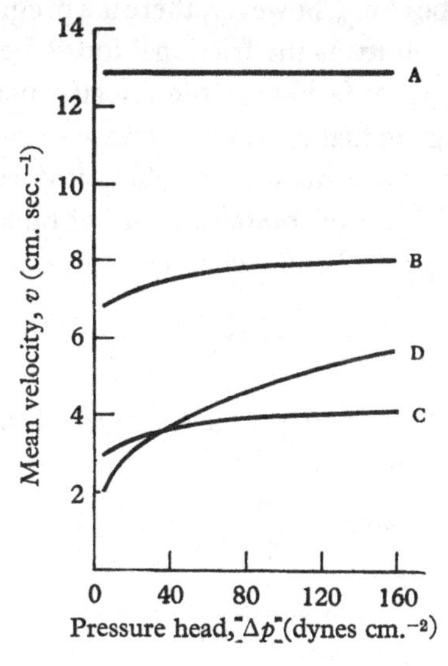


Fig. 33. Dependence of mean velocity on pressure head for flow through narrow channels. Temperature = 1·2° K. Channel widths: A, $1{\cdot}2\times10^{-5}$ cm.; B, $7{\cdot}9\times10^{-5}$ cm.; C, $3{\cdot}9\times10^{-4}$ cm.; D, $5{\cdot}0\times10^{-3}$ cm.

we talk of 'ideal superfluidity' we usually have in mind the situation that is realized in the narrowest channel of width $1{\cdot}2\times10^{-5}$ cm. The velocity is then almost independent of the pressure head and it is usually assumed that the same velocity could exist under zero pressure head. This velocity is taken to be a critical velocity $v_{s,c}$ such that when $v_s < v_{s,c}$ the flow is completely frictionless, but when $v_s \geqslant v_{s,c}$ the frictional forces are so large that $v_{s,c}$ cannot be exceeded.

As the channel width increases from $10^{-5}$ to $10^{-2}$ cm. there is a gradual transition from ideal superfluidity to the entirely different situation encountered in wide capillaries, where the mean velocity is a marked function of the pressure head. The pressure head needed to produce a given velocity is, of course, a direct measure of the frictional forces opposing the flow, the acceleration of the liquid being negligible in most experiments. Curves like those of fig. 33 may therefore be interpreted as showing how the frictional forces vary with velocity. A critical velocity is not easily defined for the wider channels, but it is possible that the curves have a finite intercept $v_{s,c}$ on the velocity axis, and velocities of flow up to $v_{s,c}$ would then be subject to no frictional retardation and would be truly superfluid. Above $v_{s,c}$, however, there is a progressive breaking down of the superfluidity as the frictional forces begin to build up in a markedly non-linear fashion as the velocity increases.

It is clear from fig. 33 that $v_{s,c}$ must decrease as the channel width $d$ increases, but the exact manner in which it does so is still undecided. The critical *rate of transfer* in cm.$^3$ of bulk liquid per sec. per cm. width of a parallel-sided channel is

$$\sigma_c = \frac{\rho_s}{\rho} v_{s,c} d. \tag{4.1}$$

Bijl, de Boer and Michels (1941) have suggested that $v_{s,c}$ is determined by a quantum relation

$$m v_{s,c} d \sim \frac{h}{4\pi}, \tag{4.2}$$

or

$$\sigma_c \sim \frac{\rho_s}{\rho} \frac{h}{4\pi m}$$

$$\sim 7 \cdot 2 \times 10^{-5} \frac{\rho_s}{\rho}. \tag{4.3}$$

Experimentally equation (4.3) is in order of magnitude agreement with the observed velocities, but Allen and Misener interpreted their results in terms of the more complicated equation

$$\sigma_c = \sigma_0 + Vd \tag{4.4}$$

with $\sigma_0 \sim 10^{-4}$ cm.$^3$ sec.$^{-1}$ cm.$^{-1}$ at 1·2° K. and $V \sim 5$ cm. sec.$^{-1}$. This point will be taken up again in connexion with the flow of the film. Allen and Misener suggested that equation (4.4) is consistent with a 'film flow' of $\sigma_0$ along the walls of the channel plus a constant velocity $V$ throughout the channel. Independently of whether this is true or not, it raises the point that $v_{s,c}$ is a mean velocity and the actual velocity may conceivably vary with distance from the wall, giving a velocity contour. The variation of $\sigma_c$ with temperature is similar to the corresponding curve for the case of the film (fig. 6) and may be due mainly to the variation of the factor $\rho_s/\rho$ in equation (4.1).

### 4.2.2. The flow of the films

The broad outlines of film flow phenomena were presented in § 1.4, and in § 7.5 we shall discuss the matter in greater detail. Here we shall concern ourselves only with those aspects of film flow which have a bearing on the nature of superfluidity. The flow of the film is frequently taken to represent the best available case of

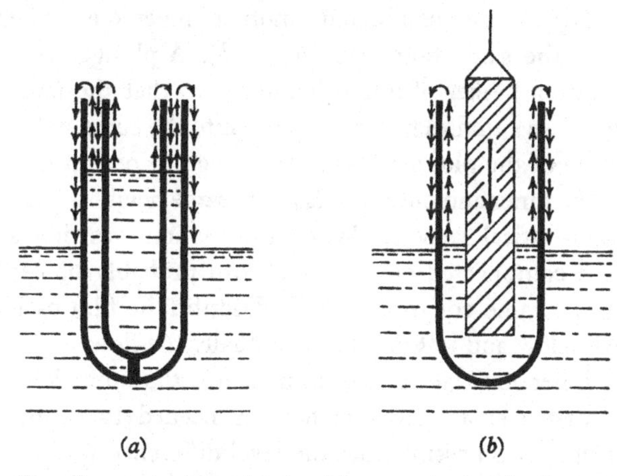


Fig. 34. Experiments to detect frictional forces opposing the flow of the film. (*a*) The double beaker experiment of Daunt and Mendelssohn (1946); (*b*) the moving plunger experiment of Picus (1954).

'ideal superfluidity' in which the flow of the superfluid component is not accompanied by viscous flow of the normal component. Several experiments have been conducted to demonstrate the absence of any frictional forces opposing this flow. The ingenious double beaker experiment of Daunt and Mendelssohn (1946) is illustrated in fig. 34*a* (see also Chandrasekhar and Mendelssohn (1955)). In such an arrangement of two coaxial beakers it is obvious that the level in the inner beaker cannot fall below the level in the annulus or there would be film flow from the annulus back into the inner beaker. What happens therefore is that the levels fall at the same rate, with the film velocity on the wall of the inner beaker needing to be only about 90 % of the critical value in order to keep up. The important question then is whether this subcritical velocity is opposed by any frictional forces, so that the inner level has to be raised above the annulus level in order to provide the pressure head needed to maintain the flow against such dissipative forces. In the experiment the two moving levels were observed to stay together to within the accuracy of measurement, which was about $10^{-2}$ cm., and it was concluded that the frictional forces must be extremely small. Similar experiments have been performed on flow through narrow channels (Bowers and White, 1951; Bowers and Mendelssohn, 1952*a*, *b*).

Picus (1954) has conducted another ingenious experiment bearing on the same point (see fig. 34*b*). A plunger was slowly lowered into a beaker of liquid helium II so that the level inside the beaker rose and film flow took place into the surrounding bath. The motion of the plunger was slow enough to permit the film to maintain the inner and outer levels at the same height while flowing with a sub-critical velocity. When the motion was first started, there were complicated effects which can probably be explained by using the same equations which describe the film oscillations discussed below and in § 7.3, but eventually the flow settled down with the inner level about 0·02 mm. above the outer level. This suggests that a small pressure head is needed to maintain the flow, but it is also possible that the level difference was a thermo-mechanical effect, consequent upon a small temperature difference of the order of $10^{-7}$ °K.

When a vessel empties through the film or a narrow channel, the rate of transfer is almost independent of the level difference and so, when the inner level reaches the outer level, the film is still moving rapidly and its momentum causes the level to overshoot and then oscillate about its equilibrium position. These oscillations were observed in narrow channels by Allen and Misener (1939) and have been used by Atkins to measure the thickness of the film (§ 7.3). The fact that these oscillations suffer very little damping enables an upper limit to be put on the frictional forces. The damping may be caused by secondary effects, not connected with the frictional forces opposing the film flow, but in any case the damping is so small that it is possible to conclude that, in experiments of the type performed by Daunt and Mendelssohn and Picus, the pressure head could not have exceeded $2 \times 10^{-4}$ cm. Another way of expressing this is that, as an element of the liquid flows along several centimetres of film, it does not lose more than 0·05 % of its kinetic energy of flow.

Although the resistance to flow is negligible at sub-critical velocities, it is clear that, once the critical velocity is reached, strong dissipative processes come into play to prevent the velocity increasing further. In this connexion it is interesting to note that the transfer rate of the film has a tendency to decrease when the level difference falls below about 0·5 cm. (see fig. 74). Eselson and Lasarew (1952) suggest that the mechanocaloric effect, consequent upon film flow, produces a temperature excess inside the beaker during emptying and a temperature defect during filling. As a consequence of the thermomechanical effect, the inner level then reaches an equilibrium position displaced slightly from the outer level and subsequently approaches the outer level at a rate depending upon inertial effects and the rapidity with which the temperature difference can be dissipated. They support this with an experiment in which the effect was made to disappear by improving the thermal contact between the inside and outside of the beaker by means of a copper bottom. However, Atkins (1950*b*) has pointed out that, even without a copper bottom, the possibility of distillation between the inner and outer liquid surfaces restricts possible temperature differences to less than $10^{-6}$ °K., which would give a negligible thermomechanical effect. The issue may be a fundamental one, because if the level difference is really a measure of the frictional forces opposing the flow, there is a suggestion that these forces vary with velocity in the region of the critical velocity. Fig. 35 shows various possibilities. Curve 1 is the ideal case of zero friction below $v_{s,c}$ and infinite friction above $v_{s,c}$. Curve 2 is the case where no velocity is truly frictionless, but there is a very marked non-linearity. Curve 3 represents a combination of frictionless flow and non-linear frictional flow.

The rate of transfer of the film is in good order of magnitude agreement with equation (4.3), although there is some evidence, as in the case of narrow channels, that $v_{s,c}d$ increases slowly with increasing $d$. The rate of flow of the film decreases slightly as the rim of the beaker is raised to greater heights above the liquid surfaces and Atkins (1950*b*) has suggested that this may be related to the fact that the film thickness at the rim decreases with height above the surface of the bulk liquid (§7.5).

Using the thin unsaturated films which exist at pressures below the saturated vapour pressure, it is possible to investigate flow in still narrower channels down to one atomic layer wide! The details are discussed in §7.6. Here we shall emphasize that the rate of transfer $(\rho_s/\rho)\, v_{s,c} d$ decreases rapidly as the film becomes thinner.

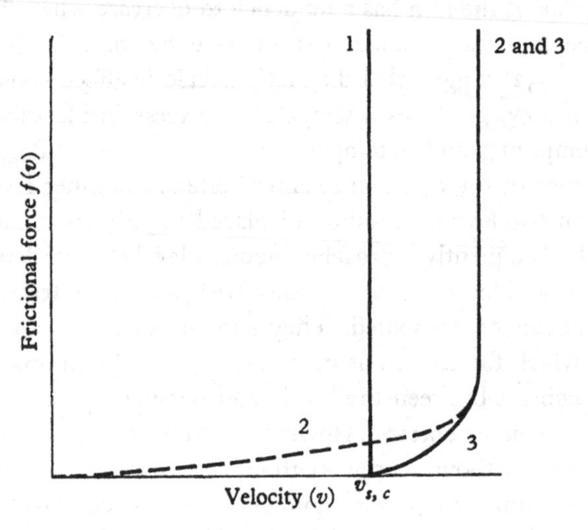


Fig. 35. Possible forms of the frictional force in film flow. 1, simple critical velocity; 2, flow which is never completely frictionless; 3, mixed flow.

However, in these extremely narrow channels the effective $\lambda$-temperature corresponding to the onset of superfluidity also decreases markedly with decreasing thickness, and the issue is therefore complicated by the fact that $\rho_s/\rho$ must be a function of thickness.

### 4.2.3. Flow through fine pores

Allen and Misener (1939) also made very narrow channels by packing jeweller's rouge inside a glass tube. The particle size was $\sim 10^{-5}$ cm. and the interstices probably had dimensions $\sim 10^{-6}$ cm. The channels available for flow were therefore very narrow and also very irregular. If we estimate that the available cross-section for flow was less than 10% of the total cross-section of the glass tube, the observed rates of flow correspond to velocities of several centimetres per second. The variation of velocity with temperature was similar to the case of the saturated film. The really novel feature was that the rate of flow was proportional to the square root of the pressure head. This result is so unlike anything else that has been observed for superfluid flow through narrow channels that it has the appearance of a valuable clue to the nature of superfluidity. One wonders if the irregular contour of the channels is the important factor. Later experiments of Chandrasekhar and Mendelssohn (1953) have confirmed these facts and have shown that the pressure gradient is present throughout the length of the column of packed powder.

A similar situation exists for flow through porous Vycor glass, which has a sponge-like structure with pore diameters in the range from $4 \times 10^{-7}$ to $10 \times 10^{-7}$ cm. These are the narrowest channels which have so far been investigated, but their cross-section varies rapidly from place to place. Liquid helium flows very readily through these narrow pores, but the rate of flow varies markedly with the pressure head and there is no clear evidence for a critical velocity (Atkins and Seki, 1957; Champeney, 1957).

### 4.2.4. Bulk liquid

If superfluid flow exists in channels wider than $10^{-2}$ cm., it breaks down at a critical velocity which is less than the velocities which have been realized in most of the experiments. All attempts to investigate the nature of superfluid flow in bulk liquid have therefore failed. To emphasize this point, three such attempts will be described.

Osborne (1950) rotated a glass vessel containing liquid helium II and plotted the contour of the resulting parabolic surface. A normal liquid rotating with an angular velocity $\Omega$ has a surface described by the equation

$$y = \frac{\Omega^2}{2g} x^2, \tag{4.5}$$

where $y$ is the height at a distance $x$ from the axis of rotation. In the case of liquid helium II, it might be suggested that only the normal component would rotate with the vessel and the superfluid component would remain stationary. The generally accepted hydrodynamical equations of the two-fluid theory would then lead to a surface of the form

$$y = \frac{\rho_n}{\rho} \frac{\Omega^2}{2g} x^2. \qquad (4.6)$$

Osborne found that liquid helium II behaved like a normal liquid and obeyed equation (4.5). The diameter of the vessel was 1·4 cm. and the velocity of the outer wall of the vessel ranged from 35 to 70 cm. sec.$^{-1}$. It is therefore possible that the experiment gave this result only because the critical velocity was greatly exceeded and a different result might be obtained if the experiment could be performed with much smaller velocities.

Andronikashvili and Kaverkin (1955) have repeated Osborne's experiment with a glass tube of internal diameter 2·7 cm., and have confirmed that the liquid has a normal parabolic surface at peripheral velocities down to 4 cm. sec.$^{-1}$. They observed that sometimes, at their top peripheral velocities of the order 40 cm. sec.$^{-1}$, the vertex of the parabola grew into a conical tip and occasionally this tip developed into a hollow vortex extending down to the bottom of the rotating vessel. Similar effects were observed by Donnelly, Chester, Walmsley and Lane (1956), but in He I as well as He II. Andronikashvili and Kaverkin inserted in their rotating liquid an apparatus for measuring the thermomechanical effect and found that, even at their top speeds, the size of this effect was the same as when the liquid was at rest. Thus, although the superfluid component is brought into rotation it still maintains its identity and the division into two components is as valid as for liquid at rest.

Andronikashvili (1952) attempted to detect the existence of a persistent current of liquid helium II. The apparatus consisted of a number of parallel disks of aluminium foil, 0·01 mm. thick and spaced 0·2 mm. apart, surrounded by a cylinder of similar foil. It was the motion of the liquid helium between the disks that was being studied. The system was brought into steady rotation just below the $\lambda$-point (at 2·17° K.) where most of the liquid could be expected to take up the motion of the disks. The arrangement was then cooled to 1·5° K. and the disks were brought to rest in about 15 sec. The next step was designed to reveal the presence of a persistent current of the superfluid component continuing to circulate even after the rotation of the disks was stopped. The temperature was suddenly raised to 1·65° K. in 15 sec. and $\rho_s$ was thereby decreased from 0·88$\rho$ to 0·78$\rho$, the idea being to see if the destroyed superfluid would give up its angular momentum to the disk system. This system was suspended from a phosphor-bronze fibre and a deflexion of about 30° might have been expected. The observed deflexion was no greater than 2° and did not depend upon the previous direction of rotation of the disks. However, the peripheral velocity was about 6·5 cm. sec.$^{-1}$, which is probably greatly in excess of the critical velocity corresponding to the dimensions of the apparatus.

## 4.3. Thermal effects

### 4.3.1. The mechanocaloric effect

The thermal effects will be treated by describing in detail the experiments of Kapitza (1942) which first put the subject on a secure quantitative foundation. The apparatus (fig. 36) was a glass Dewar vessel *A* partially immersed in a bath of liquid helium II and communicating with it through the narrow slit *B*. This slit was made by pressing together two optically polished quartz plates and the width of the slit was estimated by observing the interference fringes formed by the arrangement. The fringes were also used to align the plates parallel to one another, an arrangement of springs and rods enabling this adjustment to be made even when the apparatus was cooled down to liquid helium temperatures. The smallest gap used was about $3 \times 10^{-5}$ cm., which is presumably narrow enough to pass only the superfluid component. Heat could be supplied to the inside of the vessel by means of the heater *H*, and the temperatures outside and inside were measured on the phosphor-bronze thermometers $T_1$ and $T_2$. These two thermometers were incorporated in the same potentiometer circuit, which was designed so that the temperature difference could be measured directly. The temperature difference was assumed to be zero when there was a large gap between the two quartz plates.

Kapitza adopted two different procedures to investigate the mechanocaloric effect and the thermomechanical effect respectively. In the first procedure, which measured the mechanocaloric effect, heat was supplied to the inside of the vessel and liquid flowed in at a rate which was measured by observing the rate of rise of the inner level. The temperature difference $\Delta T$ between the inside and outside of the vessel was also measured and, to ensure that this temperature difference was not connected with the thermomechanical effect, measurements were made only when the inner level rose from about 2 to 4 mm. above the outer level. (Even when there is no flow, there is a temperature difference related to the

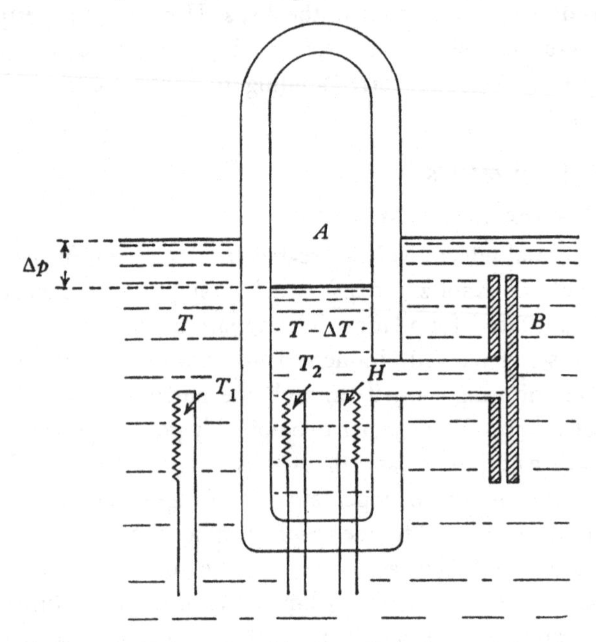


Fig. 36. Kapitza's apparatus to investigate the mechanocaloric and thermomechanical effects.

level difference by the formula for the thermomechanical effect, as we shall explain shortly. A level difference of 4 mm. corresponds to a temperature difference $\leqslant 4 \times 10^{-4}$ °K. over the temperature range covered by Kapitza.) The dependence of the flow velocity, $v$, and the temperature difference, $\Delta T$, on the rate of supply of heat, $q$, is shown in fig. 37. For small values of $q$, $\Delta T$ was zero to within the accuracy of measurement and the rate of flow was linearly proportional to $q$. Under these conditions, it was necessary to supply an amount of heat $Q^*$ to every gram of liquid emerging from the slit in order to warm it up to the same temperature as the bulk liquid at the entrance of the slit. It follows that the superfluid component has a heat deficiency of $Q^*$ cal. g.$^{-1}$ below the heat content of the bulk liquid. Between 1·3° K. and the $\lambda$-point $Q^*$ was equal to $TS$ within the accuracy of the experiments. Here $S$ is the total entropy per gram of the bulk liquid and the conclusion is that the superfluid component flowing through the slit carried no entropy with it.

At a certain critical heat input, corresponding to a critical velocity, a temperature difference suddenly appeared and the rate of flow was no longer proportional to the heat input. Below the critical velocity, however, the flow seemed to be taking place under

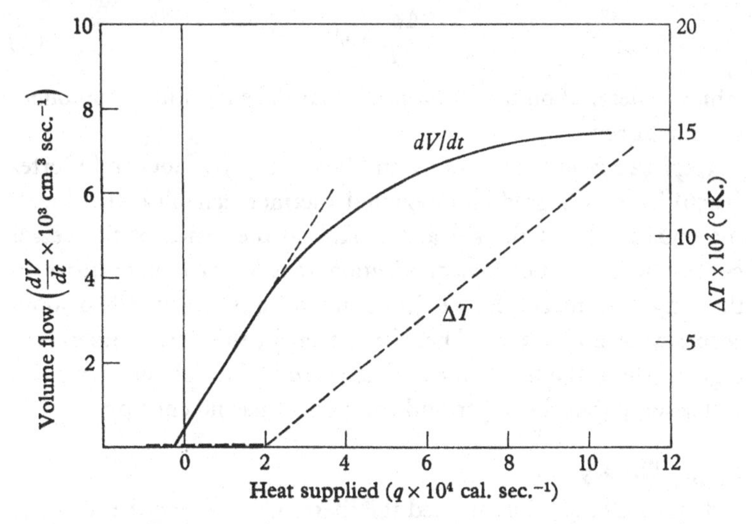


Fig. 37. Thermal initiation of superfluid flow (after Kapitza, 1942).

zero pressure head and zero temperature difference. A temperature difference therefore seems to play a similar role to a pressure difference in superfluid flow.

## 4.3.2. The thermomechanical effect

In Kapitza's second procedure no heat was supplied to the inside of the vessel. The cold inflowing superfluid component therefore lowered the inside temperature until equilibrium was reached with the temperature difference $\Delta T$ corresponding to the thermomechanical pressure head $\Delta p$, made up from the level difference $\rho gh$ and a small correction for the difference in vapour pressure inside and outside. With the correct balance between $\Delta p$ and $\Delta T$ no flow would take place into a perfectly insulated vessel, but in practice there was a small stray heat influx which slowly warmed up the inside, decreasing $\Delta T$ and allowing liquid to flow in to decrease the level difference. But this readjustment was slow and it was reasonable to assume that, at all instants, $\Delta p$ was the thermomechanical pressure head corresponding to the temperature difference $\Delta T$. In this way a range of values of $\Delta p$ and $\Delta T$ was obtained and it was found that $\Delta p$ was linearly proportional to $\Delta T$ and the slope of the curve was in agreement with H. London's equation

$$\frac{\Delta p}{\Delta T} = \rho S, \tag{4.7}$$

which is derived on the assumption that the superfluid component has no entropy.

Kapitza's results did not extend below 1·3° K. Bots and Gorter (1956) have measured an integrated thermomechanical effect with the outside end of the slit at 1·1° K. and the inside of the vessel cooled by adiabatic demagnetization to various temperatures in the range 0·1 to 1·1° K. H. London's formula was verified to an accuracy of about 5 %. Therefore, even in this low temperature region, where the thermal excitations are mainly phonons, we can still assume that the superfluid component has no entropy.

### 4.3.3. Theory

de Groot (1951) has applied the thermodynamics of irreversible processes to derive a relationship between the thermomechanical effect and the mechanocaloric effect:

$$\frac{\Delta p}{\Delta T} = \frac{\rho Q^*}{T}, \tag{4.8}$$

where $Q^*$ is the mechanocaloric heat associated with the transfer of unit mass of liquid. This relation applies to all widths of connecting channel, but we are particularly interested at present in the value assumed by $Q^*$ when the channel is so narrow that it passes only the superfluid component. The approach of H. London (1939) and, later, Landau (1941), is to assume that the superfluid component carries no entropy, so that, on arrival at the exit of the connecting channel, it must be supplied with an entropy $S$ by addition of an amount of heat $Q^* = TS$. Equation (4.8) then reduces to equation (4.7). It must be emphasized that, while (4.8) is rigorous, (4.7) is a consequence of a very special assumption about the nature of the flowing liquid.

Gorter (1949) has given a rather different formulation. He writes the entropy $S$ and the Gibbs free energy $G$ in the form $S(x, T, p)$ and $G(x, T, p)$, where $x$ is the fraction of normal component $\rho_n/\rho$. To appreciate the significance of this, consider all possible dynamical states of the liquid, some of which may be very improbable from a thermodynamical point of view, and assume that for each state it is possible to decide upon the value of $x$ by some suitable argument. Now suppose that the liquid is restricted to those states for which $x$ has a particular value $x'$ and then apply the usual arguments of statistical mechanics to derive the thermodynamic functions at $T'$ and $p'$. In this way one obtains $S(x', T', p')$ and $G(x', T', p')$ and so on for all values of $x$, $T$ and $p$. In practice, of course, at fixed $T$ and $p$, $x$ assumes the value $x_m(T, p)$ which makes the Gibbs function $G$ a minimum $\{(\partial G/\partial x)_{T,p} = 0\}$ and other values of $x$ are highly improbable. Usually $S$ is written as a function of $T$ and $p$ only: $S(T, p) = S\{x_m(T, p), T, p\}$. The procedure which has just been outlined may be familiar to the reader in the Bragg-Williams treatment of the order-disorder transition in $\beta$-brass. The degree of order is fixed at some arbitrary value and the free energy is then calculated at a particular temperature. The temperature is then fixed and the degree of order varied to minimize the free energy, the actual degree of order realized at this temperature being thus deduced.

To calculate the mechanocaloric heat $Q^*$, suppose that 1 gram of superfluid flows into a vessel containing initially $M$ grams of bulk liquid at $T$ and $p$ ($M \gg 1$). The change in $x$ is $-x/M$ and the change in entropy at constant $T$ and $p$ is $M(\partial S/\partial x)_{T,p}\,(x/M)$. To maintain normal equilibrium conditions the heat supplied must be

$$Q^* = Tx\left(\frac{\partial S}{\partial x}\right)_{T,p}.$$

The formula for the thermomechanical effect is then

$$\frac{\Delta p}{\Delta T} = \rho x\left(\frac{\partial S}{\partial x}\right)_{T,p}. \tag{4.9}$$

This reduces to H. London's formula (equation (4.7)) if

$$S(x, T, p) = xf(T, p). \tag{4.10}$$

The plausibility of this last equation can be discussed only in terms of a specific model of the liquid.

## 4.4. The thermohydrodynamical equations

Since the exact form of the thermohydrodynamical equations is still a controversial matter, it is obviously not possible to present a derivation of these equations that would be universally accepted as rigorous. We shall therefore state the equations without proof in a form based on the original ideas of Tisza and Landau. A critical analysis of their theoretical foundations has been given in the review article by Dingle (1952*a*).

The two-fluid theory requires two hydrodynamical equations, one for the superfluid component:

$$\frac{D\mathbf{v}_s}{Dt} = -\operatorname{grad} G \tag{4.11}$$

or, alternatively,

$$\rho_s \frac{D\mathbf{v}_s}{Dt} = -\frac{\rho_s}{\rho}\operatorname{grad} p + \rho_s S \operatorname{grad} T, \tag{4.12}$$

and one for the normal component:

$$\rho_n \frac{D\mathbf{v}_n}{Dt} = -\frac{\rho_n}{\rho}\operatorname{grad} p - \rho_s S \operatorname{grad} T + \eta_n(\nabla^2 \mathbf{v}_n + \tfrac{1}{3}\operatorname{grad}\operatorname{div}\mathbf{v}_n). \tag{4.13}$$

Here $\mathbf{v}_s$ and $\mathbf{v}_n$ are the velocities of the superfluid and normal components respectively, and $\eta_n$ is the viscosity associated with the normal component:

$$\frac{D}{Dt} \equiv \frac{\partial}{\partial t} + \mathbf{v}\operatorname{grad}. \tag{4.14}$$

The acceleration terms $\rho_s(D\mathbf{v}_s/Dt)$, $\rho_n(D\mathbf{v}_n/Dt)$ and the terms in grad $p$ are similar to those normally encountered in classical hydrodynamics, but there are also terms in grad $T$, expressing the fact that the superfluid component tends to move towards regions of high temperature, whereas the normal component tends to move towards lower temperatures. The significance of the terms in grad $T$ becomes apparent if we consider two vessels connected by a narrow channel. If the channel is narrow enough, the flow of the viscous normal component is negligible and $\mathbf{v}_n = 0$. In equilibrium $\mathbf{v}_s = 0$ also, and equation (4.12) then reduces to

$$\operatorname{grad} p = \rho S \operatorname{grad} T,$$

which is H. London's equation for the thermomechanical effect. Since Gorter's theory leads to a different thermomechanical equation, it is clear that in his formulation of the thermohydrodynamical equations, the term $\rho_s S \operatorname{grad} T$ must be replaced by

$$\rho(1-x)\,x\left(\frac{\partial S}{\partial x}\right) \operatorname{grad} T.$$

Equations (4.12) and (4.13) must be supplemented by two continuity equations expressing conservation of mass:

$$\frac{\partial \rho}{\partial t} + \operatorname{div}(\rho_s \mathbf{v}_s + \rho_n \mathbf{v}_n) = 0 \tag{4.15}$$

and conservation of entropy:

$$\frac{\partial}{\partial t}(\rho S) + \operatorname{div}(\rho S \mathbf{v}_n) = 0. \tag{4.16}$$

In addition, the boundary conditions play as important a role as the hydrodynamical equations in determining the nature of the flow in certain experiments. It is normally assumed that, at a wall $\mathbf{v}_n = 0$ and $\mathbf{v}_s$ is unrestricted, but there is no firm experimental foundation for these assumptions. Moreover, in some special cases the dimensions of the apparatus may be small compared with some relevant mean free path, giving rise to the type of complication that occurs in open gas phenomena. After boundary conditions and the possibility of mean free path effects have all been taken into account, the four equations (4.12), (4.13), (4.15) and (4.16) provide a complete description of the hydrodynamics, but these equations have been presented in their simplest form and we shall see later that there are additional terms which become important under certain circumstances. The theory of the attenuation of first and second sound (chapter 5) introduces terms involving the 'normal thermal conductivity' and the coefficients of second viscosity.

Moreover, the theoretical derivation of the equations ignores certain quadratic terms in $v_s^2$ and $v_n^2$, so that their validity is restricted to very small velocities. More particularly, the equations cannot be expected to apply to velocities greater than the critical velocity $v_{s,c}$. This last question of the breakdown of superfluidity at high velocities and the extra frictional forces which then appear is the subject of chapter 6.

## 4.5. The viscosity of the normal component, $\eta_n$

### 4.5.1. Experimental determination of $\eta_n$

The viscosity of the normal component was first obtained from observations of the damping of an oscillating disk, as in fig. 1*b* (Keesom and MacWood, 1938; Andronikashvili, 1948*b*; de Troyer, van Itterbeek and van den Berg, 1951; Hallett, 1952, 1955; Dash and Taylor, 1957*a*, *b*). The amplitude of oscillation was restricted to small values in order that the relevant critical velocity should not be exceeded. The damping decreased rapidly with decreasing temperature below the λ-point, and it was originally assumed that the viscosity behaves in a similar way, but this conclusion was invalidated once it was realized that only the normal component moves with the disk and the moving fluid therefore has an effective density $\rho_n$. If we solve the thermohydrodynamical equations on the assumption that $v_s = 0$, we find that the motion of the normal component falls off exponentially with distance from the disk, the 'characteristic penetration depth' being $(\eta_n T/\pi\rho_n)^{\frac{1}{2}}$, where $T$ is the period of oscillation. The velocity gradient near the disk is therefore proportional to $v(\pi\rho_n/\eta_n T)^{\frac{1}{2}}$, where $v$ is the velocity of the disk, and the viscous force on the disk is consequently proportional to $v(\pi\eta_n\rho_n/T)^{\frac{1}{2}}$. The method therefore measures the product $\eta_n\rho_n$ and the original mistake was to interpret this as $\eta\rho$. The rapid decrease of damping with falling temperature is caused principally by the factor $\rho_n$, and $\eta_n$ varies much less rapidly, as will be seen from fig. 38. In fact $\eta_n$ decreases just below the λ-point, levels out near 1·8° K. and then begins to increase again towards 1° K. There is no observable discontinuity in the damping on passing through the λ-point, but there may be a discontinuity in the slope $d\eta/dT$ (Dash and Taylor, 1957*b*).

There are several uncertainties in these oscillating disk results, apart from normal experimental errors. Below 1·5° K. the contribution of the liquid to the logarithmic decrement is comparable with the logarithmic decrement in a vacuum, and errors in the determination of this last quantity seriously affect $\eta_n$. The thermohydrodynamical equations cannot be solved exactly in the vicinity

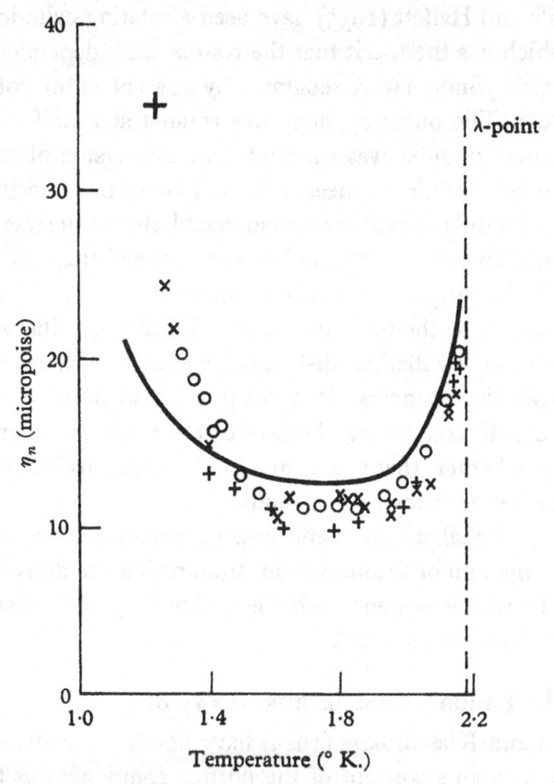


Fig. 38. The viscosity of the normal component, $\eta_n$. Oscillating disk values: O, Andronikashvili (1948*a*); ×, de Troyer *et al.* (1951); +, Hallett (1952); ——, rotating cylinder viscometer results (Heikkila and Hallett, 1955).

of the edge of the disk and there is some doubt whether the edge corrections have always been correctly assessed. Also, the contraction of the apparatus on cooling produces an uncertainty of about 2 %. The accuracy of $\eta_n$ is dependent on the values used for $\rho_n$. Andronikashvili's pile of disks experiment (fig. 2) gives good values of $\rho_n$ above 1·8° K. (Dash and Taylor, 1957*a*), but is very unreliable at lower temperatures. $\rho_n$ is therefore usually deduced from the velocity of second sound, using equation (1.6), but the results then suffer from the uncertainty in $S$. Moreover, Gorter, Kasteleijn and Mellink (1950) have suggested a slightly different equation for the velocity of second sound based on Gorter's form of the thermohydrodynamical equations.

Heikkila and Hallett (1955) have used a rotating cylinder viscometer, which has the merit that the results are independent of $\rho_n$. Two coaxial cylinders were separated by a gap of 2 mm. containing liquid He II. The outer cylinder was rotated at a uniform speed and the inner cylinder was suspended on a tungsten fibre so that the torque on it could be measured. As long as the velocity of the outer cylinder did not exceed 0·08 cm. sec.$^{-1}$, the torque was linearly proportional to the velocity, and it was assumed that under these conditions the torque was entirely due to $\eta_n$. The results, which are represented by the full curve in fig. 38, are in qualitative agreement with the oscillating disk experiments, but there are large quantitative discrepancies. It is not possible at present to decide whether the differences can be ascribed entirely to experimental errors, or whether there is some fundamental hydrodynamical distinction between the two methods.

In § 5.3.3 we shall discuss some measurements by Zinoveva (1956) on the attenuation of second sound, from which she derives values of $\eta_n$ in better agreement with the rotating cylinder viscometer than with the oscillating disk.

### 4.5.2. The Landau-Khalatnikov theory of $\eta_n$

Landau and Khalatnikov (1949) have developed a theory of $\eta_n$ based on Landau's concept of the normal component as a gas of elementary excitations (phonons and rotons). Below 1·6° K. the interactions between the elementary excitations are weak and the gas can be taken to be ideal. The viscosity of this ideal gas can then be calculated in a manner similar to the kinetic theory calculation of the viscosity of an ideal gas of atoms, the phonons and rotons being treated just as though they were ordinary particles.

Imagine the normal component flowing with a velocity $v_{n,x}$ in the direction of the $x$-axis, but with a velocity gradient $\partial v_{n,x}/\partial y$ along the $y$-axis. Then, at any point $P$ the phonons and rotons have a drift velocity $v_{n,x}$, because they *are* the normal component. The phonons and rotons travel in all directions through the liquid, carrying the drift momentum at the point $P$ through an average distance determined by their mean free paths. This mechanism gives rise to a viscosity in exactly the same way that the transfer of momentum by atoms is responsible for the viscosity of a gas. The problem, therefore, is to calculate the efficiency of the sideways transfer of drift momentum from a consideration of the collision processes which determine the mean free paths. Transfer of momentum is due to both phonons and rotons and the viscosity can conveniently be divided into a phonon viscosity and a roton viscosity

$$\eta_n = \eta_{ph} + \eta_r. \tag{4.17}$$

This assumes that phonon-roton collisions are sufficiently frequent to ensure that the phonons and rotons have the same drift velocity at all points and that they do not flow independently of one another.

Considering the roton viscosity first, analysis of a phonon-roton collision shows that the roton plays the role of a heavy particle and the phonon that of a light particle. The 'stopping power' of a roton is therefore much greater than that of a phonon, and it is therefore justifiable to assume that an overwhelming majority of roton paths are terminated by a collision with another roton, particularly above 0·6° K. where there is, in any case, a greater density of rotons than phonons. The roton viscosity can then be shown to take the form

$$\eta_r = \frac{1}{15} N_r t_r \frac{p_0^2}{\mu} \tag{4.18}$$

$$= \frac{\pi}{10} \rho_r \bar{v}_r l_r, \tag{4.19}$$

in which $t_r$ is the mean time between roton-roton collisions, $N_r$ is the number of rotons per cm.$^3$, $l_r$ is the roton mean free path and $\bar{v}_r = (2kT/\pi\mu)^{\frac{1}{2}}$ is the average velocity of the rotons (the velocity of a roton is, of course, $\partial\epsilon/\partial p$). Equation (4.19) is formally similar to the equation giving the viscosity of an ideal gas. Although $t_r$ could not be calculated directly, because insufficient is known about the interaction of two rotons, Landau and Khalatnikov took the interaction between rotons to be a $\delta$-function of their separation and showed that $N_r t_r$ is independent of temperature, which is plausible, since the time between two collisions might be expected to increase as the density of rotons decreases. The roton viscosity ought therefore to be independent of temperature. This is analogous to the fact that the viscosity of an ideal gas is independent of its density.

The phonon viscosity can also be expressed in the form

$$\eta_{ph} = \alpha \rho_{ph} c l_{ph}, \tag{4.20}$$

where $\alpha$ is a numerical factor which can be calculated approximately ($\alpha \simeq 0{\cdot}07$). The factors determining $l_{ph}$ are different above and below 0·9° K. Above 0·9° K. the phonon free paths are terminated in an overwhelming majority of cases by a collision with a roton. As the temperature decreases the density of rotons decreases rapidly and consequently the phonon mean free path increases rapidly, and the phonon viscosity also increases rapidly. After some complicated calculations, Landau and Khalatnikov find that for $T > 0{\cdot}9°$ K.

$$l_{ph} = 2{\cdot}9 \times 10^{-7} T^{-\frac{9}{2}} e^{\Delta/kT} \text{ cm.}, \tag{4.21}$$

$$\eta_{ph} = 8{\cdot}7 \times 10^{-9} T^{-\frac{1}{2}} e^{\Delta/kT} \text{ poise.} \tag{4.22}$$

These expressions are only approximate, since there is another effect influencing the transfer of momentum by the phonons (Khalatnikov, 1952*a*, 1956*b*, *c*). This is the five-phonon process in which two phonons collide and three phonons are emitted and the detailed calculations which take it into account have been described by Khalatnikov (1956*b*). Moreover, below 0·7° K. the number of rotons is so small that the mean free path of a phonon is determined also by collisions with other phonons. One then has either the above five-phonon process or an elastic four-phonon process in which two phonons collide and two phonons emerge with different momenta. In these circumstances, $T < 0{\cdot}7°$ K.

$$l_{ph} = 5{\cdot}0 \times 10^{-8} T^{-\frac{9}{2}} e^{\Delta/kT} / (1 + 2{\cdot}15 \times 10^{-5} T^{\frac{9}{2}} e^{\Delta/kT}) \text{ cm.}, \tag{4.23}$$

$$\eta_{ph} = 3{\cdot}5 \times 10^{-9} T^{-\frac{1}{2}} e^{\Delta/kT} / (1 + 2{\cdot}15 \times 10^{-5} T^{\frac{9}{2}} e^{\Delta/kT}) \text{ poise.} \tag{4.24}$$

The experimental values of $\eta_n$ can be explained rather well by combining the $\eta_{ph}$ of equation (4.22) with a constant roton viscosity $\eta_r \sim 11{\cdot}5 \times 10^{-6}$ poise. In fig. 39 we show the theoretical values of $l_r$, $l_{ph}$ and $\eta_n$. The experimental values of $\eta_n$ are taken first from the rotating cylinder viscometer experiment of Heikkila and Hallett

(1955) and secondly from the experiments by Zinoveva (1956) on the attenuation of second sound by viscous effects at the wall of a resonant cavity (§ 5.3.3). The agreement between theory and experiment may be partly fortuitous, since the values of $\eta_{ph}$ are very sensitive to the parameters $\partial^2\Delta/\partial\rho^2$ and $\partial p_0/\partial\rho$, which are not known accurately. The relevance of these parameters becomes

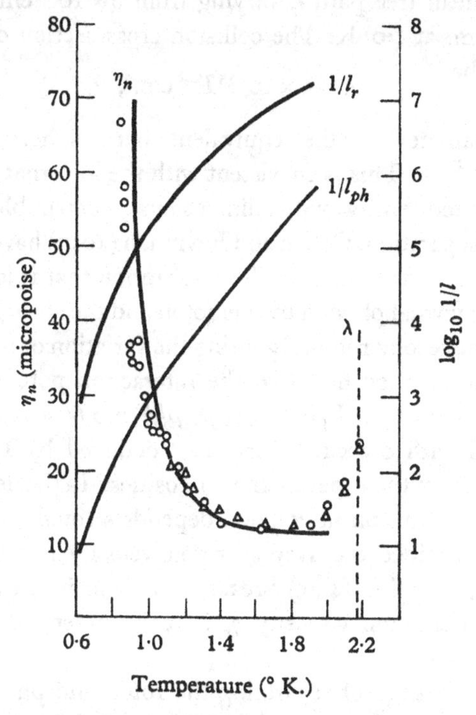


Fig. 39. The Landau-Khalatnikov theory of the viscosity of the normal component. $l_r$, mean free path of a roton in cm.; $l_{ph}$, mean free path of a phonon in cm.; $\eta_n$, viscosity of the normal component; ——, theory; △, Heikkila and Hallett (1955); ○, Zinoveva (1956).

obvious if we consider that the interaction between a phonon and a roton depends upon the fact that the phonon is a region of fluctuating density and the constants $\Delta$ and $p_0$ determining the properties of the rotons are functions of density. However, these parameters are irrelevant to the temperature variation of $\eta_n$ and there can be no doubt that the theory gives a very satisfactory explanation of the rapid increase in $\eta_n$ below 1·6° K. The theory is not applicable above 1·6° K., where the elementary excitations interact strongly, and so cannot explain the rise in $\eta_n$ just below the $\lambda$-point.

Although the theory predicts that the roton viscosity $\eta_r$ is temperature independent, the actual value is chosen empirically to fit the measurements. A value of $11{\cdot}5\times10^{-6}$ poise would correspond to a roton mean free path $l_r$ varying from $6\times10^{-8}$ cm. at 1·9° K. to $5\times10^{-6}$ cm. at 1·0° K. The collision cross-section of the roton would then be

$$\sigma_r \sim 5\times10^{-15}T^{-\frac{1}{2}}\ \text{cm.}^2, \tag{4.25}$$

and the diameter of the equivalent hard sphere would be $\sim 8\times10^{-8}T^{-\frac{1}{4}}$ cm. This is consistent with the idea that a roton is a highly localized entity with dimensions comparable with the interatomic separation. Cohen and Feynman (1957) have calculated the roton-roton interaction on the assumption that it is due to the emission of a virtual phonon by one roton and the absorption of this phonon by the second roton. Ignoring the variation of roton energy with momentum they find that the interaction potential has the form $V_0\,\delta(\mathbf{r}_1-\mathbf{r}_2)$ with $V_0=(\partial\Delta/\partial\rho)^2\,(\rho/c^2)\simeq 0{\cdot}66\times10^{-38}$ erg cm.$^3$ as compared with $0{\cdot}5\times10^{-38}$ erg cm.$^3$ deduced by Landau and Khalatnikov from the experimental viscosities. In the next approximation the roton-roton interaction depends strongly on the roton velocities and, since the average roton velocity is a function of temperature, equation (4.25) breaks down and it is by no means certain that the roton viscosity $\eta_r$ is really independent of temperature.

A possible method of separating the roton and phonon contributions is suggested by recent determinations of the viscosity from the damping of a cylindrical quartz crystal performing electrically excited torsional oscillations in the liquid (Welber and Quimby, 1957; Eisele and Hallet, 1957). The frequencies used were in the range 10 to 35 kc./s. and the penetration depth of the motion into the liquid was $\sim 5\times10^{-5}$ cm. at 1·2° K. This is larger than the roton mean free path, but smaller than the phonon mean free path, and one might therefore expect $\eta_r$ to give its full contribution to $\eta_n$ but $\eta_{ph}$ to be partially suppressed. The initial results suggest that the rapid rise in $\eta_n$ near 1·2° K. due to $\eta_{ph}$ is considerably reduced, but the experiments are not yet sufficiently developed to permit a detailed analysis. Another possible approach (Atkins, 1957) might be to study thermal conduction in channels whose width is greater than the roton mean free path, but less than the phonon mean free path (§ 6.3.2).

## 4.6. Theories of superfluidity

### 4.6.1. The nature of superfluidity

One of the principal difficulties in evaluating a theory of superfluidity is that it is not quite clear what such a theory must explain. In this section we shall therefore give a critical analysis of the various possibilities for the nature of superfluidity, while emphasizing that present experimental evidence cannot decide unambiguously between them.

It is frequently suggested (see F. London, 1954) that superfluidity is a consequence of some stringent restriction on the possible types of flow, such as the restricting equation

$$\operatorname{curl} \mathbf{v}_s = 0 \quad \text{always when} \quad v_s < v_{s,c}. \tag{4.26}$$

Applied to the case of liquid in a cylindrical vessel rotating with a sufficiently small angular velocity, this would imply that the superfluid component could never be made to rotate with the vessel. If the rotation were started above the $\lambda$-point, the whole of the liquid would rotate, but upon cooling below the $\lambda$-point the superfluid component would fall out into a state without rotation. No conclusive experiment has been performed to test this point. The experiments on rotating vessels (§ 4.2.4) merely show that this does not occur in a vessel of diameter 2·7 cm. when the peripheral velocity exceeds 4 cm. sec.$^{-1}$, which is an upper limit to the critical velocity, but what this critical velocity is and how it depends on the dimensions of the vessel is still unknown. The Andronikashvili pile of disks experiment (fig. 2) proves that the superfluid component does not move with the disks when the peripheral velocity is less than about 0·1 cm. sec.$^{-1}$ (see also § 6.5.3), but this is an oscillation experiment in which the superfluid component is given only a few seconds to take up the motion of the disks, and a different result might be obtained if the disks were set into uniform rotation for a long time. Moreover, it is not yet clear whether the critical velocity depends on the diameter of the disks or the separation between adjacent disks.

The evidence therefore still allows the possibility that the superfluid component can rotate with the vessel. If the rotation were started above the $\lambda$-point and the temperature were then lowered below the $\lambda$-point, the liquid might continue to rotate as a whole even at the smallest velocities. Blatt, Butler and Schafroth (1955) claim to have proved that the equilibrium state always requires the whole of the liquid to rotate. Allowing this, one might then argue that there is no mechanism by which the wall can drag the superfluid component into motion. This means absence of friction between the wall and the superfluid component and leads to the idea of superfluidity as a frictionless flow. The ideas of Landau and Ginsburg presented in the next section are along these lines.

Finally, one might concede that there are mechanisms which could bring the superfluid component into rotation, but that these mechanisms are so improbable and so inefficient that they have a negligible effect during the characteristic time of any experiment. To explain the results of the experiments which have been performed so far, the time to bring the superfluid component into rotation would have to be much greater than $10^2$ sec.

Similar arguments apply to the other interesting case of how the superfluid component can flow through a channel without giving up its momentum to the walls. There are three possibilities, (1) any mechanism by which the wall could bring the superfluid to rest would violate a stringent restriction such as curl $\mathbf{v}_s = 0$, or (2) there is no possible mechanism for bringing the superfluid to rest if the velocity is less than $v_{s,c}$, or (3) the processes which might bring the superfluid to rest are improbable, infrequent and inefficient.

### 4.6.2. Interaction of the superfluid component and the walls

Landau (1941) has presented the following argument. He considers the liquid at 0° K., where it consists entirely of the superfluid component, and imagines this component flowing with velocity $v_y$ through a narrow slit, and then argues that the only way the fluid can be brought to rest is by the creation of elementary excitations as the flowing liquid interacts with the walls. Suppose that a new excitation has energy $\epsilon$ and momentum $\mathbf{p}$ in the frame of reference which moves with the liquid. Then its energy in the frame of reference at rest with respect to the walls is given by (see, for instance, Dingle, 1952*a*)

$$\epsilon' = \epsilon + p_y v_y. \tag{4.27}$$

If the kinetic energy of flow is to be reduced, $\epsilon'$ must be negative, or

$$\epsilon + p_y v_y < 0,$$

$$v_y > -\epsilon/p_y. \tag{4.28}$$

Therefore, no suitable excitations can be produced and the flow is frictionless if the velocity is less than a critical value

$$v_{s,c} = \left|\frac{\epsilon}{p_y}\right|_{\text{min.}} \tag{4.29}$$

determined by the excitations which have the smallest value of $\epsilon/p_y$. For phonons $\epsilon = cp$, and $v_{s,c} = c = 2{\cdot}39 \times 10^4$ cm. sec.$^{-1}$. For rotons

$$\begin{aligned} v_{s,c} &= \left|\frac{\Delta}{p_y} + \frac{(p-p_0)^2}{2\mu p_y}\right|_{\text{min.}} \\ &= \frac{1}{\mu}[(2\mu\Delta + p_0^2)^{\frac{1}{2}} - p_0] \\ &\simeq 6 \times 10^3 \text{ cm. sec.}^{-1}. \end{aligned} \tag{4.30}$$

Both these values are many times greater than the observed critical velocities.

Ginsburg (1949) has pointed out that the critical velocity will be reduced if there are other types of excitations for which the minimum value of $\epsilon/p_y$ is less. Such excitations might not be very numerous and might make a negligible contribution to the thermodynamic functions, but they would be of major importance in dissipating the flow. He suggests that the excitations might be of the ideal gas type with

$$\epsilon = \frac{p_x^2 + p_y^2 + p_z^2}{2\mu}. \tag{4.31}$$

He then argues that, if the width of the slit is $d$ in the $z$ direction, the minimum value of $p_z$ is $\sim h/2d$ corresponding to a ground state wavelength of $2d$. $v_{s,c}$ is then the minimum value of

$$\left|\frac{p_y}{2\mu} + \frac{h^2}{8\mu d^2 p_y}\right|_{\text{min.}} = \frac{h}{2\mu d}$$

or

$$v_{s,c} d = \frac{h}{2\mu} = \frac{5 \times 10^{-4} m}{\mu}. \tag{4.32}$$

If $\mu$ has the same order of magnitude as the mass of the helium atom $m$, this equation is close to explaining the experimental results (§ 4.2.1). Ginsburg speculates that the excitations represented by equation (4.31) may be related to surface states, but independently of whether such details are correct or not, the outstanding contribution of his theory is the suggestion that we should search for new excitations which are not necessarily important to the thermodynamics but may be responsible for dissipating the flow after a certain critical velocity is exceeded.

The essence of the Landau-Ginsburg arguments may therefore be summarized as follows. The liquid flowing through the tube can be brought to rest by the creation at the walls of a large number of excitations moving in the opposite direction to the flow and with sufficient momentum to compensate exactly the flow momentum. However, if the excitations are phonons or rotons a very large energy is required to create them and the necessary energy can be derived from the kinetic energy of flow only if the flow velocity is at least $2 \cdot 39 \times 10^4$ cm. sec.$^{-1}$ for the phonons and $6 \times 10^3$ cm. sec.$^{-1}$ for the rotons. Otherwise, for the smaller flow velocities actually encountered, this large energy would have to be abstracted from the walls and the probability of this happening is comparable with the well-known example of the kettle of water boiling when placed on a block of ice. In general, for other types of excitations, the energy needed to create the excitations can be derived from the kinetic energy of flow if the velocity of flow exceeds the smallest value of $\epsilon/p$ for the excitations in question. We therefore look for suitable excitations with $\epsilon/p$ small enough to explain the observed critical velocities, and the quantized vortex lines discussed in the next section appear to satisfy this requirement. Furthermore, in order to prove that superfluidity can exist at all, we must satisfy ourselves that there are no other excitations with vanishingly small values of $\epsilon/p$. This is difficult to prove rigorously, but it will become more plausible as we proceed.

### 4.6.3. Quantized vortex lines

The treatment outlined in this section originated in a suggestion made by Onsager (1949), but the presentation is based on Feynman's (1955) more detailed development of the ideas. We shall return to Onsager's form of the theory in the next section.

If $\psi$ is a wave function describing the liquid at rest, then the wave function $\psi \exp\left[\frac{im}{\hbar}\mathbf{v}.\sum_i \mathbf{r}_i\right]$ represents a uniform translational velocity $\mathbf{v}$ of the liquid as a whole. The function

$$\psi_{\text{irrot.}} = \psi \exp\left[i \sum_i S(\mathbf{r}_i)\right], \tag{4.33}$$

where $S$ is a variable function of position and $S(\mathbf{r}_i)$ is its value at the location of the $i$th atom, represents an irrotational flow and the velocity at any point is

$$\mathbf{v} = \frac{\hbar}{m} \operatorname{grad} S. \tag{4.34}$$

In a multiply-connected region, such as the annulus between two coaxial cylinders, an irrotational flow can exist, but the wave function must be single-valued and the phase must therefore change by an integral multiple of $2\pi$ along a path which passes completely round the annulus. Hence

$$\oint \operatorname{grad} S . \mathbf{ds} = 2\pi n$$

or

$$\oint \mathbf{v} . \mathbf{ds} = 2\pi n \frac{\hbar}{m}. \tag{4.35}$$

The integral on the left-hand side is commonly called the 'circulation'. If the velocity is a function of the radius $r$ only, then

$$v = n \frac{\hbar}{m} \frac{1}{r}. \tag{4.36}$$

A velocity field of this type in an infinite liquid is the familiar vortex line of classical hydrodynamics (Lamb, 1895). It is a circular flow about a line axis, with the velocity increasing as the axis is approached in such a way that $\operatorname{curl} \mathbf{v} = 0$ everywhere except at points on the axis. It can be visualized as similar to a whirlpool on a rapidly moving river or the rotation of water as it empties down a drain, with an accompanying free surface which dips downwards towards the axis (fig. 40*a*). Notice, however, that equation (4.36) describes '*quantized* vortex lines' in which the angular momentum per helium atom about the axis is restricted to an integral multiple of $\hbar$. In what follows we shall be interested only in the 'ground state' vortex line for which $n = 1$.

The energy per unit length of a vortex line is

$$\epsilon_v = \int_a^b \tfrac{1}{2}\rho_s v^2 \, 2\pi r \, dr$$

$$= \int_a^b \tfrac{1}{2}\rho_s \left(\frac{\hbar}{mr}\right)^2 2\pi r \, dr$$

$$= \pi\rho_s \frac{\hbar^2}{m^2} \ln(b/a). \tag{4.37}$$

The upper limit $b$ is related to the size of the container. The lower limit $a$ has the order of magnitude of the interatomic distance $\delta$, but the situation near the axis is likely to be complicated by the high velocities there and the tendency for the centrifugal force to make the atoms fly apart, decreasing the density near the axis and changing the local values of the potential and zero-point energies.

If the axis of a vortex line is bent round into a large circle of radius $R$, one obtains a vortex ring. Its energy is

$$\epsilon_{vr} \approx 2\pi^2 R \rho_s \frac{\hbar^2}{m^2} \ln\left(\frac{R}{a}\right) \tag{4.38}$$

and its momentum perpendicular to the plane of the ring can be shown to be

$$p_{vr} \approx 2\pi^2 R^2 \rho_s \frac{\hbar}{m}. \tag{4.39}$$

It therefore provides us with exactly the type of excitation we need to explain the destruction of superfluid flow in narrow channels, because

$$\frac{\epsilon_{vr}}{p_{vr}} \approx \frac{\hbar}{mR} \ln\left(\frac{R}{a}\right). \tag{4.40}$$

In a channel of width $d$, the maximum value of $R$ would be $\sim \frac{1}{2}d$ and the corresponding critical velocity would be

$$v_{s,c} \approx \frac{2\hbar}{md} \ln\left(\frac{d}{2a}\right). \tag{4.41}$$

This is a very satisfactory result, because it not only makes $v_{s,c}$ increase with decreasing $d$, but it also makes $v_{s,c}d$ a slowly increasing function of $d$ in agreement with experiment. The numerical magnitude of the critical velocity is still too large by a factor of about ten when applied to the film, but this might reasonably be reduced by a more detailed analysis. Feynman (1955) obtained (4.41) by considering the formation of vortex lines at the exit of the channel and has discussed in some detail the complications which may arise.

### 4.6.4. Liquid in a rotating cylindrical vessel

If a liquid is contained in a rotating cylindrical vessel subject to no external couple, then the motion of the liquid can be obtained by applying a thermodynamic argument under conditions of constant temperature $T$, pressure $p$ and total angular momentum $M$. If the vessel has radius $R$, height $L$, moment of inertia $I_v$ and an angular velocity $\Omega$, and the angular velocity of the liquid at a distance $r$ from the axis is $\omega(r)$, then equilibrium corresponds to a minimum value of the energy

$$E = \tfrac{1}{2} I_v \Omega^2 + \pi L \rho \int_0^R \omega^2 r^3 \, dr \tag{4.42}$$

subject to the condition that the total angular momentum is constant:

$$M = I_v \Omega + 2\pi L \rho \int_0^R \omega r^3 \, dr. \tag{4.43}$$

Introducing a Lagrangian multiplier $\lambda$ and varying $\Omega$, we have

$$\frac{\partial}{\partial \Omega}(E - \lambda M) = I_v(\Omega - \lambda) = 0,$$

whence

$$\lambda = \Omega. \tag{4.44}$$

We must then choose the function $\omega(r)$ to give a minimum value of

$$E - \Omega M = -\tfrac{1}{2} I_v \Omega^2 + \pi L \rho \int_0^R (\omega^2 - 2\omega\Omega) r^3 \, dr. \tag{4.45}$$

The solution is obviously

$$\omega = \Omega \quad \text{for all } r. \tag{4.46}$$

The liquid therefore rotates like a solid body with a uniform angular velocity. The velocity at any point is $v = \Omega r$ and $\operatorname{curl} \mathbf{v} = 2\Omega$ everywhere. The free surface is a paraboloid, as shown in fig. 40*b*. The same result is obtained by solving the Navier-Stokes equation for a viscous liquid.

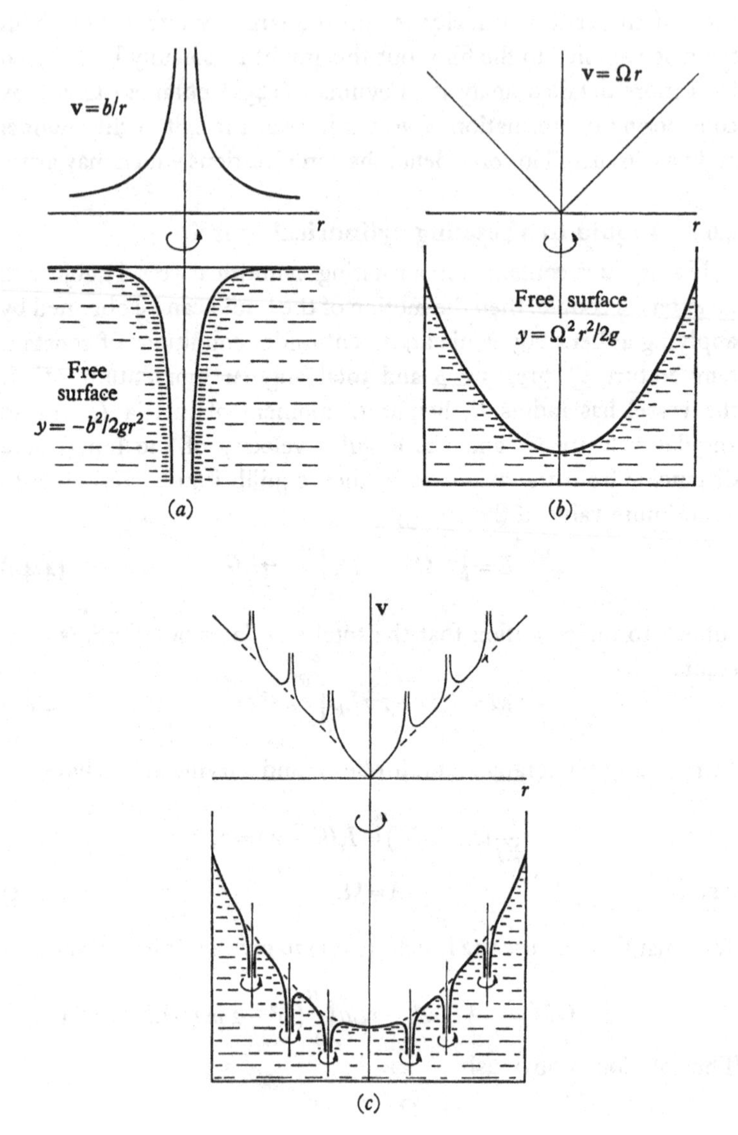


Fig. 40. Possible rotational motions of a liquid and the corresponding form of the free surface. (*a*) A vortex line; (*b*) normal liquid in a rotating vessel; (*c*) liquid helium II in a rotating vessel.

Feynman (1955) maintains that this is not a permissible solution for liquid He II, because, as a consequence of the weakness of the interatomic forces and the looseness of the structure, a region of the liquid containing a few atoms must rotate independently and its angular momentum must be either zero or of the order of $\hbar$. Therefore, if we require that $\text{curl}\,\mathbf{v}=2\Omega$ everywhere as for a rotating solid body, the smallest permissible value of $\Omega$ is of the order $\hbar/m\delta^2$, which is very large ($\delta$ is the interatomic distance). How then can we bring the liquid into rotation to achieve a minimum value of $E-\Omega M$, while maintaining $\text{curl}\,\mathbf{v}_s=0$ throughout most of the liquid? First, of course, the normal component $\rho_n$ has viscosity and will be brought into rotation like a normal liquid, so we need consider only the superfluid component. The total density of the liquid could be brought into rotation by the creation of phonons and rotons at the walls, but this would require an enormous expenditure of energy for the same reasons that we considered in §4.6.2. Feynman's suggestion is that we should embed in the liquid a few quantized vortex lines with their axes parallel to the axis of rotation of the vessel. Then $\text{curl}\,\mathbf{v}_s=0$ everywhere except in the immediate vicinity of the vortex axes, where there are a few atoms which rotate around one another in a manner reminiscent of the rotational states of a polyatomic molecule. The circulation around any curve enclosing $N_v$ vortex lines is

$$\oint \mathbf{v}.\mathbf{ds}=N_v\frac{2\pi\hbar}{m} \tag{4.47}$$

which, on a macroscopic scale, is almost the same as saying that the average curl in the region enclosed by the curve is

$$[\text{curl}\,\mathbf{v}]_{Av}=n_v\frac{2\pi\hbar}{m}, \tag{4.48}$$

where $n_v$ is the number of vortex lines per unit area. Therefore, in the rotating vessel experiment, if we have a uniform density of vortex lines given by

$$2\Omega=n_v\frac{2\pi\hbar}{m}, \tag{4.49}$$

the net result is very similar to rotation like a solid body and the free surface assumes the form of a paraboloid with small indentations near the axes of the vortex lines (fig. 40*c*). However, it can easily be shown that the indentations have a diameter of about $10^{-4}$ cm. and would not have been observed in the actual experiments (§ 4.2.4) which produced an apparently normal paraboloidal surface. Moreover, the energy needed to form the vortex lines can be deduced from equation (4.37) and shown to be only slightly greater than the energy of the solid-like rotation provided that all vortex lines are in their ground state. For example, in Osborne's experiment the angular velocity was about 100 radians/sec., so the vortex lines were spaced about $2 \times 10^{-4}$ cm. apart and their energy was only about 1·0001 times the energy for the solid-like rotation. However, it is possible that by the exercise of a little more ingenuity we could devise a flow pattern with still lower energy.

In later sections we shall see that good evidence for the existence of the vortex lines is provided by the attenuation of second sound in the rotating liquid (§ 6.1) and by certain details of the breakdown of superfluidity above the critical velocity (§ 6.4).

In general a complicated rotational flow may be produced by a complex arrangement of vortex lines which are not necessarily straight but can twist and turn amongst one another. Some lines end on the boundaries but others close up on themselves. In a small region containing $n_v^+$ anticlockwise lines and $n_v^-$ clockwise lines per unit area, the average value of the curl is

$$[\text{curl}\,\mathbf{v}]_{Av} = (n_v^+ - n_v^-)\frac{2\pi\hbar}{m}. \tag{4.50}$$

In the limit, when the density of lines is high and they interweave and move about in a confused tangle, we have the case of turbulence.

In a cylindrical vessel of radius $R$, the smallest possible angular momentum corresponds to the existence of a single vortex line and a peripheral velocity given by

$$v = \frac{\hbar}{m}\frac{1}{R}. \tag{4.51}$$

For smaller peripheral velocities no rotation at all would be permissible, unless we are prepared to allow the single vortex line to be intermittently created and destroyed, so that it is present for only a fraction of the time. Equation (4.51) gives us something very much like the critical velocity we have been discussing, but perhaps we should emphasize that the experimental evidence on this type of situation is far from definite.

Returning to the more general case of several vortex lines, the free energy can probably be lowered still further by the superposition of several states of the type we have been considering, and the correct description would then involve assigning to each region of the liquid a probability that a vortex line is contained within that region. Onsager's (1949) treatment is equivalent to replacing the vortex lines by vortex sheets. He imagines the liquid divided up by coaxial cylindrical surfaces of radii $r_1, r_2, \ldots, r_k, \ldots$ in increasing order of magnitude, and assumes that there is an irrotational flow $v_k = k\hbar/mr$ in the annular region between $r_k$ and $r_{k+1}$, but that the surfaces themselves are vortex sheets across which there is a discontinuity in the velocity. Adjusting the $r_k$ to minimize the energy at constant angular momentum (see F. London, 1954, p. 151) one obtains

$$r_k = \sqrt{\frac{(2k-1)\hbar}{2m\Omega}}, \tag{4.52}$$

where $\Omega$ is the angular velocity of the container. For large values of $k$ the velocity of the liquid converges on the value $\Omega r$, which is characteristic of a normal liquid.

Landau and Lifshitz (1955) also divided the liquid up with cylindrical vortex sheets, but made no attempt to quantize the circulation between the sheets. The velocity between the sheets was taken to be $v_k = b_k/r$ and $b_k$ was determined by minimizing the energy at constant angular momentum as all the $b_k$'s and all the $r_k$'s were varied. Each vortex sheet was assumed to have the same surface energy $\alpha$ per unit area, which was taken to be $\sim \rho_s(kT_\lambda u_2^4/\rho)^{\frac{1}{3}}$ on the basis of a dimensional argument, and later suggested by Ginsburg (1956) to be

$$\sim \frac{\hbar^2}{2m\delta^4} \sim 5 \times 10^{-2}\,\text{erg cm.}^{-2},$$

$\delta$ being the interatomic distance. At high rotational speeds $(\rho_s \Omega^2 R^3/8\alpha \gg 1)$, $b \to \Omega r^2$ and $v \to \Omega r$ as for a normal liquid. However, for slowly rotating liquid this method does not give a few vortex sheets, as in Onsager's treatment, but an infinite number of sheets becoming more densely packed as the axis of rotation is approached. Lifshitz and Kaganov (1956) have calculated the total angular momentum of the liquid under these circumstances and have deduced the effective density as a function of the angular velocity of the vessel. At the angular velocities used in Andronikashvili's (1946) pile of disks experiment the effective density is little larger than $\rho_s$, but at higher angular velocities (when $\rho_s\Omega^2R^3/8\alpha > 1$) the effective density tends towards $\rho$. The numerical predictions of this theory are sufficiently different from those of the Onsager-Feynman theory that one might hope to devise an experiment to decide between them and settle the important question of whether the circulation should be quantized. This point will be taken up again in §6.1, where we shall present the experimental evidence and theoretical arguments advanced by Hall and Vinen (1956) in favour of vortex lines rather than vortex sheets.

To conclude this section, let us emphasize that the flow patterns we have been discussing are stable modes of motion which allow the liquid to rotate without dissipation. The formation of vortex lines as a first stage in the dissipation of the energy of flow is a separate issue which applies to the case when the motion of the liquid is inconsistent with the motion of the boundaries.

CHAPTER 5

# FIRST AND SECOND SOUND

## 5.1. The two types of wave propagation

### 5.1.1. The two wave velocities

If we ignore certain second order terms and irreversible effects, the thermohydrodynamical equations may be written as

$$\rho_s \frac{\partial \mathbf{v}_s}{\partial t} = -\frac{\rho_s}{\rho}\operatorname{grad} p + \rho_s S \operatorname{grad} T, \tag{5.1}$$

$$\rho_n \frac{\partial \mathbf{v}_n}{\partial t} = -\frac{\rho_n}{\rho}\operatorname{grad} p - \rho_s S \operatorname{grad} T, \tag{5.2}$$

$$\frac{\partial \rho}{\partial t} + \operatorname{div}(\rho_s \mathbf{v}_s + \rho_n \mathbf{v}_n) = 0, \tag{5.3}$$

$$\frac{\partial}{\partial t}(\rho S) + \operatorname{div}(\rho S \mathbf{v}_n) = 0. \tag{5.4}$$

Eliminating $\mathbf{v}_s$ and $\mathbf{v}_n$ and dropping in the process the remaining second order terms such as $(\partial \rho_s/\partial t)\, v_s$, one obtains

$$\frac{\partial^2 \rho}{\partial t^2} = \nabla^2 p, \tag{5.5}$$

$$\frac{\partial^2 S}{\partial t^2} = \frac{\rho_s}{\rho_n} S^2 \nabla^2 T. \tag{5.6}$$

The conditions at any point in the liquid may be completely specified by giving the deviations from equilibrium of any two thermodynamic quantities, and it will be convenient to use the entropy $S$ and density $\rho$:

$$\Delta p = \left(\frac{\partial p}{\partial \rho}\right)_S \Delta \rho + \left(\frac{\partial p}{\partial S}\right)_\rho \Delta S, \tag{5.7}$$

$$\Delta T = \left(\frac{\partial T}{\partial \rho}\right)_S \Delta \rho + \left(\frac{\partial T}{\partial S}\right)_\rho \Delta S. \tag{5.8}$$

Equations (5.5) and (5.6) then transform to

$$\frac{\partial^2\rho}{\partial t^2}=\left(\frac{\partial p}{\partial \rho}\right)_S \nabla^2\rho+\left(\frac{\partial p}{\partial S}\right)_\rho \nabla^2 S, \tag{5.9}$$

$$\frac{\partial^2 S}{\partial t^2}=\frac{\rho_s}{\rho_n}S^2\left[\left(\frac{\partial T}{\partial \rho}\right)_S \nabla^2\rho+\left(\frac{\partial T}{\partial S}\right)_\rho \nabla^2 S\right]. \tag{5.10}$$

Let us look for a solution of these equations in the form of a plane wave with a velocity $u$ along the $z$-axis:

$$\rho=\rho_0+\rho'\,e^{i\omega(t-z/u)}, \tag{5.11}$$

$$S=S_0+S'\,e^{i\omega(t-z/u)}. \tag{5.12}$$

Substituting in equations (5.9) and (5.10),

$$\left[\left(\frac{u}{u_1}\right)^2-1\right]\rho'+\left(\frac{\partial p}{\partial S}\right)_\rho\left(\frac{\partial \rho}{\partial p}\right)_S S'=0, \tag{5.13}$$

$$\left(\frac{\partial T}{\partial \rho}\right)_S\left(\frac{\partial S}{\partial T}\right)_\rho \rho'+\left[\left(\frac{u}{u_2}\right)^2-1\right]S'=0, \tag{5.14}$$

where

$$u_1^2=\left(\frac{\partial p}{\partial \rho}\right)_S, \tag{5.15}$$

$$u_2^2=\frac{\rho_s}{\rho_n}S^2\left(\frac{\partial T}{\partial S}\right)_\rho. \tag{5.16}$$

Equations (5.13) and (5.14) are compatible if

$$\left[\left(\frac{u}{u_1}\right)^2-1\right]\left[\left(\frac{u}{u_2}\right)^2-1\right]=\left(\frac{\partial p}{\partial S}\right)_\rho\left(\frac{\partial \rho}{\partial p}\right)_S\left(\frac{\partial T}{\partial \rho}\right)_S\left(\frac{\partial S}{\partial T}\right)_\rho$$

$$=(C_p-C_v)/C_p. \tag{5.17}$$

In liquid helium II, the difference between $C_p$ and $C_v$ is so small that it can be neglected and it is then obvious that there are two solutions for the wave velocity:

$$u\simeq u_1,$$

$$u_1^2=\left(\frac{\partial p}{\partial \rho}\right)_S$$

$$=\frac{\gamma V}{K_T} \tag{5.18}$$

and
$$u \simeq u_2,$$
$$u_2^2 = \frac{\rho_s}{\rho_n} S^2 \left(\frac{\partial T}{\partial S}\right)_\rho$$
$$= \frac{\rho_s}{\rho_n} \frac{TS^2}{C_v}. \tag{5.19}$$

If $u \simeq u_1$, then it is clear from equation (5.13) that $S' \simeq 0$ and this solution therefore corresponds to first sound, in which there are density oscillations at approximately constant entropy. However, several approximations have been made and actual first sound is accompanied by small variations in entropy and temperature, although a more complete analysis (Dingle, 1950) shows that the exact solution corresponds more closely to adiabatic than to isothermal conditions, and so liquid helium II resembles ordinary substances in this respect in spite of its anomalously large thermal conductivity. Similarly, if $u \simeq u_2$, equation (5.14) indicates that $\rho' \simeq 0$ and this second solution corresponds to second sound, in which entropy and temperature fluctuations occur at approximately constant density. Again, however, we must realize that small density and pressure fluctuations do occur in real second sound. Let us now consider the behaviour of $\mathbf{v}_s$ and $\mathbf{v}_n$ in the two types of sound by going back again to the fundamental equations (5.1) and (5.2). In first sound the term in grad $T$ may be ignored and so

$$\mathbf{v}_s = \mathbf{v}_n \quad \text{(first sound)}, \tag{5.20}$$

which implies that the two components move together in phase. In second sound it is the term in grad $p$ which can be ignored and then

$$\rho_s \mathbf{v}_s + \rho_n \mathbf{v}_n = 0 \quad \text{(second sound)}, \tag{5.21}$$

implying that the two components move in opposite directions out of phase in such a way that the net flow of matter is zero. Of course, these phase relationships have to be slightly modified when we consider the exact solutions of the equations, or when we introduce irreversible effects.

### 5.1.2. Irreversible effects and attenuation

The classical effects which attenuate first sound in ordinary substances are viscosity and thermal conductivity. Introducing these factors into the thermohydrodynamical equations, we have

$$\rho_s \frac{\partial \mathbf{v}_s}{\partial t} = -\frac{\rho_s}{\rho} \operatorname{grad} p + \rho_s S \operatorname{grad} T, \tag{5.22}$$

$$\rho_n \frac{\partial \mathbf{v}_n}{\partial t} = \frac{\rho_n}{\rho} \operatorname{grad} p - \rho_s S \operatorname{grad} T + \eta_n(\nabla^2 \mathbf{v}_n + \tfrac{1}{3} \operatorname{grad} \operatorname{div} \mathbf{v}_n), \tag{5.23}$$

$$\frac{\partial \rho}{\partial t} + \operatorname{div}(\rho_s \mathbf{v}_s + \rho_n \mathbf{v}_n) = 0, \tag{5.24}$$

$$\frac{\partial}{\partial t}(\rho S) + \operatorname{div}(\rho S \mathbf{v}_n) = \frac{\chi}{T} \nabla^2 T. \tag{5.25}$$

Apart from the term $\operatorname{div} \rho S \mathbf{v}_n$, the last equation is the same as the equation of heat conduction for ordinary substances and $\chi$ is analogous to the normal thermal conductivity, as distinct from the abnormal thermal conductivity due to the counterflow of the two components. Let us look for a solution corresponding to a first sound wave:

$$\rho = \rho_0 + \rho' \, e^{i\omega(t - z/u)}, \tag{5.26}$$

$$S' = 0, \tag{5.27}$$

$$v_s = v_n = v_0 \, e^{i\omega(t - z/u)}. \tag{5.28}$$

Equation (5.24) then reduces to

$$v_0 = \frac{u}{\rho} \rho'. \tag{5.29}$$

Instead of equation (5.9) one obtains

$$\frac{\partial^2 \rho}{\partial t^2} = \left(\frac{\partial p}{\partial \rho}\right)_S \nabla^2 \rho - \eta_n \operatorname{div} [\nabla^2 \mathbf{v}_n + \tfrac{1}{3} \operatorname{grad} \operatorname{div} \mathbf{v}_n], \tag{5.30}$$

which reduces to

$$u^2 = \left(\frac{\partial p}{\partial \rho}\right)_S + i \frac{4}{3} \frac{\eta_n \omega}{\rho}. \tag{5.31}$$

The solution is complex and the imaginary part of $u$ corresponds to an attenuation $\alpha_1$:

$$\rho = \rho_0 + \rho' \, e^{-\alpha_1 z} \, e^{i\omega(t - z/u)}. \tag{5.32}$$

Putting in the actual figures, the dispersion is found to be negligible and $u_1^2 = (\partial p/\partial \rho)_S$ to a sufficient approximation. The attenuation coefficient is readily shown to be

$$\alpha_1 = \frac{2}{3}\frac{\eta_n \omega^2}{\rho u_1^3}$$

$$= \frac{8\pi^2}{3}\frac{\eta_n \nu^2}{\rho u_1^3}, \tag{5.33}$$

if $\nu$ is the frequency. As we shall see later, the thermal conductivity also gives rise to a term in the attenuation, but this depends upon certain second order terms which were omitted in the preceding analysis.

Now let us consider a solution corresponding to a second sound wave:

$$S = S_0 + S' e^{i\omega(t-z/u)}, \tag{5.34}$$

$$\rho' = 0, \tag{5.35}$$

$$v_n = -\frac{\rho_s}{\rho_n} v_s = v_0 e^{i\omega(t-z/u)}. \tag{5.36}$$

From equation (5.25), ignoring the small correction term involving $\chi$, we obtain

$$v_0 = \frac{uS'}{S}. \tag{5.37}$$

The equivalent of equation (5.6) is

$$\frac{\partial^2 S}{\partial t^2} = \frac{\rho_s}{\rho_n} S^2 \left(\frac{\partial T}{\partial S}\right)_\rho \nabla^2 S$$

$$-\eta_n \frac{\rho_s S}{\rho \rho_n} \operatorname{div}\left[\nabla^2 \mathbf{v}_n + \tfrac{1}{3} \operatorname{grad} \operatorname{div} \mathbf{v}_n\right]$$

$$+\frac{\chi}{\rho T}\left(\frac{\partial T}{\partial S}\right)_\rho \frac{\partial}{\partial t}(\nabla^2 S). \tag{5.38}$$

Inserting the second sound plane wave solution (5.34), this eventually reduces to

$$u^2 = \frac{\rho_s}{\rho_n}\frac{TS^2}{C} + \frac{i\omega}{\rho}\left(\frac{\rho_s}{\rho_n}\frac{4}{3}\eta_n + \frac{\chi}{C}\right). \tag{5.39}$$

The velocity is approximately the same as before $\left(u_2^2=\frac{\rho_s}{\rho_n}\frac{TS^2}{C}\right)$ and the attenuation coefficient is

$$\alpha_2=\frac{\omega^2}{2\rho u_2^3}\left(\frac{\rho_s}{\rho_n}\frac{4}{3}\eta_n+\frac{\chi}{C}\right). \tag{5·40}$$

Later in this chapter we shall see that these are not the only effects which attenuate first and second sound.

## 5.2. First sound

### 5.2.1. The velocity of first sound

The velocity of first sound in liquid helium has been measured by several investigators at various frequencies. The first measurements were made by Findlay, Pitt, Grayson-Smith and Wilhelm (1938) using a standing wave technique at 1·338 Mc./s. The standing waves were formed between a quartz crystal transducer and a plane reflector and the signal at the crystal was followed through nodes and antinodes as the reflector was moved away. More recently, van Itterbeek and Forrez (1954) have used an improved version of this technique at frequencies of 0·220, 0·420, 0·512 and 0·800 Mc./s. and estimate that the accuracy of their velocity measurements is better than 0·1 %. A radar-type pulse technique has been used by Pellam and Squire (1947) at a carrier frequency of 15 Mc./s. and by Atkins and Chase (1951 *a*) at a carrier frequency of 14 Mc./s. In order to determine the limiting value of the velocity at 0° K., Chase (1953) extended the pulse method down to 0·85° K., using frequencies of 2, 6 and 12 Mc./s. and later Chase and Herlin (1955) extended the 12 Mc./s. results still further down to 0·1° K. The optical method of Debye and Biquard has been used by van Itterbeek, van den Berg and Limburg (1954) and by van den Berg, van Itterbeek, van Aardenne and Herfkens (1955), in both cases at a frequency of 1 Mc./s. The principle of this method is that a system of standing acoustical waves involves a periodic spatial variation in density which can be used as an optical diffraction grating, and the nature of the diffraction pattern gives the grating spacing, which is one-half of the wavelength. All these various measurements agree to within their experimental errors, proving that there is less than $\frac{1}{2}$% dispersion over the frequency range 0·22 to 15 Mc./s. Fig. 41 shows how the velocity varies with temperature. The extrapolated value of the velocity at 0° K. is $239 \pm 2$ m. sec.$^{-1}$.

The dip in velocity at the $\lambda$-point is interesting. In liquid helium II there is some effect which makes the velocity of sound decrease as the $\lambda$-point is approached, even though the density is

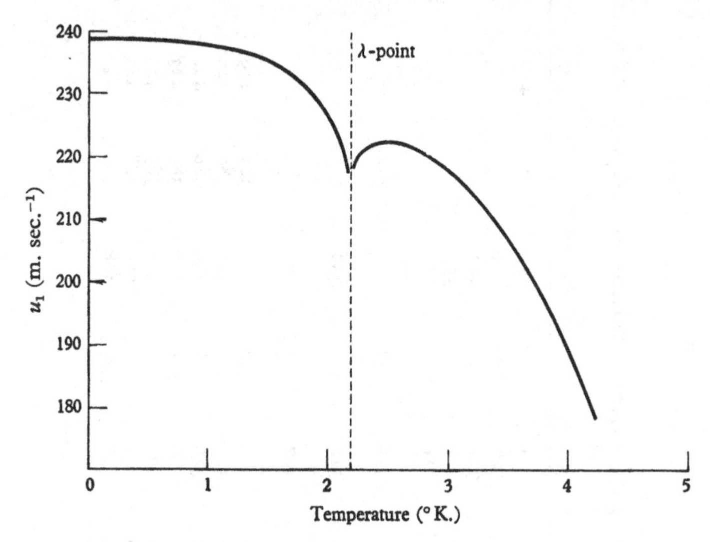


Fig. 41. The velocity of first sound as a function of temperature at the vapour pressure.

increasing. In the immediate vicinity of the $\lambda$-point there are strong attenuation and dispersion effects which will be described in § 5.2.2.

The velocity at high pressures has been measured by Findlay, Pitt, Grayson-Smith and Wilhelm (1939) and by Atkins and Stasior (1953). The results are shown in fig. 42 and are tabulated in Table IV. Findlay *et al.* report a discontinuity at the $\lambda$-point at high pressures, but Atkins and Stasior found no evidence for this. The measured velocities agree satisfactorily with the values which would be predicted from direct measurements of the compressibility (Keesom and Keesom, 1933).

TABLE IV. *The velocity of first sound in liquid helium at various temperatures and pressures (smoothed values)*

| $T$° K. \ $p$ | Vapour pressure | Atmospheres 2·5 | 5·0 | 10·0 | 15·0 | 20·0 | 25·0 | 30·0 | 40·0 | 50·0 | 60·0 | 70·0 |
|---|---|---|---|---|---|---|---|---|---|---|---|---|
| 1·25 | 237 | 257 | 273 | 300 | 326 | 346 | 365 | | | | | |
| 1·50 | 235 | 256 | 272 | 299 | 325 | 345 | 362 | | | | | |
| 1·75 | 233 | 252 | 270 | 298 | 323 | 342 | 355 | | | | | |
| 1·80 | 232 | 251 | 269 | 297 | 321 | 339 | 352 | | | | | |
| 1·90 | 229 | 249 | 267 | 295 | 318 | 333 | 348 | 372 | | | | |
| 2·00 | 227 | 247 | 265 | 292 | 312 | 336 | 358 | 379 | | | | |
| 2·10 | 222 | 240 | 259 | 288 | 317 | 340 | 361 | 382 | | | | |
| 2·20 | 219 | 240 | 259 | 293 | 322 | 344 | 366 | 385 | 419 | | | |
| 2·25 | 220 | 242 | 261 | 295 | 323 | 345 | 367 | 386 | 420 | | | |
| 2·50 | 222 | 244 | 265 | 298 | 326 | 348 | 369 | 388 | 422 | 451 | | |
| 3·00 | 218 | 242 | 264 | 298 | 327 | 349 | 370 | 389 | 423 | 452 | 481 | 510 |
| 3·50 | 206 | 230 | 256 | 296 | 325 | 349 | 370 | 389 | 423 | 452 | 481 | 510 |
| 4·00 | 190 | 216 | 246 | 290 | 321 | 347 | 369 | 388 | 423 | 452 | 481 | 510 |
| 4·20 | 180 | 206 | 241 | 285 | 318 | 345 | 368 | 387 | 422 | 452 | 481 | 510 |

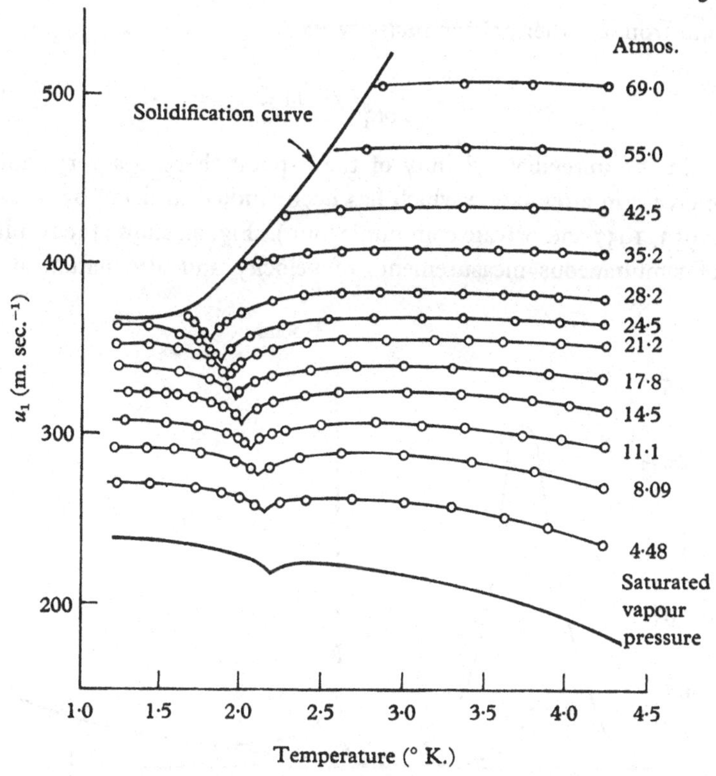


Fig. 42. The velocity of first sound at high pressures (Atkins and Stasior, 1953).

## 5.2.2. The attenuation of first sound near the λ-point

The attenuation at 12 Mc./s. is shown schematically in fig. 43 for the whole temperature range from 0·1 to 4·5° K. The various investigations have all employed the pulse technique at frequencies ranging from 2 to 15 Mc./s. Above 1·2° K. the attenuation varies as the square of the frequency, but at lower temperatures the situation is more complicated for reasons which will be apparent later.

The pioneer investigation was made by Pellam and Squire (1947) at 15 Mc./s. from 4·5 down to 1·57° K. They found that the attenuation in liquid helium I above 3° K. can be adequately explained as the sum of two classical effects arising from the viscosity $\eta$ (see equation (5·33)):

$$\alpha_v = \frac{2}{3}\frac{\omega^2}{\rho u_1^3}\eta \tag{5·41}$$

and from the thermal conductivity $\chi$:

$$\alpha_T = \frac{\omega^2}{2\rho u_1^3}(\gamma - 1)\frac{\chi}{C_p}. \tag{5.42}$$

In the immediate vicinity of the $\lambda$-point there is a very rapid increase in attenuation which has been studied in detail by Chase (1953, 1957 and private communication). Fig. 44 shows the results of simultaneous measurements of velocity and attenuation at a

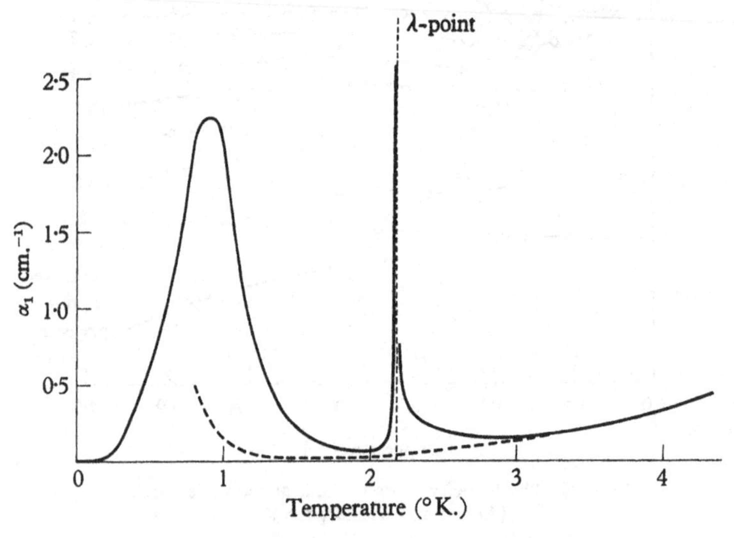


Fig. 43. The attenuation of first sound at 12 Mc./s. ----, classical viscous attenuation.

frequency of 0·985 Mc./s. Notice that the attenuation has a maximum value at about $2 \times 10^{-4}$ °K. below the $\lambda$-point and the velocity has a minimum value at about $3 \times 10^{-4}$ °K. below the $\lambda$-point. There was an uncertainty of about $10^{-4}$ °K. in the exact position of the $\lambda$-point, but this does not affect the result that the minimum in the velocity occurred at a lower temperature than the maximum in the attenuation. The form of the two curves strongly suggests that the attenuation is due to some relaxation time $\tau$ which varies rapidly with temperature and passes through the value $1/\omega$ at the temperature of maximum attenuation.

Landau and Khalatnikov (1954) have given a general treatment of the attenuation of sound in the vicinity of a second order phase transition, with particular emphasis on the case of liquid helium. Their formal presentation can be most easily understood by thinking in terms of the Bragg-Williams theory of the order-disorder transition in alloys. In that case, the Gibbs function of the

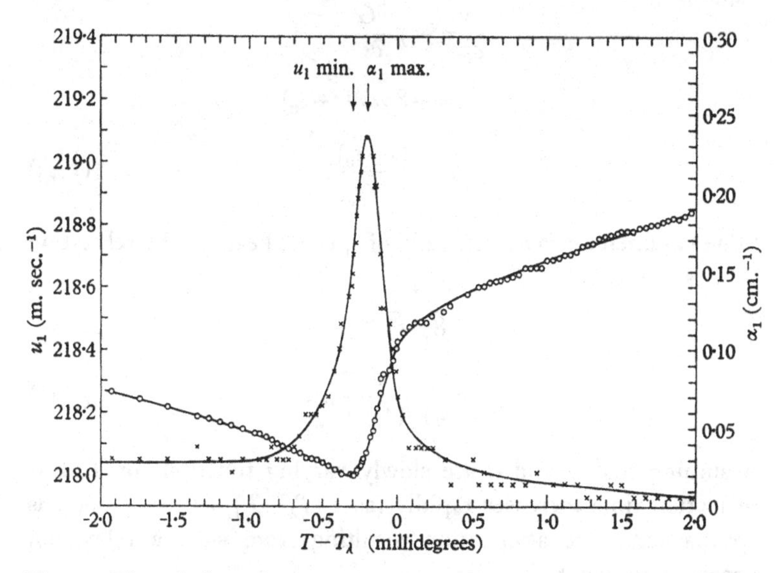


Fig. 44. The velocity and attenuation of first sound in the immediate vicinity of the $\lambda$-point at a frequency of 0·985 Mc./s. (Chase, private communication). $\times$, attenuation $\alpha_1$; $\bigcirc$, velocity $u_1$.

low temperature modification just below the $\lambda$-temperature has the form

$$G_{\text{II}}(p,T,s) = G_{\text{I}}(p,T) - a(p)(T_\lambda - T)s^2 + c(p,T)s^4, \quad (5{\cdot}43)$$

where $G_{\text{I}}$ is the Gibbs function which would be applicable to the high temperature modification if it could exist below the transition temperature and $s$ is the parameter defining the long-range order. The observed value of $s$ is found from the condition $\partial G_{\text{II}}/\partial s = 0$, which gives

$$s_0^2 = \frac{a}{2c}(T_\lambda - T). \quad (5{\cdot}44)$$

For liquid helium the equivalent of $s_0^2$ is $\rho_s/\rho$. In a first sound wave $\mathbf{v}_s \simeq \mathbf{v}_n$ (equation (5.20)), and so $\rho_s/\rho_n$ remains almost unaltered by the motion. Actually, however, in order to achieve true thermodynamic equilibrium, the ratio $\rho_s/\rho$ ought to adjust itself to conform with the small changes in pressure and temperature occurring in the sound wave. The rate of approach of $s$ to its true equilibrium value is

$$\begin{aligned}\frac{ds}{dt} &= -\gamma \frac{\partial G}{\partial s} \\ &= -8\gamma c s_0^2 (s - s_0) \\ &= -\frac{(s - s_0)}{\tau}.\end{aligned} \tag{5.45}$$

(The parameter $\gamma$ is *not* the ratio of specific heats.) The relaxation time is therefore

$$\begin{aligned}\tau &= \frac{1}{8\gamma c s_0^2} \\ &= \frac{1}{4\gamma a (T_\lambda - T)}.\end{aligned} \tag{5.46}$$

Assuming that $\gamma$ and $a$ are slowly varying functions of $T$, this relaxation time increases rapidly (as $1/(T_\lambda - T)$) as the $\lambda$-point is approached. The attenuation resulting from such a relaxation process is known to be

$$\alpha = \frac{\omega^2 \tau}{1 + \omega^2 \tau^2} \frac{u_{\mathrm{I}}^2 - u_{\mathrm{II}}^2}{2u_{\mathrm{I}}^3}, \tag{5.47}$$

where $u_{\mathrm{II}}$ is the velocity at frequencies so low that $s$ has plenty of time to adjust to its equilibrium value before the phase of the wave changes, and $u_{\mathrm{I}}$ is the velocity at very high frequencies when $s$ has no time to change significantly during one period. Applying equation (5.47) to the data of fig. 44, at the temperature of maximum attenuation the relaxation time $\tau$ is $1{\cdot}62 \times 10^{-7}$ sec. and $u_{\mathrm{I}} - u_{\mathrm{II}} = 0{\cdot}356$ m. sec.$^{-1}$. A possible method of estimating $u_{\mathrm{I}}$ might be to extrapolate the velocities in He I from temperatures just above the region of high attenuation and similarly $u_{\mathrm{II}}$ might be estimated by extrapolating the velocities in He II from temperatures just below the region of high attenuation. In this way one obtains $u_{\mathrm{I}} - u_{\mathrm{II}} \sim 0{\cdot}5$ m. sec.$^{-1}$. The data are not adequate for a determination of how $\tau$ varies with $(T_\lambda - T)$, principally because of the uncertainty in the exact position of $T_\lambda$. The Landau-Khalatnikov theory predicts zero attenuation above the $\lambda$-point, whereas in fact the attenuation is anomalously high up to 3° K.

It is still an open question whether the velocity would tend to zero at the $\lambda$-point if the dispersion effects could be eliminated by working at vanishingly small frequencies. A possible difficulty in approaching very close to the $\lambda$-point is that first sound is adiabatic and, in a practical case, the temperature swing might be at least $10^{-5}$° K. and the pressure swing might easily be sufficient to change the $\lambda$-point by at least $10^{-5}$° K. in accordance with the slope of the $\lambda$-curve.

Pippard (1951) has proposed a specific mechanism to explain this anomalously high attenuation near the $\lambda$-point. He postulates that, as a consequence of statistical fluctuations, the liquid just above the $\lambda$-point should be visualized as a matrix of He I containing small inclusions of He II. Because the coefficient of expansion is positive for He I but negative for He II, the sound wave compressions warm the matrix but cool the inclusions. Heat is then transferred from the matrix to the inclusions with a characteristic relaxation time and there results the type of attenuation which is always associated with relaxation effects. Moreover, the cooling effect of the inclusions implies that the temperature rise of the matrix will be reduced, the situation will be intermediate between the ideal isothermal and adiabatic cases and the effective value of $\gamma$ in equation (5.18) will be decreased. This is advanced as the explanation of the sharp decrease in the velocity as the $\lambda$-point is approached from above. Pippard is able to show that this mechanism predicts the correct order of magnitude for the anomalous attenuation above the $\lambda$-point, but an exact calculation proves to be difficult. Just below the $\lambda$-point, the same considerations apply to a matrix of He II containing inclusions of He I, but the situation is then more complicated and not even an order of magnitude calculation is attempted. An essential assumption in the theory is that the anomalous behaviour of the specific heat just above the $\lambda$-point is also due to fluctuations, and this in itself is of sufficient importance to $\lambda$-transitions in general to merit careful consideration.

### 5.2.3. The attenuation of first sound below 2° K.

In their pioneer investigations Pellam and Squire (1947) observed that the attenuation begins to increase again below 2° K. We have already shown (equation (5.33)) that the classical viscous attenuation $\alpha_v$ is present in He II if the coefficient of viscosity is taken to be $\eta_n$, and it can also be shown that the thermal attenuation $\alpha_T$ of equation (5.42) is also applicable if $\chi$ is carefully interpreted as a type of 'normal' thermal conductivity which does not involve counterflow of the two components (Dingle, 1950; Kronig and Thellung, 1950). Above 1·1° K., $\alpha_T$ is negligible compared with $\alpha_v$. The classical viscous attenuation has been plotted on fig. 43 and is seen to be an order of magnitude smaller than the observed attenuation. This is a situation very commonly encountered in other media and it is usually ascribed to a coefficient of second viscosity, $\zeta_2$, which is almost always a consequence of some type of relaxation effect. Then

$$\alpha_1 = \frac{\omega^2}{2\rho u_1^3}\left(\tfrac{4}{3}\eta_n + \zeta_2\right). \tag{5.48}$$

A well-known example is the relaxation effect which gives rise to the anomalous absorption of sound in a diatomic gas when the interchange of energy between the translational motion and the rotational and vibrational modes is not fast enough to keep up with the changes in temperature accompanying the adiabatic compressions and rarefactions of the sound wave.

Khalatnikov (1950, 1952*b*) has postulated a similar mechanism in He II. When a small element of the liquid is adiabatically compressed, as in a sound wave, the phonon-roton gas must adjust itself to the new density and temperature, and the time taken is assumed to be the relaxation time relevant to the absorption of sound. First, the relative numbers of high and low energy excitations must change, but this turns out not to be important because it takes place very rapidly via the elastic collision processes which have already been considered in connexion with the viscosity (§ 4.5.2, Landau and Khalatnikov, 1949). Secondly, the total number of phonons and rotons must also change, and this is a slower process and the determining one. Therefore, instead of the elastic collisions which determine the viscosity, we must now consider inelastic collisions during which phonons or rotons are created or annihilated. Certain simple types of process can be immediately excluded. The three-phonon process, in which a single phonon divides into two phonons, cannot simultaneously satisfy the laws of conservation of energy and momentum. A single roton can split into two rotons only if it has a total energy greater than $2\Delta$, but there are very few such rotons, since most rotons have an energy of the order of $\Delta + kT$. The same objection applies to any purely roton process in which the total number of rotons changes. The important process adjusting the number of phonons present is the five phonon process, in which two colliding phonons produce three scattered phonons, and this gives rise to the relaxation time $\theta_{ph}$. The process adjusting the number of rotons is one in which a roton collides with a phonon of sufficiently high energy to make the emission of two rotons possible, and the characteristic relaxation time of this process is called $\theta_{rph}$. Both these processes can, of course, proceed in the opposite direction and decrease the number of phonons and rotons present. Khalatnikov was able to determine the functional form of the variation of $\theta_{ph}$ and $\theta_{rph}$ with temperature

$$\left.\begin{aligned} \frac{1}{\theta_{ph}} &= a\left(\frac{\partial \mu_{ph}}{\partial N_{ph}}\right)_{\rho,S} T^{11}, \\ \frac{1}{\theta_{rph}} &= b\left(\frac{\partial \mu_r}{\partial N_r} + \frac{\partial \mu_{ph}}{\partial N_{ph}}\right)_{\rho,S} e^{-2\Delta/T}, \end{aligned}\right\} \tag{5.49}$$

$\mu_{ph}$ and $\mu_r$ are the chemical potentials of the phonons and rotons, and $N_{ph}$ and $N_r$ represent the number of phonons and rotons per unit volume. The coefficient of second viscosity then becomes

$$\zeta_2 = \theta_{ph}\left(\frac{\partial p}{\partial \mu_r} + \frac{\partial p}{\partial \mu_{ph}}\right)^2 \frac{\partial \mu_{ph}}{\partial N_{ph}} + \theta_{rph}\left(\frac{\partial p}{\partial \mu_r}\right)^2 \left(\frac{\partial \mu_r}{\partial N_r} + \frac{\partial \mu_{ph}}{\partial N_{ph}}\right), \tag{5.50}$$

all partial derivatives being taken at constant $\rho$ and $S$. Since the nature of the interactions between elementary excitations is not well understood, the parameters $a$ and $b$ cannot be calculated, so Khalatnikov chose them to fit the experimental data of Pellam and Squire. The resulting relaxation times are shown in fig. 45, but they are probably not accurate to better than a factor of two or three. A more recent discussion of this point has been given by

Arkhipov (1954), who obtains $a=1\cdot2\times10^{43}$ and $b=4\cdot3\times10^{50}$. Khalatnikov showed that the relaxation effects do not lead to an appreciable dispersion, the difference between the velocities at zero and infinite frequencies being about 1 % at 2° K. and decreasing rapidly with decreasing temperature.

At the time this theory was advanced the only available experimental results were those of Pellam and Squire extending down to 1·57° K., which is just below the minimum in the attenuation curve. When the measurements were extended down to 1·2° K. by Atkins

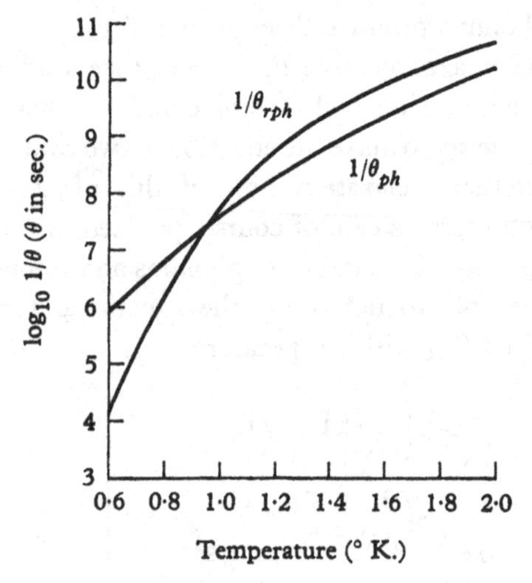


Fig. 45. The relaxation times in Khalatnikov's theory of the absorption of first sound.

and Chase (1951 *b*) they found an extremely rapid rise in attenuation in striking qualitative agreement with Khalatnikov's prediction. In these latter experiments, with $\omega\sim10^8$ c./s. and $T>1\cdot2°$ K., the conditions $\omega\theta_{ph}\ll1$ and $\omega\theta_{rph}\ll1$ were always satisfied, so that equation (5.48) was valid and the attenuation varied as the square of the frequency. But the relaxation times increase rapidly with decreasing temperature and near 1° K. $\omega\theta\sim1$. The attenuation would then be expected to pass through a maximum and to cease to be proportional to $\omega^2$. The experiments of Chase (1953) extending down to 0·85° K. and using frequencies of 2, 6 and 12 Mc./s. confirmed this expectation (fig. 46). The 12 Mc./s. results were extended still further right down to 0·1° K. by Chase and Herlin (1955) using a magnetic cooling technique (fig. 47). They obtained a double peak near 1° K., but this can probably be explained as an error in measurement (Chase, 1956) and does not appear in the later experiments of Whitney (1957) and Newell and Wilks (1956).

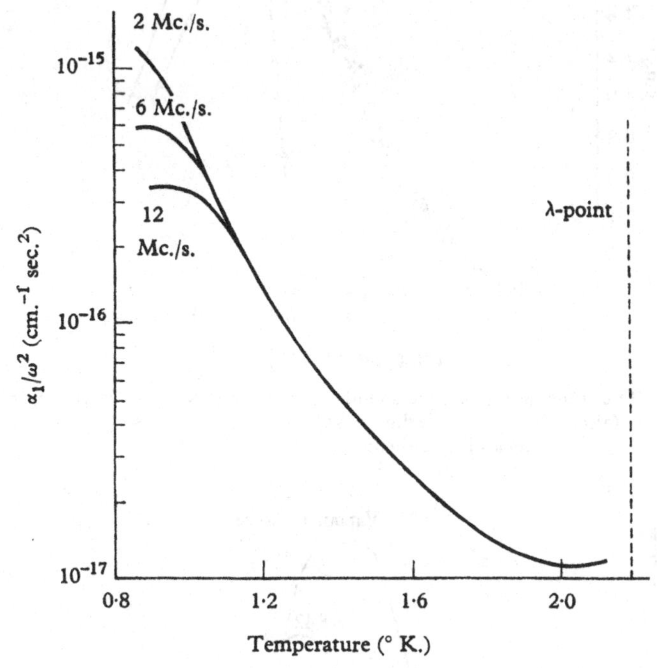


Fig. 46. The attenuation of first sound just below 1° K. (after Chase, 1953).

The detailed interpretation of the temperature region near 1° K. should be attempted with caution, because the classical viscous and thermal attenuations (equations (5.41) and (5.42)) each contribute about 20 % to the total attenuation. The normal viscosity increases rapidly with decreasing temperature as the mean free path for elastic collisions increases (§4.5.2, fig. 39), but the situation soon becomes complicated when the mean free path becomes comparable in magnitude with the wavelength of the sound ($\sim 2 \times 10^{-3}$ cm.) and for this reason the viscous attenuation probably also has a maximum near 1° K. The same considerations apply to the thermal conductivity $\chi$, which is discussed more fully in §5.3.3.

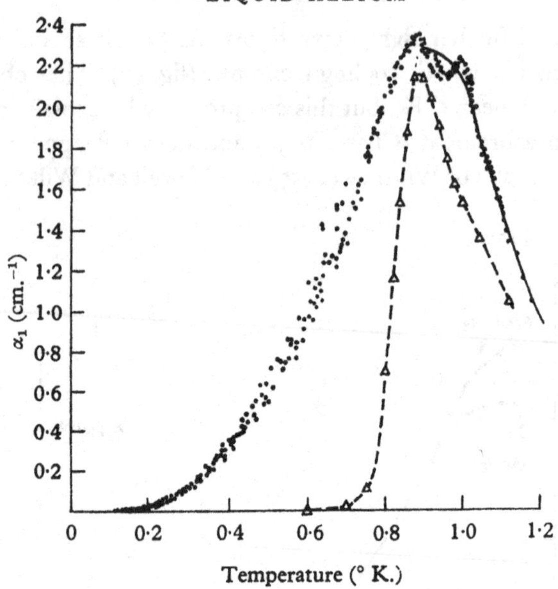


Fig. 47. The attenuation of first sound down to 0·1° K. at a frequency of 12·1 Mc./s. (after Chase and Herlin, 1955). ——, earlier results shown in fig. 46; ---△---, Khalatnikov's theory.

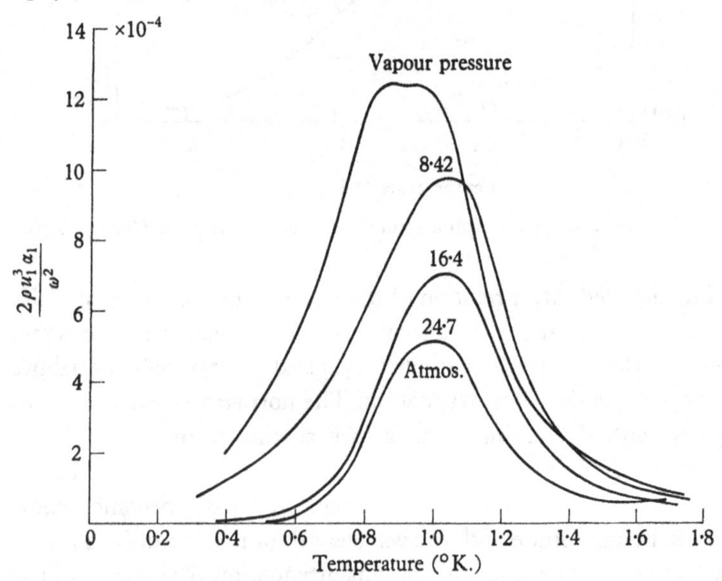


Fig. 48. The attenuation of first sound at high pressures (after Newell and Wilks, 1956). Frequency = 14·4 Mc./s. The number adjacent to the peak of each curve is the pressure in atmospheres.

In addition, note that $\gamma - 1$, which appears in equation (5.42), is zero at 1·15° K. and has a maximum near 0·9° K.

The attenuation below 0·8° K. is too large to be explained by extrapolating any of the theories applicable at higher temperatures. Between 0·8 and 0·3° K. it is proportional to $T^{2\cdot 8}$, but below 0·3° K. it varies as a still higher power of $T$.

Newell (1955), Newell and Wilks (1956) and Whitney (1957) have measured the attenuation at higher pressures. Some of their results are shown in fig. 48. Referring to equation (5.48), much of the variation of $\alpha_1$ with pressure comes from variations in $u_1$ and $\rho$, so the quantity plotted in fig. 48 is

$$\frac{2\rho u_1^3 \alpha_1}{\omega^2} = \tfrac{4}{3}\eta_n + \zeta. \tag{5.51}$$

## 5.3. Second sound

### 5.3.1. The nature of second sound

Since second sound is unique to liquid helium, some of its fundamental aspects will be described in detail. Much of this section is based on papers by Peshkov (1948*b*), Dingle (1948, 1950), Pellam (1949) and Osborne (1948, 1951).

Consider a plane wave of second sound travelling in the $z$-direction

$$T = T_0 + T' \, e^{i\omega(t - z/u_2)}. \tag{5.52}$$

A more complicated second sound disturbance, such as a pulse, may be Fourier analysed into plane waves of this type. The velocity of the normal component is in the $z$-direction.

$$v_{nz} = v_{nz,0} \, e^{i\omega(t - z/u_2)}. \tag{5.53}$$

Inserting these two equations into equation (5.2) and remembering that grad $p$ is zero to first order,

$$v_{nz} = \frac{\rho_s}{\rho_n} \frac{S}{u_2} T'. \tag{5.54}$$

The heat flow is

$$W_z = \rho S T v_{nz}$$

$$= \rho \frac{\rho_s}{\rho_n} \frac{T S^2}{u_2} T' \tag{5.55}$$

or

$$W_z = \rho C u_2 T'. \tag{5.56}$$

The heat flow is therefore in phase with the temperature $T'$ and is 90° out of phase with the temperature gradient $\partial T'/\partial z$. It is possible to make an analogy with an electrical transmission line: $W_z$ is the analogue of the electric current, $T'$ is the analogue of the voltage and the thermal impedance $(\rho C u_2)^{-1}$ is the analogue of the characteristic impedance of the transmission line. Since the specific heat $C$ decreases rapidly with decreasing temperature, the excitation of second sound becomes easier at lower temperatures, because a larger temperature swing $T'$ is obtained for the same heat input (Osborne, 1948).

The kinetic energy per unit volume is

$$\text{K.E.} = \tfrac{1}{2}\rho_n v_n^2 + \tfrac{1}{2}\rho_s v_s^2$$

$$= \frac{1}{2}\frac{\rho\rho_n}{\rho_s} v_n^2 \qquad (5.57)$$

since $\rho_s v_s + \rho_n v_n = 0$. The potential energy is the work done to produce the existing fluctuations in the thermodynamic quantities

$$\text{P.E.} = \delta U + p\delta V - T\delta S.$$

This is zero to first order, but to second order it is (Landau and Lifshitz, 1938, p. 114)

$$\text{P.E.} = \tfrac{1}{2}(\delta S . \delta T - \delta p . \delta V).$$

In second sound the variations in $p$ and $V$ may be ignored, so the potential energy per unit volume is

$$\text{P.E.} = \tfrac{1}{2}\delta S . \delta T$$

$$= \tfrac{1}{2}\rho \frac{C}{T}(\delta T)^2.$$

Using equation (5.54) to replace $\delta T = T'$ by $v_n$

$$\text{P.E.} = \tfrac{1}{2}\rho \frac{C}{T}\left(\frac{\rho_n u_2}{\rho_s S}\right)^2 v_n^2$$

$$= \frac{1}{2}\frac{\rho\rho_n}{\rho_s} v_n^2. \qquad (5.58)$$

The density of potential energy is therefore equal to the density of kinetic energy, as it should be in wave propagation. The total density of mechanical energy is

$$\text{T.M.E.} = \frac{\rho\rho_n}{\rho_s} v_n^2. \qquad (5.59)$$

The instantaneous flow of mechanical energy per unit area is easily found if one remembers that $v_n$ varies with $t$ and $z$ in accordance with equation (5.53)

$$P = u_2 \frac{\rho \rho_n}{\rho_s} v_n^2 \tag{5.60}$$

or, using equations (5.54) and (5.55),

$$P = \frac{T'W}{T}. \tag{5.61}$$

$P$ can be considered to be the analogue of the Poynting vector in electrodynamics (Peshkov, 1948*b*). The reader should distinguish carefully between the heat current $W$ and the flow of mechanical energy $P$. A more complete treatment of these matters, which does not ignore the small variations in density, has been given by Dingle (1948).

Now let us consider the factors involved in the transmission and reception of second sound. Fig. 49*a* represents a typical transmitter, which might consist of a thin heating layer of carbon deposited on a plastic backing. The heater is usually so thin compared with the wavelength of a thermal wave in it that its temperature may be considered to be the same at all points, whereas the backing is usually so thick that the thermal waves have effectively died out before they reach the other side of it and so it may be considered as infinite in extent. If the current per unit width of the heater is

$$i = i_0 \cos \tfrac{1}{2}\omega t \tag{5.62}$$

the instantaneous Joule heat per unit area is

$$\begin{aligned} q &= r i_0^2 \cos^2 \tfrac{1}{2}\omega t \\ &= \tfrac{1}{2} r i_0^2 (1 + \cos \omega t), \end{aligned} \tag{5.63}$$

where $r$ is the electrical resistance per unit area. The constant term in this equation represents a steady heat input and does not interest us. The second sound arises from the oscillating term and therefore has a frequency which is double that of the alternating current fed into the heater. The heat input can therefore be represented by the real part of

$$q = q_0 e^{i\omega t}. \tag{5.64}$$

If the temperature of the heater is $T_0 + T_h e^{i\omega t}$ and its thermal capacity per unit area is $\Gamma_h$, the instantaneous supply of heat to the heater is

$$q_h = i\omega\Gamma_h T_h e^{i\omega t} \tag{5.65}$$

and its equivalent thermal impedance is $1/i\omega\Gamma_h$ per unit area. The heat current through the liquid in the immediate vicinity of the heater is

$$q_l = \rho C u_2 T_l e^{i\omega t}. \tag{5.66}$$

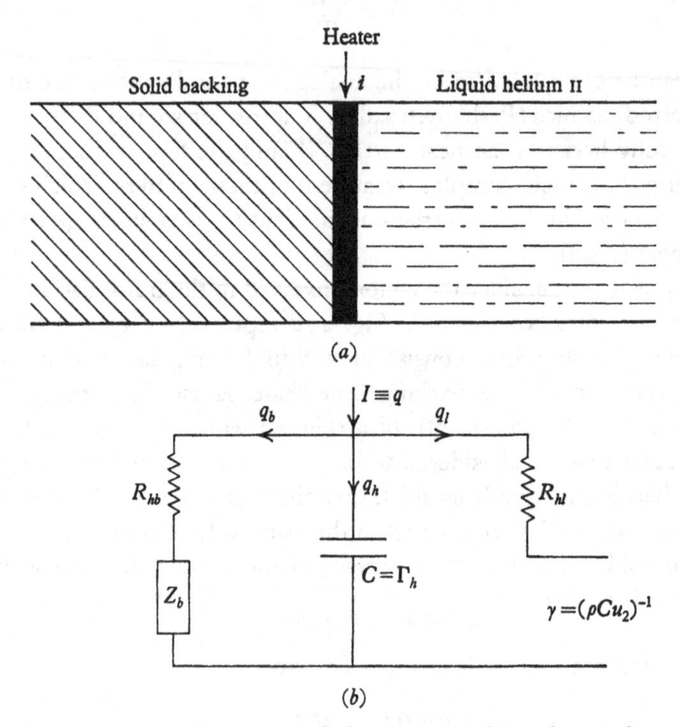


Fig. 49. The electric circuit equivalent to a second sound transmitter. (a) Typical transmitter; (b) equivalent electrical circuit.

The characteristic impedance of the liquid is

$$\gamma = (\rho C u_2)^{-1}. \tag{5.67}$$

As we shall explain more fully in §6.7, a heat flow from the heater to the liquid is accompanied by a temperature discontinuity at the surface known as the Kapitza boundary effect:

$$q_l = (T_h - T_l)\, e^{i\omega t}/R_{hl}, \tag{5.68}$$

where $R_{hl}$ is the Kapitza thermal impedance. A similar effect probably exists between the heater and its backing and the heat current into the backing is

$$q_b = \frac{(T_h - T_b)}{R_{hb}} e^{i\omega t}. \tag{5.69}$$

If the solid backing has a density $\rho_b$, a specific heat $C_b$ per g. and a thermal conductivity $\chi_b$, the thermal wave in the solid is known to have the form

$$T_0 + T_b \exp\left[ z(1+i)\sqrt{\frac{\rho_b C_b \omega}{\chi_b}} \right]. \tag{5.70}$$

The heat current through the backing in the vicinity of the heater is

$$\begin{aligned} q_b &= \chi\left(\frac{\partial T}{\partial z}\right)_{z=0} \\ &= (1+i)(\rho_b C_b \chi_b \omega)^{\frac{1}{2}} T_b e^{i\omega t}. \end{aligned} \tag{5.71}$$

The thermal impedance of the solid backing is therefore

$$Z_b = \tfrac{1}{2}(1-i)(\rho_b C_b \chi_b \omega)^{-\frac{1}{2}}. \tag{5.72}$$

The total heat supplied is used in part to raise the temperature of the heater, but some of it flows across the boundaries into the liquid and the backing

$$q = q_h + q_l + q_b. \tag{5.73}$$

These equations will readily be seen to be consistent with the equivalent electrical circuit of fig. 49*b*. The heat $q$ is equivalent to a current $I$ fed into three circuits in parallel. The heater is represented by a capacitor $C = \Gamma_h$. The boundary effects are represented by the resistances $R_{hl}$ and $R_{hb}$, although we ought not to exclude the possibility that these quantities are complex at sufficiently high frequencies. The liquid helium is replaced by a transmission line of characteristic impedance $\gamma$ and the backing by a complex impedance $Z_b$. The quantity of most interest is the temperature in the liquid, which is seen to be

$$T_l = \tfrac{1}{2} i_0^2 r \frac{\gamma}{R_{hl} + \gamma} \left[ \frac{1}{R_{hb} + Z_b} + \frac{1}{R_{hl} + \gamma} + i\Gamma_h \omega \right]^{-1}. \tag{5.74}$$

Except in the immediate vicinity of the $\lambda$-point $R_{hl} \gg \gamma$, and at low frequencies $\Gamma_h \omega$ and $1/Z_b$ may be neglected, so the equation simplifies to

$$T_l = \tfrac{1}{2} i_0^2 r \gamma = \tfrac{1}{2} i_0^2 r (\rho C u_2)^{-1}. \tag{5.75}$$

At higher frequencies $T_l$ is reduced because $\Gamma_h$ becomes important, as does $Z_b$ if we can assume $R_{hb}$ is not much greater than $R_{hl}$. Osborne (1951) experimented with a transmitter and receiver consisting of graphite deposited on a plastic backing, and found that his square pulses were received with a rise time of the order of 100–200 $\mu$sec., suggesting a fall off in the efficiency of the transducers at frequencies of the order of 1 kc./s.

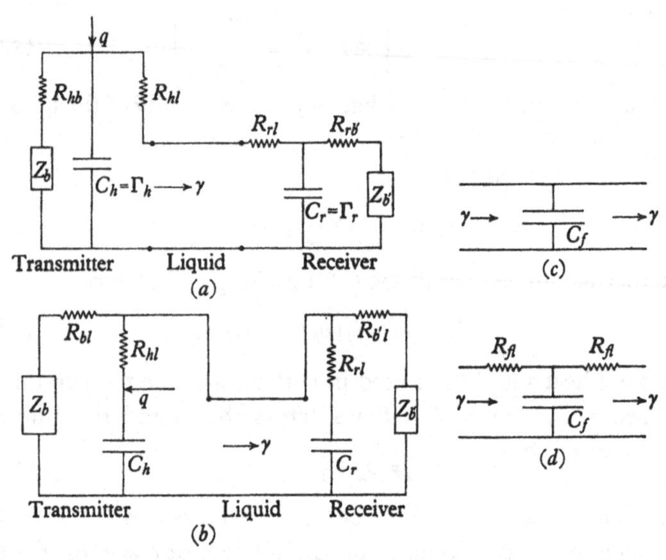


Fig. 50. Equivalent electric circuits of various situations encountered in second sound work. (*a*) Graphite transmitter and receiver; (*b*) wire transmitter and receiver; (*c*) copper foil without Kapitza boundary effect; (*d*) copper foil with Kapitza boundary effect.

The receiver is always a resistance thermometer, usually a layer of graphite or a phosphor-bronze wire. It may be represented by terminating the transmission line with a suitable circuit element. The case of a graphite transmitter and a graphite receiver is shown in fig. 50*a*. At low frequencies $1/\Gamma_r\omega$ and $Z_{b'}$ are large compared with the Kapitza resistance $R_{rl}$ and the full temperature swing in the liquid appears in the thermometer. In fact the temperature amplitude has twice the value given by equation (5.75) because of the reflexion at the receiver. At higher frequencies there is some loss of temperature swing across $R_{rl}$. Osborne found that his pulses had a much shorter rise time if the heater was a constantan wire and the receiver a phosphor-bronze wire. When the wires are wound on an open frame this is obviously a consequence of the elimination of the backing, but there is also an improvement even when the wire is placed immediately in front of a solid backing, presumably because there is not good thermal contact between the wire and the backing, which is therefore coupled directly to the liquid through a Kapitza boundary resistance. This may be better understood by considering the equivalent circuit of fig. 50*b*.

The Kapitza boundary resistance plays a significant part in these considerations, so Osborne investigated its importance by placing a thin copper foil in the direct path of the second sound beam. The thickness of the foil was very small compared with the penetration of thermal waves into it, so, in the absence of a Kapitza resistance, it could have been represented by a capacitor $C_f$ across the transmission line as in fig. 50*c*. Since $1/\omega C_f$ was large compared with $\gamma$, this capacitor would have had a negligible effect and the second sound would have passed through the foil with little loss in intensity. In practice no second sound appeared on the other side of the foil, which is readily understandable in terms of the circuit of fig. 50*d* which includes the Kapitza boundary resistance $R_{fl} \gg \gamma$.

An alternative method of detecting second sound was suggested by Onsager and used by Lane, Fairbank, Schultz and Fairbank (1946, 1947). When second sound is reflected from the liquid-vapour interface, the temperature oscillations of the liquid surface produce a periodic vaporization and the resulting puffs of vapour give rise to first sound in the vapour phase which can be detected by an ordinary microphone. Dingle (1948, 1950) has discussed the theory of this conversion.

Second sound is generated at one face of a porous partition when first sound is incident on the opposite face (Peshkov, 1948*a*). The pressure variations of the first sound force the superfluid component backwards and forwards through the fine pores to produce a periodic heating and cooling on the exit side.

Kurti and McIntosh (1955) have developed a second sound transmitter consisting of a block of paramagnetic salt, the temperature of which is made to oscillate by the application of an alternating magnetic field. This has the advantage that the temperature swing is equally positive and negative and there is no superimposed steady heat flow as in most of the other methods. The method could also be used to produce a negative temperature pulse.

### 5.3.2. The velocity of second sound

The pioneer experiments in which Peshkov (1944, 1946) discovered second sound and measured its velocity made use of continuous waves and a resonant cavity of variable length. His results were confirmed by Lane, Fairbank, Schultz and Fairbank (1946, 1947) who generated continuous waves of second sound at a heater immersed in the liquid, allowed the second sound to impinge on the liquid surface and become converted into first sound in the vapour, and then detected this with a microphone hung above the surface. During their experiment the surface slowly fell as the liquid evaporated away, and they observed a resonance whenever the height of the surface above the heater was an exact number of half wavelengths. The pulse technique (Pellam, 1949; Osborne, 1951) also gives velocities in agreement with these earlier results. Combining these investigations we may conclude that there is no measurable dispersion between $10^2$ and $10^4$ c./s. in the temperature range from 1·2° K. to the $\lambda$-point. As the result of a careful investigation conducted at a temperature of 1·63° K., Peshkov (1949) found that the velocity was also constant to within the experimental accuracy of 2 % at lower frequencies down to 10 c./s. Below 1° K. the situation is not so simple, as we shall see later.

These initial measurements above 1·2° K. are in very satisfactory agreement with the formula

$$u_2^2 = \frac{\rho_s}{\rho_n} \frac{TS^2}{C}, \tag{5.76}$$

but we should remember that the values of $\rho_s$ and $\rho_n$ given by the Andronikashvili experiment (§ 1.3) are not very accurate below 1·5° K. Frequently equation (5.76) is accepted and used to calculate $\rho_n$ from the observed velocities of second sound (de Klerk, Hudson and Pellam, 1954).

In accordance with equation (5.76) the velocity is zero at the $\lambda$-point, where $\rho_s = 0$, and rises to a flat maximum near 1·65° K. (fig. 51). The subsequent behaviour at lower temperatures was once a subject of controversy, since Tisza (1947) had predicted that $u_2$

would fall to zero at 0° K., whereas Landau (1941) had predicted that $u_2 = u_1/\sqrt{3}$ at 0° K. This difference arose, not so much because of the distinction between the Landau theory and the theory of Bose-Einstein condensation, but rather because Tisza associated the Debye waves in the liquid with the superfluid component, while in Landau's theory they are inevitably associated with the

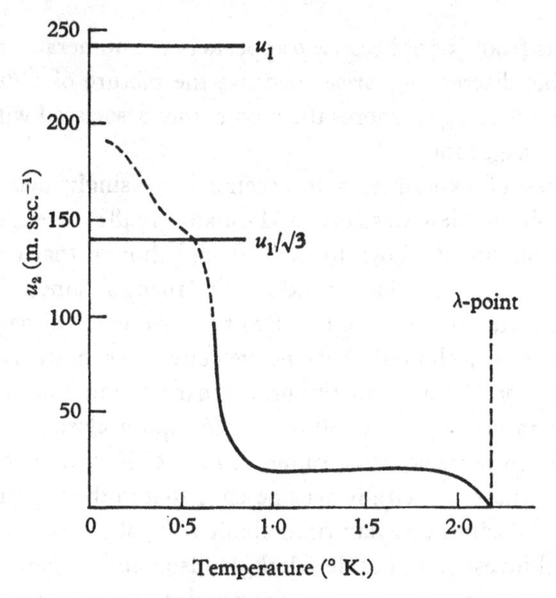


Fig. 51. The velocity of second sound at the vapour pressure.

normal component. Near 0° K., the normal component in Landau's theory is composed entirely of phonons and so

$$S = \alpha T^3,$$

$$C = 3\alpha T^3,$$

$$E = \tfrac{3}{4}\alpha T^4,$$

$$\rho_n = \tfrac{4}{3}\rho\frac{E}{u_1^2} = \frac{\rho\alpha T^4}{u_1^2},$$

$$\rho_s \simeq \rho. \tag{5.77}$$

Inserting these values into equation (5.76), one readily obtains

$$u_2 = u_1/\sqrt{3} \quad \text{as} \quad T \to 0^\circ \text{K} \tag{5.78}$$

Ward and Wilks (1951, 1952) have derived this equation independently of the two fluid theory by considering second sound as a density wave in a phonon gas. Dingle (1952*a*) has pointed out, though, that this approach leads to the formula

$$u_2^2 = \frac{\rho}{\rho_n}\frac{TS^2}{C}, \tag{5.79}$$

which differs from (5.76) because $\rho$ appears in the numerator rather than $\rho_s$. This discrepancy arises because the picture of a density wave in the phonon gas ignores the momentum associated with the superfluid background.

In a series of experiments it became increasingly clear that Landau's point of view was correct. Peshkov (1948*c*) extended his velocity measurements down to 1·03° K. and showed that there is a minimum at 1·12° K. Maurer and Herlin (1949) obtained similar results independently. Pellam and Scott (1949), using a magnetic cooling technique, showed that the velocity rises quite rapidly below 1° K. and Atkins and Osborne (1950) found that at still lower temperatures $u_2$ levels off at a value quite close to $u_1/\sqrt{3}$. However, when more accurate values of $u_1$ at 0° K. were available (Atkins and Chase, 1951*a*), it became clear that in the experiment of Atkins and Osborne $u_2$ had risen about 10% above $u_1/\sqrt{3}$. The more careful investigation of de Klerk, Hudson and Pellam (1954) showed that near 0·5° K. $u_2$ has a plateau close to the value $u_1/\sqrt{3}$ but rises again at still lower temperatures and tends towards $u_1$ at 0° K. (fig. 51). We shall now show that this increase from $u_1/\sqrt{3}$ to $u_1$ is probably due to mean free path effects.

All these investigations below 1° K. used the pulse technique and it was observed that the pulse begins to broaden at 0·9° K. and at 0·1° K. is violently distorted to many times its original length (fig. 52). Just below 1° K. some of the distortion may be caused by dispersion of the frequencies constituting the pulse (Dingle, 1952*b*), but the major effect at lower temperatures is undoubtedly that the mean free path of the phonons becomes comparable with the wavelengths of the pulse components and even with the dimensions of the apparatus itself (Gorter, 1952; Atkins, 1953*a*; Mayper and Herlin, 1953). The predictions of Landau and Khalatnikov embodied in fig. 39 suggest that the mean free path is about 0·1 mm.

at 0·8° K., but is greater than 1 cm. below 0·5° K. If we visualize second sound as a density wave in the phonon gas, it is obvious that it cannot exist unless the phonons establish contact with one another by collisions in a region of dimensions less than the wavelength.

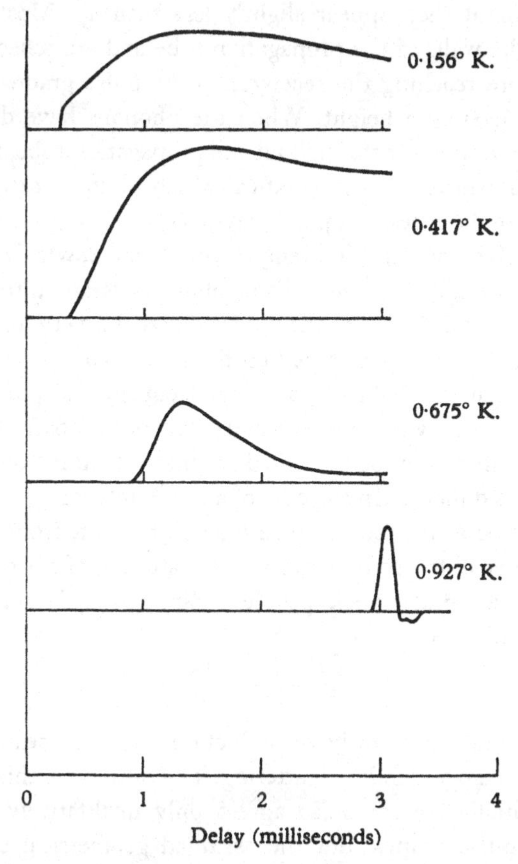


Fig. 52. Second sound pulses below 1° K. Width of input pulse = 20 μsec. at 0·156° K. and 40 μsec. at all other temperatures. Path length = 6·25 cm. (after Kramers, van Peski-Tinbergen, Wiebes, van den Burg and Gorter, 1954).

At 0·1° K., where the mean free path of the phonons is probably many times larger than the dimensions of the apparatus, the propagation of a square heat pulse may be pictured as follows. The heat pulse generates phonons at the transmitter and these phonons subsequently describe paths which may involve many collisions with the walls of the propagation tube, but negligibly few collisions with other phonons. Some phonons travel directly to the receiver and form the rising edge of the pulse, which therefore corresponds to the phonon velocity $u_1$, except that the arrival of the first few phonons might be masked by the receiver noise and the apparent velocity might then appear slightly less than $u_1$. Most phonons travel to the walls of the propagation tube and are reflected many times before reaching the receiver, so the pulse gradually builds up to its maximum height. When the phonons have distributed themselves uniformly throughout the propagation tube, the liquid is at a uniform excess temperature which in some cases may be smaller than the excess temperature corresponding to the maximum of the received signal. The temperature then slowly falls as heat leaks out through the tube walls or phonons escape through holes in the walls, and this gives the received signal a very long tail. In practice the pulse tail observed on the oscilloscope is often determined principally by the poor low frequency response of the receiver. In this way the very elongated pulse observed at very low temperatures can be explained qualitatively, but a quantitative treatment is difficult. In the case of an infinitely long, very narrow tube of radius $R$, the temperature at a point remote from the transmitter can be obtained by treating the scattering of the phonons at the walls as a 'random walk' problem (Ziman, 1953*b*) and this leads to the diffusion equation

$$\frac{\partial T}{\partial t} = \tfrac{2}{3} u_1 R \frac{\partial^2 T}{\partial x^2}, \tag{5.80}$$

which is the equation for heat conduction when the mean free path is determined by boundary scattering. The solution of this equation for an initially square pulse agrees only qualitatively with the observed pulse shapes, but the assumed geometrical conditions correspond poorly with the experimental ones.

Kramers, van Peski-Tinbergen, Wiebes, van den Burg and Gorter (1954; see also Kramers, 1955) have made a very thorough investigation of the propagation of heat pulses below 1° K. As might be expected from the above considerations, the shape and arrival time of the received signal were found to vary with the length and diameter of the propagation tube, even at temperatures as high as 0·9° K. Between 0·7 and 0·9° K. the observations could be explained simply in terms of ordinary dispersion of the pulse frequencies (Dingle, 1952*b*) and in particular, as this treatment predicts, the velocity derived from the arrival time of the pulse maximum was independent of the tube geometry. At the very lowest temperatures, near 0·1° K., the phenomena were in general agreement with the picture of phonon diffusion given above, and the arrival time of the foot of the pulse corresponded to the velocity of first sound $u_1$ for all the tubes if it was assumed that the Kapitza boundary effects at the surfaces of the transmitter and receiver introduced a constant time delay of $8 \pm 2\,\mu$sec. Variation of the pulse amplitude by a factor of 100 had only a small effect which could be explained entirely in terms of masking of the foot of the pulse by the receiver noise.

At 0·1° K. the rising edge of a sufficiently large received pulse commenced with a small, almost vertical portion, which could be ascribed to those phonons travelling directly from the transmitter to the receiver, particularly since the variation of its height among the different tubes was almost exactly proportional to the solid angle subtended by the receiver at the transmitter. This steep front would be expected to disappear upon raising the temperature to the point where the phonon mean free path becomes comparable with the length of the tube, so that the phonon is then not able to reach the receiver before colliding with another phonon. The steep front was found to disappear at 0·2° K. for a tube 6 cm. long and at 0·34° K. for a tube 3 cm. long. Khalatnikov predicts much longer mean free paths than this at these temperatures and Kramers (1955) therefore suggests that the phonon mean free paths may be shortened by collisions with the few $He^3$ atoms which are always present. In a later chapter we shall see that second sound propagation in a $He^3$-$He^4$ solution is free from pulse distortion effects for this very reason (King and Fairbank, 1954). Kramers *et al.* repeated their heat pulse experiments with atmospheric helium, which is richer in $He^3$, and found that the apparent velocity did not start to rise towards $u_1$ until much lower temperatures than with well helium. When considering the scattering of phonons by $He^3$ atoms at extremely low temperatures, we should remember that the phonon wavelength is longer than the average distance between scattering centres, so there are interference effects and the scattering must therefore be caused by local fluctuations in the $He^3$ density.

Osborne (1956) has investigated sinusoidal signals in resonant cavities. No sharp resonances were observed below 0·6° K., but the phase and amplitude at the receiver were measured as functions of the frequency (60 to 2000 c./s.) and the length of the cavity. The results were expressed formally as a variation of velocity and

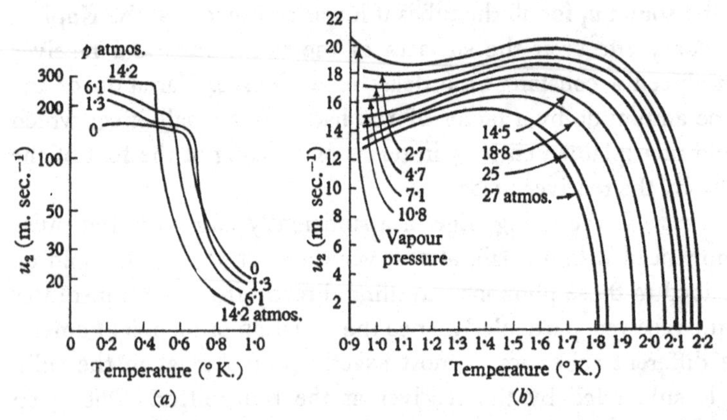


Fig. 53. The velocity of second sound at high pressures. (*a*) is based on Mayper and Herlin (1953); (*b*) is based on Peshkov and Zinoveva (1948) and Maurer and Herlin (1951).

attenuation with frequency. The interpretation of these results is discussed in detail in the original paper. The important conclusions are that the mean free path of the phonons is about 0·04 cm. at 0·58° K., about 0·08 cm. at 0·54° K. and greater than 2 cm. below 0·4° K. At the quartz wall of the cavity about one-third of the phonons were diffusely reflected and the remaining two-thirds specularly reflected.

The velocity of second sound at higher pressures has been measured above 1° K. by Peshkov and Zinoveva (1948) and by Maurer and Herlin (1951). Below 1° K. (Mayper and Herlin, 1953) the phenomena described above are again observed, but the apparent velocity at the lowest temperatures now tends towards the velocity of first sound characteristic of the increased density. Fig. 53 summarizes these various investigations. From these data one may obtain values of $\rho_n/\rho$ at various pressures and hence deduce the important quantity $\partial p_0/\partial \rho$.

### 5.3.3. The attenuation of second sound

Although the attenuation of second sound is much greater than the attenuation of first sound at the same frequency, it was not so readily observed because of the difficulty of exciting second sound at a frequency greater than a few kilocycles. Atkins and Hart (1954) investigated pulsed continuous waves at carrier frequencies of 10 and 20 kc./s., using a constantan wire transmitter and a phosphor-bronze wire receiver in order to get good high frequency response. They dispensed with the customary propagation tube in order to avoid boundary effects and therefore had to apply a small correction for beam spreading. The attenuation was found to increase with increasing amplitude of the second sound signal; this may be a consequence of the large amplitude effects discussed in the next section. The attenuation extrapolated to zero amplitude is shown in fig. 54 together with the results of a simultaneous investigation conducted by Hanson and Pellam (1954). These latter investigators employed continuous waves in the frequency range from 100 to 350 kc./s. They used carbon transducers and at these very high frequencies their signal amplitudes were very small, but they were able to get good signal to noise ratios by using a receiver with a very narrow pass-band of about 4 c./s. Their beam spreading correction was always less than 5 % and the attenuation was so large that the part of the received signal which had experienced multiple reflexions between the transmitter and the receiver was never more than a few per cent of the whole. At the very small amplitude used by Hanson and Pellam the attenuation was independent of amplitude.

Fig. 54 also shows values of the attenuation between 0·7 and 1·3° K. obtained by Kramers (1955) and Zinoveva (1956). Kramers estimated his values from the distortion of square second sound pulses. Zinoveva used continuous waves in a resonant cavity at frequencies between 200 and 4000 c./s. and deduced the attenuation from the width of the resonance peaks. The attenuation could be divided into two parts, a part varying as the square root of the frequency $\nu^{\frac{1}{2}}$, ascribable to surface losses, and a part varying as the square of the frequency $\nu^2$, ascribable to volume losses. The attenuation shown in fig. 54 corresponds to the volume losses. The surface losses had previously been discovered by Peshkov

(1948*a*) and have been treated theoretically by Dingle (1948, 1950) and by Khalatnikov (1952*a*, 1956*c*). From them Zinoveva was able to deduce the values of $\eta_n$ which have already been presented in fig. 39.

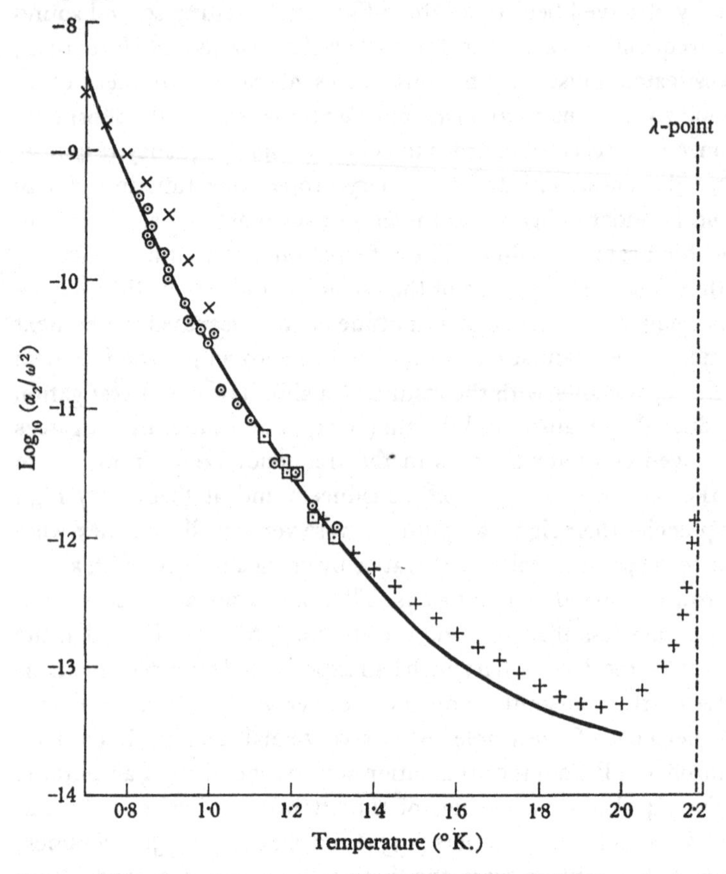


Fig. 54. The attenuation of second sound. ⊡, Atkins and Hart (1954); +, Hanson and Pellam (1954); ×, Kramers *et al.* (1954); ⊙, Zinoveva (1956); ——, theory (Khalatnikov, 1952*b*).

Just before these various measurements were made, Khalatnikov (1952*b*) had predicted the results below 2° K. with remarkable success. His expression for the attenuation of second sound is

$$\alpha_2 = \frac{\omega^2}{2\rho u_2^3}\left[\frac{\rho_s}{\rho_n}\{\tfrac{4}{3}\eta_n + \zeta_2 - \rho(\zeta_1+\zeta_4) + \rho^2\zeta_3\} + \frac{\chi}{C}\right]. \quad (5.81)$$

$\zeta_2$ is the coefficient of second viscosity which appeared in the expression for the attenuation of *first* sound (equation (5.48)) and $\zeta_1$, $\zeta_3$ and $\zeta_4$ are similar coefficients related to the same relaxation effects. These relaxation effects now make a contribution which is comparable in magnitude to that from $\eta_n$, but the largest contribution is from $\chi$, which is analogous to a coefficient of thermal conductivity. Khalatnikov obtains $\chi$ in a fundamental way by considering the transport equation for the elementary excitations when there is a small space-dependent departure from equilibrium. The resulting expression for $\chi$ reduces to zero for the special case of an ideal gas of phonons, but in practice the presence of even a very small number of rotons, with which the phonons can collide, gives $\chi$ a finite value, which actually increases rapidly as the temperature is lowered. Above 0·9° K. Khalatnikov's formulae can be expressed in a way which emphasizes that $\chi$ is a type of thermal conductivity. There are separate roton and phonon contributions

$$\chi = \chi_r + \chi_{ph} \tag{5.82}$$

corresponding to the transport of heat by the diffusion of phonons and rotons separately

$$\chi_r = \frac{\pi}{6}\rho \bar{v}_r l'_r C_r, \tag{5.83}$$

$$\chi_{ph} = \rho c l'_{ph} C_{ph}. \tag{5.84}$$

The mean free paths $l'_r$ and $l'_{ph}$ are almost the same as the ones appearing in the expressions for the normal viscosity $\eta_n$ (§4.5.2). The theoretical values of $\chi$ are shown in fig. 55 and are compared with the experimental values deduced from $\alpha_2$.

The Khalatnikov theory breaks down near the $\lambda$-point, where the results of Hanson and Pellam (1954) indicate that there is a rapid rise in attenuation very similar to the one encountered in first sound.

### 5.3.4. Large amplitude effects and shock waves

As in the case of ordinary sound, the non-linearity of the thermohydrodynamical equations governing second sound results in a variation of velocity with amplitude and the development of shock waves at large amplitudes. These effects were discovered by Osborne (1951) and have been discussed theoretically by Tem-

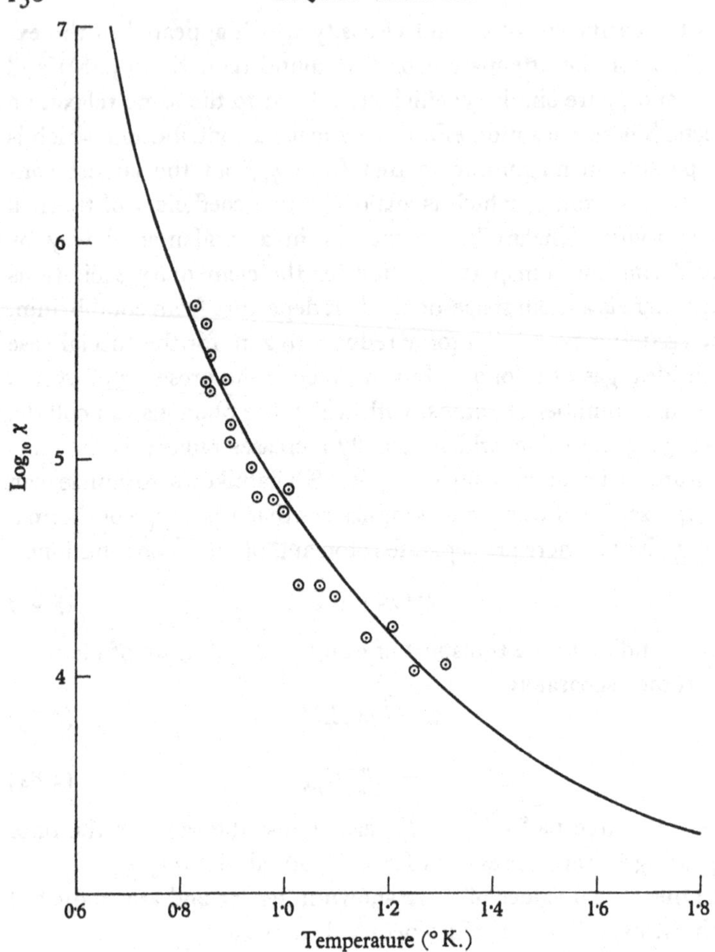


Fig. 55. The thermal conductivity $\chi$ in units of erg sec.$^{-1}$ cm.$^{-1}$ deg.$^{-1}$. ——, theory (Khalatnikov, 1952*b*); ⊙, experiment (Zinoveva, 1956).

perley (1951) and by Khalatnikov (1951, 1952*d*). Khalatnikov solves the thermohydrodynamical equations to second order and finds that the local value of the velocity of second sound, $u_2$, at a point where the velocity of the normal component is $v_n$, differs from the velocity $u_{2,0}$ for vanishingly small amplitudes in the following way

$$u_2 = u_{2,0} + \tau_2 v_n, \tag{5.85}$$

$$\tau_2 = \frac{TS}{C}\frac{\partial}{\partial T}\left[\ln\left(u_{2,0}^3 \frac{C}{T}\right)\right]. \tag{5.86}$$

The variation of $\tau_2$ with temperature is shown in fig. 56 and it is seen to be negative between 0·4 and 0·9° K. and also between 1·9° K. and the $\lambda$-point, so that in these temperature ranges the local value of the second sound velocity is *less* than the small amplitude value, a novel situation not encountered in the case of first sound.

Osborne's (1951) original observations were concerned with the progressive change in shape of a second sound pulse of large

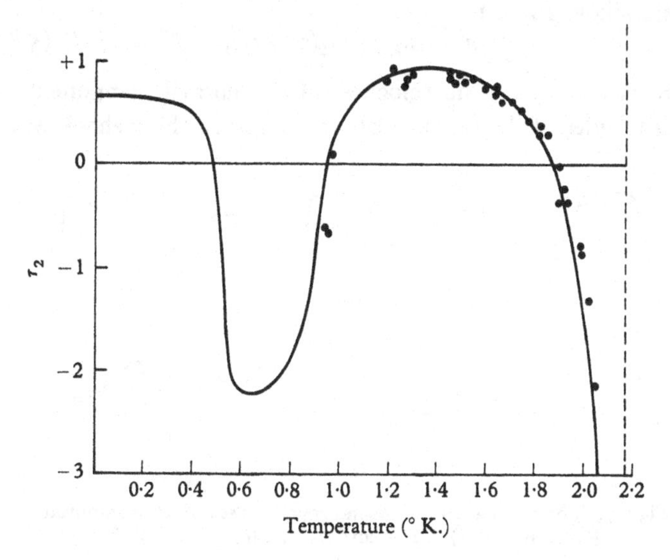


Fig. 56. The quantity $\tau_2$ in Khalatnikov's theory of the velocity of second sound at large amplitudes. ——, theory (Khalatnikov, 1951); ●, experiment (Dessler and Fairbank, 1956).

amplitude. Consider the propagation of a pulse with a finite time of rise and fall as shown in fig. 57, which is a picture in space of the pulse as it travels through the liquid. At those temperatures where $\tau_2$ is positive (1·05 to 1·9° K. in Osborne's experiments) the top of the pulse *CD* has a velocity greater than the leading foot *A* or the trailing tail *B*, and so the leading edge *AC* becomes steeper while the tail *DB* lengthens. Osborne was able to demonstrate this change in shape as the receiver was withdrawn from the transmitter and also at a fixed transmitter-receiver separation as the amplitude of the pulse was increased. Above 1·9° K., in the region of negative $\tau_2$, the effect is reversed. The top of the pulse *CD* then travels more slowly than $A$ and $B$, so the leading edge is lengthened whereas the rear of the pulse becomes steeper.

At sufficiently large amplitudes and sufficiently large distances from the transmitter, the steep edge of the pulse becomes practically vertical and a shock wave front is formed (at the forward edge of the pulse when $\tau_2$ is positive and at the rear of the pulse when $\tau_2$ is negative). The velocity of this shock front can be shown to be (Khalatnikov, 1952*d*)

$$u_{2,s} = u_{2,0} + \tfrac{1}{2}\tau_2(v_1 + v_2), \tag{5.87}$$

where $v_1$ and $v_2$ are the velocities of the normal component on opposite sides of the shock. Osborne observed these shock waves

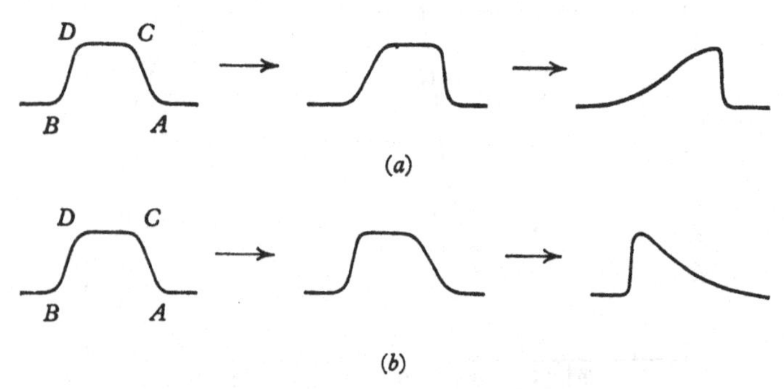


Fig. 57. The propagation of second sound pulses of large amplitude (Osborne, 1951). (*a*) Below 1·9° K.; (*b*) above 1·9° K.

and showed that they travelled with an increased velocity at 1·05° K., in order of magnitude agreement with equation (5.87).

The increase in attenuation with amplitude found by Atkins and Hart (see previous section) may be related to these finite amplitude effects.

Dessler and Fairbank (1956) have verified equation (5.86) directly by measuring the velocity of a small second sound pulse riding on top of a larger pulse of variable amplitude. Their results are included in fig. 56 and are seen to be in good agreement with Khalatnikov's formula. The change in velocity in this experiment can be conveniently divided into two parts. First, the large pulse increases the local temperature and the velocity of the small pulse corresponds to a temperature slightly higher than the temperature outside the large pulse. Secondly, the large pulse superimposes on the small pulse a thermal current which affects its velocity through the intermediary of the second order terms in the thermohydrodynamical equations. Khalatnikov (1956 *a*) has shown that, in a region where the flow velocities of the two components are $v_s$ and $v_n$ parallel to the direction of propagation, the velocity of first sound is

$$u_1 = u_{1,0} + \frac{\rho_n v_n + \rho_s v_s}{\rho} \tag{5.88}$$

and the velocity of second sound is

$$u_2 = u_{2,0} + \frac{\rho_n v_n + \rho_s v_s}{\rho} + \left(\frac{2\rho_s}{\rho} - \frac{TS}{\rho_n C}\frac{\partial \rho_n}{\partial T}\right)(v_n - v_s). \tag{5.89}$$

This last equation is equivalent to equation (5.86) if $\rho_n v_n + \rho_s v_s = 0$ and we add on the change in velocity due to the local rise in temperature.

### 5.3.5. The Pitot tube, the Rayleigh disk and the radiation balance

From the thermohydrodynamical equations one can readily derive an equation which is the equivalent of the Bernoulli equation of classical hydrodynamics (Landau, 1941):

$$p + \tfrac{1}{2}\rho_n v_n^2 + \tfrac{1}{2}\rho_s v_s^2 = \text{constant along a stream line.} \tag{5.90}$$

It follows that an increase in $v_s^2$ or $v_n^2$ is accompanied by a decrease in pressure, which is the classical Bernoulli effect, except that it can now occur even if the net mass flow $\rho_n \mathbf{v}_n + \rho_s \mathbf{v}_s$ is zero, as in a second sound wave. Pellam and his co-workers have performed an interesting series of experiments which depend on this Bernoulli effect and which confirm the value of the two-fluid approach to second sound phenomena.

In a Pitot tube type of experiment, Pellam (1950) used a liquid helium manometer to measure the pressure difference between the centre and end of a resonant second sound cavity half a wavelength long. In accordance with the above Bernoulli equation, the pressure was lower at the antinode near the centre of the cavity than at the node near the end.

The Rayleigh disk is a classical device for detecting the mechanical energy in a sound wave. The disk is suspended with the normal to its plane horizontal and at an angle of 45° to the direction of propagation of the sound. The velocity distribution around the disk gives rise to a variation in Bernoulli pressure and a resultant torque which, in the case of ordinary sound, can be shown to be

$$\tau = \tfrac{4}{3}\rho a^3 v^2, \tag{5.91}$$

where $a$ is the radius of the disk and $v$ is the local value of the undisturbed particle velocity. Because the torque varies as $v^2$ it always acts in the same direction and its average value can be obtained by inserting the root mean square value of $v$. For a second sound wave one can readily prove that the two components will act separately to give a total torque (Usui, 1951)

$$\tau' = \tfrac{4}{3}a^3(\rho_n v_n^2 + \rho_s v_s^2). \tag{5.92}$$

The torque is therefore a direct measure of the total mechanical energy $(\rho_n v_n^2 + \rho_s v_s^2)$ associated with the wave. Pellam and Morse (1950) and Pellam and Hanson (1952) have measured this torque and have found it to be in good agreement with equation (5.92). A rearrangement of this equation gives

$$\tau' = \frac{4}{3}\frac{a^3\overline{W}^2}{\rho C T u_2^2}, \tag{5.93}$$

where $\overline{W}$ is the root mean square heat current density. The experiment may therefore be used to derive values of the specific heat $C$ in good agreement with other methods of measurement. A lucid review of these researches has been given by Pellam (1955*b*).

Herrey (1955) has performed a somewhat similar experiment in which a beam of 100 kc./s. second sound was wholly intercepted by a disk at 45° to its direction of propagation and the force on the disk was measured. This differs from the Rayleigh disk inasmuch as the frequency is much higher and the flow does not completely encircle the disk, so the *force* is a measure of the radiation *pressure* $\rho_n v_n^2 + \rho_s v_s^2$, but this is exactly the quantity which was measured by the *torque* on the Rayleigh disk. The experimental results with this radiation balance were also in approximate agreement with the two fluid theory, except for an anomalous variation of the force with the heat current density, but this may have been caused by the presence of a steady heat current superimposed on the alternating one. A thin copper foil disk proved to be a perfect reflector of second sound, but a porous copper membrane gave only partial reflexion.

## 5·4. The thermohydrodynamical equations

The various effects which produce attenuation of first and second sound correspond to extra terms in the thermohydrodynamical equations. With the addition of these terms (Khalatnikov, 1952*b*) these equations become:

$$\rho_s \frac{D\mathbf{v}_s}{Dt} = -\frac{\rho_s}{\rho}\operatorname{grad} p + \rho_s S \operatorname{grad} T$$
$$+\rho_s \operatorname{grad}[\zeta_3 \operatorname{div} \rho_s(\mathbf{v}_s - \mathbf{v}_n) + \zeta_4 \operatorname{div} \mathbf{v}_n], \quad (5{\cdot}94)$$

$$\frac{D}{Dt}(\rho_n \mathbf{v}_n + \rho_s \mathbf{v}_s) = -\operatorname{grad} p + \eta_n(\nabla^2 \mathbf{v}_n + \tfrac{1}{3}\operatorname{grad}\operatorname{div}\mathbf{v}_n)$$
$$+\operatorname{grad}[\zeta_1 \operatorname{div} \rho_s(\mathbf{v}_s - \mathbf{v}_n) + \zeta_2 \operatorname{div} \mathbf{v}_n], \quad (5{\cdot}95)$$

$$\frac{\partial \rho}{\partial t} + \operatorname{div}(\rho_n \mathbf{v}_n + \rho_s \mathbf{v}_s) = 0, \quad (5{\cdot}96)$$

$$\frac{\partial}{\partial t}(\rho S) + \operatorname{div}(\rho S \mathbf{v}_n) = \frac{\chi}{T}\nabla^2 T. \quad (5{\cdot}97)$$

The coefficients of first viscosity, $\eta_n$, and thermal conductivity, $\chi$, arise from transfer of momentum and energy respectively by diffusion of the elementary excitations. The coefficients of second viscosity, $\zeta_1$, $\zeta_2$, $\zeta_3$, $\zeta_4$, are related to the relaxation times involved as the number density of elementary excitations returns to its equilibrium value. From Onsager's reciprocal relations, it can be shown that $\zeta_1 = \zeta_4$. The exact form of the thermohydrodynamical equations, particularly with regard to higher order terms, is, however, still a subject of controversy; see, for example, Dingle's review article (1952*a*) and F. London's book (1954). A very detailed set of thermohydrodynamical equations has been presented by Khalatnikov (1956*c*).

CHAPTER 6

# FURTHER ASPECTS OF THE THERMOHYDRODYNAMICS

## 6.1. Second sound in uniformly rotating liquid

### 6.1.1. The experiments and their interpretation in terms of vortex lines

When second sound is propagated in rotating liquid helium it experiences an extra attenuation. This effect was first reported by Wheeler, Blakewood and Lane (1955) who investigated d.c. pulses travelling in an axial direction in the annulus between a fixed outer cylinder and a rotating inner cylinder. Turbulence is very likely to occur in such an arrangement and it is probable that the increased attenuation was caused by scattering of the second sound from eddies. We shall therefore devote most of the present discussion to the experiments of Hall and Vinen (1956) who designed their apparatus to study stable flow subject to no dissipation processes.

Hall and Vinen deduced the attenuation from the bandwidth of a second sound resonator. In the non-rotating condition the bandwidth was measured either by plotting a resonance curve or by switching off the second sound transmitter power and noting the characteristic decay time of the amplitude at the receiver. The increase in attenuation upon starting up the rotation was then very conveniently followed by setting the frequency on resonance and recording photographically the variations in resonant amplitude. When the rotation was first started there was a large transient attenuation of the same order of magnitude as that observed by Wheeler, Blakewood and Lane, which strongly suggests that the liquid is first brought into rotation by the development of turbulence. This transient died down after about 30 sec. and the attenuation then settled down to a steady value slightly higher than in the non-rotating state. When the rotation was stopped a large transient attenuation again appeared and lasted about 200 sec.

Principal interest is centred on the steady excess attenuation accompanying uniform stable rotation. The resonators (fig. 58) were designed to rotate as a whole with the liquid, including the transmitter and receiver, and the only contact between the rotating liquid inside the can and the outside liquid was via the vapour and the film. The apparatus therefore came very close to satisfying the requirement that, if the can and its contents were isolated and subjected to no external couple, the liquid would continue to rotate unchanged with no dissipative process slowing it down.

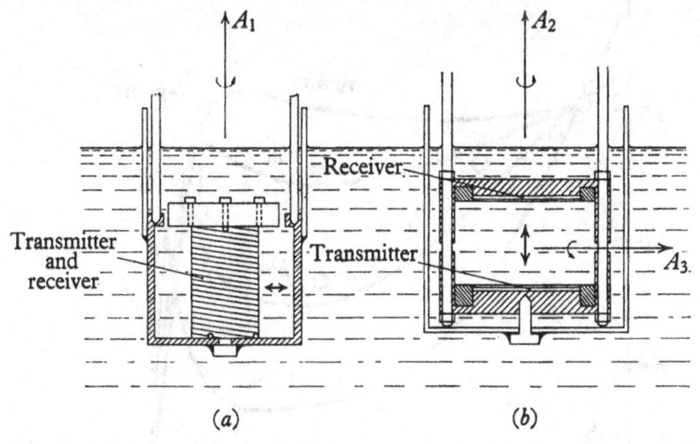


Fig. 58. The second sound resonators used by Hall and Vinen (1956). (*a*) Radial mode resonator; (*b*) axial mode resonator. The small arrows in the liquid inside the resonators indicate the direction in which the second sound causes the normal component to oscillate.

First consider the radial mode resonator of fig. 58*a* in which the direction of propagation is radial and perpendicular to the axis of rotation $A_1$. When this resonator was rotated at speeds of 0·46, 0·65 and 1·22 revolutions per sec. the velocity of the second sound changed by less than 0·1 %, proving that the two-fluid theory is still valid in rotating liquid with the values of $\rho_s$ and $\rho_n$ unchanged. Except near the $\lambda$-point, where the situation is complicated and not yet understood, the excess attenuation was linearly proportional to the angular velocity, $\Omega$, independent of the amplitude of the second sound and independent of frequency. The total attenuation can therefore be expressed as

$$\alpha = \alpha_0 + \frac{B\Omega}{2u_2}. \qquad (6.1)$$

Within the accuracy of these experiments, $B$ is a function of temperature only and its experimental values are shown in fig. 59.

This extra attenuation is readily understandable if the rotating superfluid component contains vortex lines or vortex sheets (§ 4.6.4). Second sound is an oscillatory motion of the rotons and phonons relative to the superfluid and this motion would be hindered by collisions of the phonons and rotons with the vortex lines or sheets. An attenuation of the type given by equation (6.1)

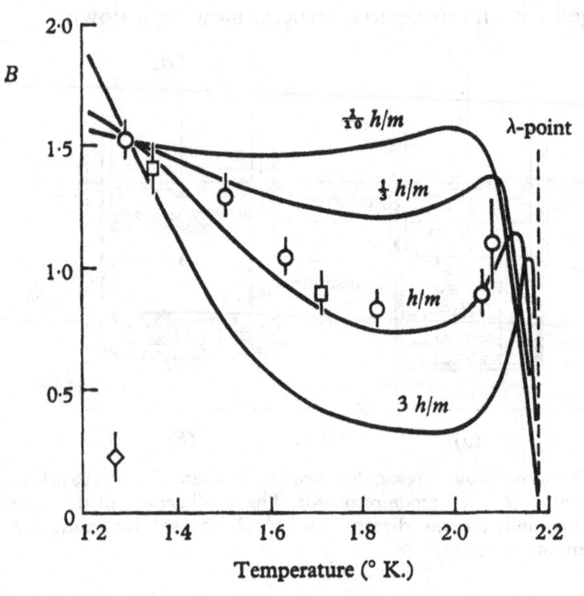


Fig. 59. The parameter $B$ as a function of temperature. ○, radial mode resonator rotating about axis $A_1$; ◇, axial mode resonator rotating about axis $A_2$; □, axial mode resonator rotating about axis $A_3$. The full curves are derived from the theory discussed in the text and the figure adjacent to each curve represents the value assumed for the circulation around a vortex line. In each case $\alpha\sigma_0$ has been chosen to fit the experimental results at 1·3° K. (after Hall and Vinen, 1956).

would result if these collisions were equivalent to a force of mutual friction between the superfluid and normal components of the form

$$\mathbf{f}_{sn} = B\frac{\rho_s \rho_n}{\rho}\Omega(\mathbf{v}_s - \mathbf{v}_n). \tag{6.2}$$

The justification of this equation will be considered in the next section, but the proportionality to $\Omega$ is immediately obvious in Feynman's theory, since the number of quantized vortex lines per unit area is $m\Omega/\pi\hbar$.

If this explanation is correct, there should be a much smaller excess attenuation when the direction of propagation is parallel to the axis of rotation, for the collisions of the phonons and rotons with the vortex lines cannot result in a large transfer of momentum parallel to the axis of the line. This is the situation realized in the axial mode resonator of fig. 58*b* rotating about the axis $A_2$, and the rate of increase in attenuation with rotation was in fact much smaller in this case, the corresponding value of $B$ being no greater than one-fifth of the value for the radial mode resonator. A finite effect is still possible, since the lines probably wobble due to thermal agitation and zero-point motion and thereby present a collision cross-section to excitations moving parallel to their axes.

When the axial mode resonator was turned on its side and rotated about the axis $A_3$, the value of $B$ was exactly the same as for the radial mode resonator. This is what would be expected from the vortex *line* model, since the motion of the normal component is perpendicular to the lines in both cases. The model of co-axial cylindrical vortex *sheets*, however, would predict a lower attenuation in the case of the axial mode resonator, because there would then be regions of the liquid with the motion of the normal component at grazing incidence to the sheets.

These experiments therefore provide impressive evidence for the existence of vortex lines in uniformly rotating superfluid, and indicate that an arrangement of vortex lines is to be preferred to an arrangement of co-axial vortex sheets. Hall and Vinen have made a critical analysis of the thermodynamics of rotation and the conditions under which $E-\Omega M$ is a minimum. They conclude that an arrangement of vortex lines is always energetically favourable compared with an arrangement of vortex sheets. Moreover, the lines must all be in their ground state, must be distributed with approximately uniform density and must rotate with the vessel. This applies not only to the quantized models of Onsager (1949), London (1954) and Feynman (1955), but also to the non-quantized model of Landau and Lifshitz (1955). If we ignore quantization, but introduce a surface energy $\alpha$ per unit area of vortex sheet, instead of an arrangement of co-axial cylindrical vortex sheets, it is preferable to have a number of vortex tubes with their axes uniformly distributed throughout the liquid. Moreover, if

$\alpha \sim (\hbar^2/2m\delta^4)$ (Ginsburg, 1956), the optimum diameter of the vortex tube is of the order of the interatomic distance $\delta$, the optimum circulation around a tube is of the order of $h/m$ and the number of tubes per unit area is of the order of $m\Omega/\pi\hbar$. The Landau-Lifshitz theory therefore leads to a system of vortex lines very similar to the quantized vortex lines of Onsager and Feynman. Ginsburg's justification of $\alpha \sim (\hbar^2/2m\delta^4)$ is that the wave function has a node at the vortex sheet and varies rapidly in a distance of the order of $\delta$. In Feynman's theory there is a similar node along the axis of the vortex line. The point at issue, then, is whether a rigorous analysis would lead to a strict quantization of the form

$$mvr = \hbar, \tag{6.3}$$

or whether the situation is more complicated and this equation is true only in order of magnitude.

### 6.1.2. Calculation of the magnitude of the extra attenuation

On the assumption that the extra attenuation is caused by collisions between rotons and vortex lines, Hall and Vinen were able to calculate the magnitude of the parameter $B$ appearing in equations (6.1) and (6.2). Phonons were ignored because they make a negligible contribution to $\rho_n$ in the temperature range covered by the experiments.

The first step is to calculate the collision diameter $\sigma_{rv}$ for scattering of rotons by a quantized vortex line. The principal part of the interaction was found to arise from the fact that the energy of a roton in the velocity field of a vortex line is

$$\epsilon' = \epsilon(p) + \mathbf{p}.\mathbf{v}_s. \tag{6.4}$$

To allow for the atomic structure in the vicinity of the axis of the vortex line, the superfluid velocity was taken to be

$$v_s = \frac{\hbar}{mr}(1 - e^{-r/a_0}), \tag{6.5}$$

where the cut-off distance $a_0$ is of the order of the atomic diameter. Using a first order Born approximation, the collision diameter was found to be

$$\sigma_{rv} = \frac{\pi^3 \mu \hbar^3}{2m^2 k T a_0^2 p_0}\left[\left(1 + \frac{4a_0^2 p_0^2}{\hbar^2}\right)^{\frac{1}{2}} - 1\right] \tag{6.6}$$

$$= \frac{\sigma_0}{T}. \tag{6.7}$$

With $a_0 \sim 1$ Å., $\sigma_0 \sim 75$ Å. deg., which is a very reasonable value, but the exact value is very sensitive to $a_0$ and cannot be estimated precisely. In the course of the calculation, it emerges that the scattering is approximately symmetrical in the forward direction and that the roton can exchange momentum only in a direction normal to the vortex line (ignoring wobbling).

A calculation of the type familiar in kinetic theory then gives the force exerted on unit length of the vortex line by the collisions of the rotons in the form

$$\mathbf{f} = D(\mathbf{v}_R - \mathbf{v}_L), \tag{6.8}$$

$$D = \alpha \rho_n \sigma_{rv} \bar{v}_r. \tag{6.9}$$

$\mathbf{v}_R$ is the drift velocity of the rotons in the vicinity of the vortex line and $\mathbf{v}_L$ is the drift velocity of the vortex line. $\bar{v}_r = \sqrt{(2kT/\pi\mu)}$ is the average thermal velocity of the rotons at the temperature in question. The parameter $\alpha$ is of the order of unity, but cannot be calculated precisely, because the motion of the rotons in the immediate vicinity of the axis of the vortex line is likely to be very complicated.

A vortex line tends to drag along the rotons in its vicinity and so $\mathbf{v}_R$ is not the same as $\mathbf{v}_n$, which is the velocity of the normal component averaged over a region containing several vortex lines. A rather laborious treatment of this drag gives

$$(\mathbf{v}_n - \mathbf{v}_R) = \frac{\mathbf{f}}{E}, \tag{6.10}$$

where

$$E = -\frac{4\pi\eta_n(\ln \frac{1}{2}\lambda l_r + 1)}{(\ln \frac{1}{2}\lambda l_r + 1)^2 + \frac{\pi^2}{16}}, \tag{6.11}$$

$l_r$ is the mean free path for roton-roton collisions and

$$\lambda = \left(\frac{\rho_n \omega}{\eta_n}\right)^{\frac{1}{2}}, \tag{6.12}$$

$\omega$ being the angular frequency of the second second. Eliminating $\mathbf{v}_R$ from (6.8) and (6.10), we get

$$\mathbf{f} = C(\mathbf{v}_n - \mathbf{v}_L), \tag{6.13}$$

$$C = \frac{ED}{E+D}. \tag{6.14}$$

Similarly, the vortex lines may have a drift velocity relative to the superfluid component and $\mathbf{v}_L$ is not necessarily the same as $\mathbf{v}_s$. In fact, a classical result known as the Magnus effect tells us that, when the vortex lines are subject to a force $\mathbf{f}$ per unit length, their velocity is given by

$$\mathbf{f}=\rho_s(\mathbf{v}_L-\mathbf{v}_s)\times\mathbf{K}, \tag{6.15}$$

$\mathbf{K}$ represents the circulation around the vortex line. Its direction is, of course, parallel to the axis of the line and in Feynman's theory its magnitude is $2\pi\hbar/m$. Eliminating $\mathbf{v}_L$ from (6.13) and (6.15), we obtain $\mathbf{f}$ in terms of $(\mathbf{v}_s-\mathbf{v}_n)$:

$$\mathbf{f}=\frac{C}{1+\dfrac{C^2}{\rho_s^2K^2}}\frac{\mathbf{K}\times[\mathbf{K}\times(\mathbf{v}_s-\mathbf{v}_n)]}{K^2}+\frac{\rho_s}{1+\dfrac{C^2}{\rho_s^2K^2}}[\mathbf{K}\times(\mathbf{v}_s-\mathbf{v}_n)]. \tag{6.16}$$

Remembering that there are $2\Omega/K$ vortex lines per unit area and that $\mathbf{K}$ is in the same direction as $\boldsymbol{\Omega}$, we finally obtain for the force of mutual friction per unit volume exerted by the superfluid on the normal component

$$\mathbf{f}_{sn}=-B\frac{\rho_s\rho_n}{\rho}\frac{\boldsymbol{\Omega}\times[\boldsymbol{\Omega}\times(\mathbf{v}_s-\mathbf{v}_n)]}{\Omega}-B'\frac{\rho_s\rho_n}{\rho}[\boldsymbol{\Omega}\times(\mathbf{v}_s-\mathbf{v}_n)], \tag{6.17}$$

where

$$B=\frac{C}{1+\dfrac{C^2}{\rho_s^2K^2}}\frac{2\rho}{\rho_n\rho_s}\frac{1}{K} \tag{6.18}$$

and

$$B'=\frac{2\rho}{\rho_n}\frac{1}{1+\dfrac{\rho_s^2K^2}{C^2}}. \tag{6.19}$$

The first term is parallel to $\mathbf{v}_s-\mathbf{v}_n$ and has the form stated in equation (6.2) to be necessary to explain the experimental results. The second term is perpendicular to $(\mathbf{v}_s-\mathbf{v}_n)$ but is too small to have been detected in the experiments. As will be seen from fig. 59, the experimental values of $B$ can be explained very satisfactorily if $\alpha\sigma_0$ is chosen to be 14·8 Å. deg. and the circulation $K$ is put equal to $2\pi\hbar/m$, as in Feynman's theory. This value of $\alpha\sigma_0$ is consistent with the calculation of the collision diameter (equation (6.6)) if $a_0\sim 1$ Å. If different values of the circulation $K$ are chosen, but $\alpha\sigma_0$ is adjusted in each case to fit the experimental value of $B$ at

1·3° K., the variation of $B$ with temperature no longer agrees with experiment. Poor agreement with experiment also results if we neglect the tendency for the vortex lines to drag the rotons, but the value of $B$ is not very sensitive to the motion of the vortex lines through the superfluid component, except near the $\lambda$-point where the experimental results are not reliable.

These researches of Hall and Vinen therefore inspire confidence in the Onsager-Feynman theory of quantized vortex lines in rotating superfluid, and also indicate that, even if the circulation is not strictly quantized according to equation (6.3), this equation must be a close approximation to the actual situation.

## 6.2. The breakdown of superfluidity

When the velocity of flow through a channel exceeds the critical value dissipative processes appear and the velocity varies in a markedly non-linear way with the pressure head, as shown in fig. 33. A phenomenological description of this situation might be attempted by introducing additional non-linear frictional terms into the hydrodynamical equations. Ignoring terms involving the coefficients of second viscosity, these equations would take the form

$$\rho_s \frac{D\mathbf{v}_s}{Dt} = -\frac{\rho_s}{\rho} \operatorname{grad} p + \rho_s S \operatorname{grad} T - F_{sn}(\mathbf{v}_s - \mathbf{v}_n) - F_s(\mathbf{v}_s), \quad (6.20)$$

$$\rho_n \frac{D\mathbf{v}_n}{Dt} = -\frac{\rho_n}{\rho} \operatorname{grad} p - \rho_s S \operatorname{grad} T + F_{sn}(\mathbf{v}_s - \mathbf{v}_n) - F_n(\mathbf{v}_n) + \eta_n(\nabla^2 \mathbf{v}_n + \tfrac{1}{3} \operatorname{grad} \operatorname{div} \mathbf{v}_n). \quad (6.21)$$

The frictional terms $F_{sn}$, $F_s$ and $F_n$ may be complicated functions involving the operators grad, div and curl.

$F_{sn}$ is a force of mutual friction between the two components and it therefore appears with opposite sign in the two equations. The idea of a mutual friction was first introduced by Gorter and Mellink (1949) in order to explain certain details of thermal conduction in moderately narrow channels, and they suggested that it should vary as the cube of the relative velocity of the two components

$$\mathbf{F}_{sn} = A\rho_s \rho_n |\mathbf{v}_s - \mathbf{v}_n|^2 (\mathbf{v}_s - \mathbf{v}_n). \quad (6.22)$$

Since the phonons and rotons constitute the normal component and since vortex lines may be considered as part of the superfluid component, the force $\mathbf{f}_{sn}$ of equation (6.17), which arises from collisions between rotons and vortex lines, is a type of mutual friction. It seems very plausible that $\mathbf{F}_{sn}$ might arise from the development of vorticity in the superfluid component, followed by collisions between the resulting eddies and the phonons and rotons. Similarly, the term $F_s(\mathbf{v}_s)$ may be at least partly due to the presence in the superfluid component of an 'eddy viscosity', in which the eddies play the same role as do the atoms in producing the viscosity of a gas. However, since we are now concerned with dissipative processes, the system of vortex lines is probably highly disordered and very similar to turbulence, so we cannot expect the situation to be as simple as it was for the stable flow of the uniformly rotating liquid.

In the next few sections we shall present the evidence for the Gorter-Mellink mutual friction varying as the cube of the relative velocity, and we shall then discuss the experiments of Vinen which indicate that this mutual friction is intimately related to the formation of vorticity. We shall then consider the experiments which indicate that the mutual friction alone is not sufficient to explain all the non-linear dissipative effects which occur. At present, the addition of extra frictional forces to the hydrodynamical equations must be regarded as a tentative formalism which may give a satisfactory phenomenological explanation of the flow under certain circumstances, but may be totally inapplicable under other circumstances. (In classical hydrodynamics the phenomenon of turbulence does not require a revision of the fundamental hydrodynamical equations, but rather a consideration of special types of solutions of those equations). It is possible, for example, that the functional form of the frictional forces depends upon the dimensions and shape of the channel.

## 6.3. Mutual friction

### 6.3.1. The thermomechanical effect

In an extremely narrow channel connecting two chambers at different temperatures, $v_n$ is necessarily zero and $v_s$ must also be zero when there is no net transfer of matter. $F_s$, $F_n$, $F_{sn}$, $Dv_s/Dt$ and $Dv_n/Dt$ then also vanish and equation (6.20) reduces to

$$\operatorname{grad} p = \rho S \operatorname{grad} T, \tag{6.23}$$

which is H. London's equation for the thermomechanical effect. In a channel of finite width, however, when $v_n$ has a non-zero value, the non-linear frictional forces become important, the thermomechanical pressure head falls below the value given by equation (6.23) and is no longer a linear function of the temperature difference. In these circumstances there is a counterflow of the two components in the channel, but no net transfer of matter ($\rho_n v_n + \rho_s v_s = 0$) and the heat flow per unit area of cross-section of the channel can readily be shown to be

$$\mathbf{W} = \rho S T \mathbf{v}_n = \rho_s S T (\mathbf{v}_n - \mathbf{v}_s). \tag{6.24}$$

In the steady state with $D\mathbf{v}_s/Dt = D\mathbf{v}_n/Dt = 0$, equations (6.20) and (6.21) can be added to give

$$\operatorname{grad} p = \eta_n \nabla^2 \mathbf{v}_n - \mathbf{F}_s - \mathbf{F}_n. \tag{6.25}$$

The early experiments of Allen and Reekie (1939) revealed that, even in the wider channels when the pressure head is a markedly non-linear function of the temperature difference, it still varies linearly with the heat current $W$. Gorter and Mellink (1949) explain this by assuming that $F_s$ and $F_n$ can be ignored, so that

$$\operatorname{grad} p = \eta_n \nabla^2 \mathbf{v}_n, \tag{6.26}$$

which, for a parallel-sided channel of width $d$, leads to a mean velocity

$$\bar{\mathbf{v}}_n = -\frac{d^2}{12\eta_n} \operatorname{grad} p \tag{6.27}$$

and an average heat current proportional to the pressure head

$$\overline{\mathbf{W}} = -\frac{\rho S T d^2}{12\eta_n} \operatorname{grad} p. \tag{6.28}$$

If we make allowance for the fact that the channel width $d$ can never be measured with great accuracy, this equation agrees very satisfactorily with the measurements of Duyckaerts (1943) and Mellink (1947) for channel widths of 19, 10·5 and $5 \times 10^{-4}$ cm.

The term $\eta_n \nabla^2 \mathbf{v}_n$ in equation (6.25) is proportional to $1/d^2$ and therefore may mask $F_s$ and $F_n$ only when $d$ is sufficiently small. Winkel, Broese van Groenou and Gorter (1955) report that very accurate measurements with a channel width of $6 \times 10^{-4}$ cm. give

$$\Delta p = \alpha_1 W + \alpha_2 W^2, \tag{6.29}$$

which suggests a finite value of $\mathbf{F}_s + \mathbf{F}_n$ varying as the square of one of the velocities.

### 6.3.2. Thermal conduction

If we provisionally ignore $F_s$ and $F_n$ and substitute the value of grad $p$ from equation (6.28) into equation (6.20), we obtain a relationship between the heat current and the temperature gradient:

$$\rho_s S \operatorname{grad} T = -\frac{12 \rho_s \eta_n}{\rho^2 S T d^2} \overline{\mathbf{W}} + F_{sn}\left(\frac{-\overline{\mathbf{W}}}{\rho_s S T}\right). \tag{6.30}$$

The second term on the right is independent of the channel width $d$, but the first term varies as $1/d^2$ and therefore becomes unimportant in sufficiently wide channels, so that the equation can then be written approximately as

$$\rho_s S \operatorname{grad} T = F_{sn}\left(\frac{-\overline{W}}{\rho_s S T}\right). \tag{6.31}$$

Fig. 60 shows the experimental results of Keesom, Saris and Meyer (1940) for thermal conduction in glass capillaries of various diameters between 0·03 and 0·16 cm. In accordance with equation (6.31), the heat current is independent of the diameter of the capillary. The temperature gradient varies as the cube of the heat current, and Gorter and Mellink (1949) therefore postulated that $F_{sn}$ should have the form $A \rho_s \rho_n (v_s - v_n)^3$, as we have already stated in equation (6.22). $A$ is about 50 cm. sec. g.$^{-1}$, varies only slowly with temperature and is independent of channel width in these wider channels.

With this form for $F_{sn}$, equation (6.30) may be rewritten as

$$\rho_s S \operatorname{grad} T = -\frac{12 \rho_s \eta_n W}{\rho^2 S T d^2} - \frac{A \rho_s \rho_n W^3}{(\rho_s S T)^3}. \tag{6.32}$$

In narrower channels the linear term in $W$ has to be considered and the equation predicts that grad $T$ should vary linearly with $W$ near the origin, but that the cubic variation should take over at higher values of $W$. Such a behaviour had already been observed by Mellink (1947), using an apparatus similar to the one shown in fig. 36 with slit widths in the range $2 \times 10^{-3}$ to $2 \times 10^{-5}$ cm. More detailed investigations (Hung, Hunt and Winkel, 1952; Winkel, Broese van Groenou and Gorter, 1955) have indicated that the mutual friction term in $W^3$ is completely absent near the origin and does not put in an appearance until a critical value of $v_s$ is

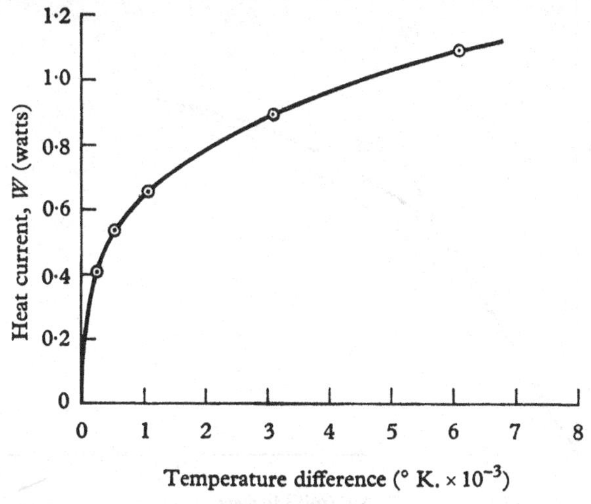


Fig. 60. Thermal conduction in wide glass capillaries. Temperature = 1·60° K. Within experimental error this curve describes the results for three capillaries with diameters 0·033, 0·0705 and 0·157 cm. (after Keesom, Saris and Meyer, 1940).

exceeded. For the curve of fig. 61, the linear term may be estimated from the tangent at the origin and the value of $A$ then plotted as a function of $v_s$. $A$ is found to be zero up to $v_s \simeq 14$ cm. sec.$^{-1}$ and then to rise rapidly towards the value given by other methods. This critical velocity agrees with the critical velocities obtained by the methods we shall describe in later sections only on the assumption that it is the superfluid velocity $v_s$ which becomes critical. In some instances, at velocities below the critical velocity, there was evidence for the existence of a small frictional term varying as the square of the velocity and agreeing in magnitude with the deviations from the linear relationship between $\Delta p$ and $W$ in the thermomechanical effect (equation (6.29)).

From the initial slope of the grad $T$ versus $W$ curve the normal viscosity $\eta_n$ can be deduced. After due allowance has been made for the heat conducted along the glass walls of the channel, $\eta_n$ is found to be independent of the channel width above 1·5° K. and its magnitude lies within 50% of the values measured by other methods. Between 1·5 and 1·1° K., however, it increases by only about 30% and the very rapid rise indicated by oscillating disk and

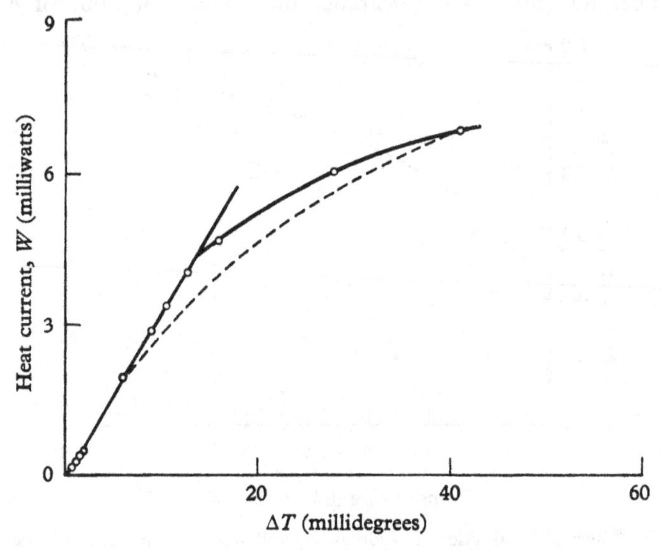


Fig. 61. Thermal conduction in a narrow channel of width $2{\cdot}4 \times 10^{-4}$ cm. Temperature = 1·72° K.; ----, $\Delta T = aW + bW^3$, with $a$ chosen to give the correct slope at the origin and $b$ chosen to fit the curve at the highest value of $W$ (after Winkel, Broese van Groenou and Gorter, 1955).

rotating cylinder viscometer experiments (fig. 38) is not reproduced (Broese van Groenou, Poll, Delsing and Gorter, 1956). According to the Landau-Khalatnikov theory (§ 4.5.2 and fig. 39) the mean free path of the phonons at 1·1° K. is about $5 \times 10^{-4}$ cm., whereas the mean free path of the rotons is about $2 \times 10^{-6}$ cm. The channel widths in the experiments lay in the range $7 \times 10^{-5}$ to $6 \times 10^{-4}$ cm., so it is possible that the phonon contribution to $\eta_n$ might have been reduced by Knudsen effects, while the roton contribution remained unaltered. Since the rapid rise below 1·5° K. is attributed to the phonon contribution, this would explain the results qualitatively (Atkins, 1957).

Below 0·6° K. there is very good evidence that thermal conduction is influenced by mean free path effects. The rotons have then fallen out of the picture and the phonon mean free path has probably increased considerably. Heat flow through the liquid in a channel is therefore best visualized as a diffusion of phonons which collide with the walls, but rarely with one another, so that their effective mean free path is determined by the width of the channel. Fairbank and Wilks (1955) investigated thermal conduction in two german silver tubes with internal diameters of 0·029 and 0·080 cm., and found that the heat flow was proportional to the temperature difference below 0·6° K. It was therefore possible to define a thermal conductivity $\chi_t$ and a mean free path $l_t$:

$$\chi_t = \tfrac{1}{3} c C_{ph} l_t. \tag{6.33}$$

At 0·6° K. $l_t$ was almost exactly equal to the diameter of the tube, but was about 30% higher at 0·2° K. This increase at lower temperatures may correspond to specular reflexion of a finite fraction of the phonons at the walls, since the phonon wavelength of $5 \times 10^{-6}$ cm. at 0·2° K. was probably comparable in magnitude with the size of the irregularities of the wall. Whitworth (1955 and private communication) has extended these measurements to wider tubes with diameters ranging from 0·15 to 1·01 cm.; he found that the wider the tube the lower was the temperature at which the effective mean free path $l_t$ became constant at a value close to the tube diameter. At 0·58° K. the variation of conductivity with tube size was consistent with a viscous flow of the phonons down the tube with a slip correction corresponding to a mean free path of 0·038 cm. This agrees approximately with the value deduced by Osborne (1956) from second sound measurements (see §5.3.2). The increase in $l_t$ at the lowest temperatures was reduced in the case of a glass tube by etching its inner surface, which supports the specular reflexion hypothesis.

### 6.3.3. Flow through narrow channels

When there is a net transfer of liquid through the channel under the combined influence of a pressure difference and a temperature difference, equation (6.20) becomes

$$\rho S \operatorname{grad} T - \operatorname{grad} p = \frac{\rho}{\rho_s} F_{sn}(\mathbf{v}_s - \mathbf{v}_n) + \frac{\rho}{\rho_s} F_s(\mathbf{v}_s). \tag{6.34}$$

Neglecting $F_s$ and $F_n$ as the thermomechanical effect in narrow channels suggests, $v_n$ can be calculated from equation (6.27) using the values of $\eta_n$ found from thermal conduction in the same channel. Since $\rho_s v_s + \rho_n v_n$ can be derived from the net transfer of matter through the slit, $v_s$ and $v_s - v_n$ can then be deduced, and in this type of experiment $v_s \gg v_n$, so the uncertainty in $v_n$ is not very important.

Hung, Hunt and Winkel (1952) and Winkel, Delsing and Gorter (1955) have performed this type of experiment with channel widths ranging from $4 \times 10^{-5}$ to $7 \times 10^{-4}$ cm. At velocities well above the critical velocity, the results could always be explained in terms of equation (6.34) by ignoring $F_s$ and putting

$$F_{sn} = A\rho_s \rho_n (v_s - v_n)^m. \qquad (6.35)$$

The driving force $\rho S \Delta T - \Delta p$ was found to be proportional to the length of the slit and the frictional forces cannot therefore be mainly due to end effects. For channel widths greater than $10^{-4}$ cm., $m$ was almost exactly 3 in accordance with the original Gorter-Mellink hypothesis, but for smaller channel widths $m$ increased to a value between 4 and 5 at $d = 0{\cdot}4 \times 10^{-4}$ cm. Moreover, $A$ was not independent of channel width but varied approximately as $1/d^{\frac{1}{2}}$. The mutual friction must therefore be more complicated than was originally suspected.

Measurements of isothermal flow at high pressure heads give similar results. Swim and Rorschach (1955) used channel widths of 2·4 and $4{\cdot}3 \times 10^{-4}$ cm. and pressure heads up to $2{\cdot}3 \times 10^{3}$ dyne cm.$^{-2}$ (15 cm. head of liquid helium). Their results are explicable in terms of a mutual friction with $m = 3{\cdot}0$ for $d = 4{\cdot}3 \times 10^{-4}$ cm. and $m = 3{\cdot}7$ for $d = 2{\cdot}4 \times 10^{-4}$ cm. Wansink, Taconis, Staas and Reuss (1955) produced a channel of width about $2 \times 10^{-5}$ cm. and 3 cm. long by sealing a gold wire into soft glass and studied flow through it at pressure heads up to $10^{6}$ dyne cm.$^{-2}$ (7500 cm. head of liquid helium). Even up to these high pressure gradients, the results are still explicable in terms of a mutual friction with $m = 3{\cdot}9$. It is interesting to notice that the mean velocity varied by a factor of more than two over the range of pressure heads used. The apparent independence of pressure head in curves such as the one for $d = 1{\cdot}2 \times 10^{-5}$ cm. in fig. 33 is therefore merely a consequence of working with small pressure heads.

Similar experiments were performed by Bowers and Mendelssohn (1952 *a*, *b*) with a slit width of $1{\cdot}2 \times 10^{-4}$ cm. and the apparatus of fig. 62, which incorporated a tube $X$ to measure the pressure at an intermediate point along the slit. As there was no attempt to produce perfect thermal isolation of the inside of the vessel, the investigation was only semi-quantitative, but it yielded qualitative results of considerable interest. When the vessel was raised or lowered and flow took place under gravitation alone, the level in

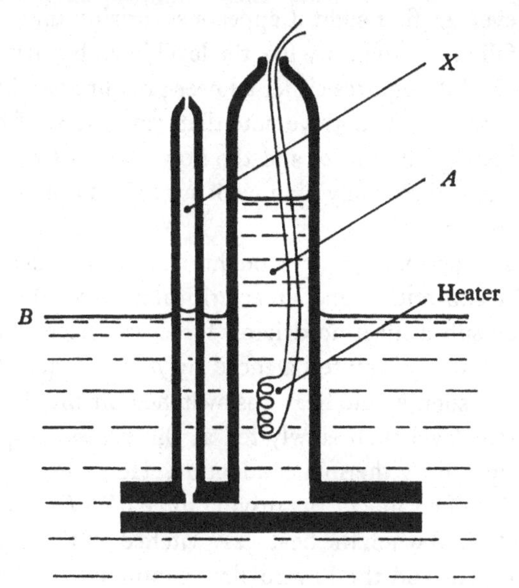


Fig. 62. The apparatus of Bowers and Mendelssohn (1952*a*, *b*).

$X$ rapidly settled down to a position very close to the level of the outer bath $B$, indicating that the flow between $X$ and $B$ was frictionless and required no pressure gradient. Because of the annular shape of the slit, the velocity was greatest at the inner radius and decreased with increasing radius, so that the critical velocity was exceeded only over some region between $A$ and $X$ where all the pressure drop was developed, whereas the velocity between $X$ and $B$ was always less than critical and no pressure drop was produced there. This experiment is therefore comparable with a similar experiment on the film described in §4.2.2 and fig. 34*a*, and it clearly demonstrates the existence of a critical velocity and of frictionless flow in channels of this size. When heat was supplied to $A$, the level there rose because of the thermomechanical effect, but the level in $X$ again stayed near the bath level, indicating that sub-critical thermal flow takes place under zero temperature gradient as well as under zero pressure gradient. At a certain critical heat input, however, the critical velocity was exceeded and frictional effects appeared. The rate of flow was then no longer proportional to the heat input and the level in $X$ was *depressed below* the bath level. At first sight it appears surprising that the level in $X$ should fall rather than rise like the level in $A$, but frictional flow between $B$ and $X$ requires either a lower pressure or a higher temperature or both in $X$, and we note that, since most of the flow in this experiment is due to $\rho_s$ and the flow of $\rho_n$ is small, the temperature in $X$ can rise only by flow of superfluid out of $X$, causing an initial depression.

The following procedure brought out clearly the difference between the sub-critical and super-critical regions. When the flow was entirely sub-critical no driving force was needed in the slit and $\Delta p$ was exactly counterbalanced by $\rho S \Delta T$ (equation (6.34)). Therefore, as soon as the heat was switched off the flow stopped, and the inner level then slowly fell as the excess temperature in $A$ was dissipated by thermal conduction. However, in the super-critical region the value of the driving force $\rho S \Delta T - \Delta p$ across the slit was finite, and when the heat was switched off this driving force was still present and the inward flow continued until the inside temperature had fallen to a value corresponding to thermomechanical equilibrium. The critical velocity determined by the onset of this 'overshoot' agreed with the value corresponding to the appearance of a level difference between $X$ and $B$ and was of the order of 25 cm. sec.$^{-1}$.

Hung, Hunt and Winkel (1952) and Winkel, Delsing and Poll (1955) have developed the overshoot procedure to the point of obtaining very accurate measurements of the frictional forces from the volume of liquid flowing into the vessel after the heat is switched off (for details of the calculations see Winkel, 1955). Above the critical velocity the non-linear frictional force obtained in this way agrees very well with the results of the other methods that have already been described. Below the critical velocity there is a slight indication of a very small frictional force varying linearly with the velocity, which, if genuine, might imply that $F_s = kv_s$. In this type of experiment $v_n$ is small compared with $v_s$, but in thermal conduction experiments $v_n$ can be comparable with or greater than $v_s$. The critical velocities obtained in the two cases agree only on the assumption that $v_s$, rather than $v_n$ or $v_s - v_n$, becomes critical. The critical velocity $v_{s,c}$ then lies in the range 8 to 20 cm. sec.$^{-1}$ and increases with decreasing channel width $d$, but less rapidly than $1/d$.

Chandrasekhar and Mendelssohn (1953) have performed an interesting series of experiments in which the slit of fig. 62 was replaced by a tube packed with jeweller's rouge. The phenomena accompanying thermal flow in this case were exactly similar to those observed with the slit. The level in $X$ remained at the same height as in $B$ until a critical heat input was reached, after which the level in $X$ was slightly depressed. This critical heat input was also accompanied by departures from the linear relationship between heat input and rate of flow and by the onset of the overshoot phenomenon. On the other hand, gravitational flow seemed to have an entirely different character. In agreement with the experiments of Allen and Misener (§4.2.3) the rate of flow varied as the square root of the pressure head and the level in $X$ always took up an intermediate position between the levels in $A$ and $B$, giving no evidence for frictionless flow under any conditions. This contrast between thermal and gravitational flow is intriguing and may be connected with the irregular shape of the channels between the rouge particles.

## 6.4. Mutual friction and vorticity

### 6.4.1. Second sound in a heat current

In an elegant series of experiments Vinen (1957*a*, *b*, *c*, 1958) has established the connexion between mutual friction and vorticity. He first investigated heat flow through a tube 10 cm. long with a rectangular cross-section either 0·240 by 0·645 cm. (case 1) or 0·400 by 0·783 cm. (case 2). Emphasis was placed on smaller heat currents, corresponding to values of $v_s - v_n$ in the range from 0 to 15 cm. sec.$^{-1}$, compared with the range from 10 to 100 cm. sec.$^{-1}$

in previous investigations. Below 1·8° K., the relationship between heat current and temperature gradient was found to be

$$\operatorname{grad} T = \frac{A\rho_n}{\rho_s^3 S^4 T^3}(W - W_0)^3 \tag{6.36}$$

in accordance with a mutual friction of the form

$$F_{sn} = A\rho_s\rho_n(|v_s - v_n| - v_0)^3. \tag{6.37}$$

The values of $A$, shown as a function of temperature in fig. 63, agreed in magnitude with the rather scattered values calculated

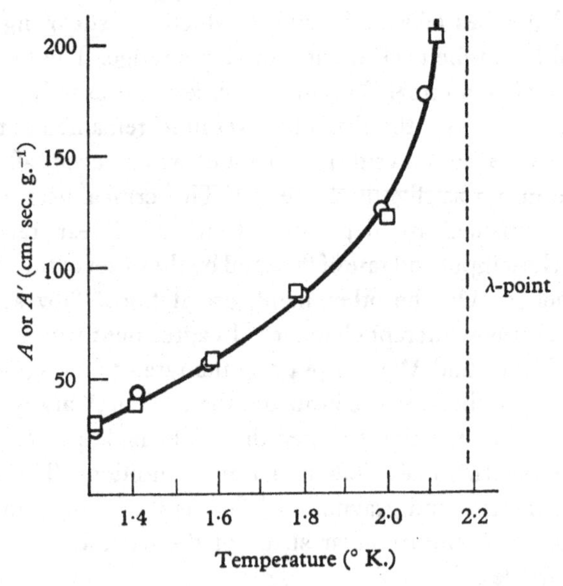


Fig. 63. The constant $A$ determining the magnitude of the mutual friction deduced by Vinen from: □, thermal conduction; ○, the excess attenuation of second sound propagated perpendicularly to a heat current.

by Gorter and Mellink from previous measurements, and $v_0$ was approximately 0·4 cm. sec.$^{-1}$ and varied only slowly with temperature. When a thin mica plate was placed down the centre of the smaller tube (case 1) effectively halving its width, $A$ was unaltered. There was also no change when the mica plate was roughened or replaced by a coil of wire, and it therefore seems that the boundary conditions are unimportant and that $A$ is a bulk effect independent of channel width in these wider channels.

The tube was then used as a second sound resonator with the direction of propagation perpendicular to the heat current and parallel to the larger side of the cross-section. The attenuation of the second sound was deduced from the bandwidth of the resonator and the resonant amplitude in the manner already explained in § 6.1.1. The velocity of the second sound was not influenced by the presence of the heat current (within $\pm\frac{1}{2}\%$) proving that the two-fluid theory is still applicable with unaltered values of $\rho_s$ and $\rho_n$. There was, however, a large increase in attenuation varying as $(W-W_0)^2$ below 1·8° K.:

$$\alpha=\alpha_0+C(W-W_0)^2. \tag{6.38}$$

The $W_0$ of this equation is identical with the $W_0$ of equation (6.36).

Phenomenologically, this excess attenuation can be explained by assuming that the mutual friction is instantaneously

$$F_{sn}=A'\rho_s\rho_n\,|\,\overline{\mathbf{v}_s-\mathbf{v}_n}\,|^2(\mathbf{v}_s-\mathbf{v}_n). \tag{6.39}$$

$(\mathbf{v}_s-\mathbf{v}_n)$ is the instantaneous relative velocity of the two components and is the sum of two perpendicular vectors corresponding to the heat current and the second sound. $\overline{\mathbf{v}_s-\mathbf{v}_n}$ is the time-average of $(\mathbf{v}_s-\mathbf{v}_n)$ averaged over one period of the second sound, and is therefore the relative velocity produced by the heat current alone in the absence of the second sound $(W=\rho_s ST(\overline{\mathbf{v}_n-\mathbf{v}_s}))$. Calculating the energy dissipation per cycle due to this frictional force, the extra attenuation can be shown to be

$$\alpha'=\frac{A'\rho}{2\rho_s^2S^2T^2u_2}W^2, \tag{6.40}$$

which agrees with equation (6.38) if $W_0$ is ignored. The experimental values of $A'$ agree exactly with the values of $A$ deduced from the thermal conduction experiments (fig. 63). The same values of $A'$ can be used to explain the results in the two tubes of different cross-sections.

The mutual friction can therefore be imagined to originate in the following way. In supercritical heat currents turbulence is developed in the superfluid component which therefore contains a tangled mass of vortex lines, the total length of the lines per unit volume being proportional to $(\overline{\mathbf{v}_s-\mathbf{v}_n})^2$. The mutual friction can then be seen to be a consequence of collisions between the vortex lines and the elementary excitations, which are mainly rotons above 1° K. As in § 6.1.2 the force per unit length of vortex line is proportional to $(\mathbf{v}_s - \mathbf{v}_n)$ and the total force on all the vortex lines in unit volume is proportional to

$$\overline{(\mathbf{v}_s - \mathbf{v}_n)^2}(\mathbf{v}_s - \mathbf{v}_n),$$

as in equation (6.39). For a heat current in the absence of second sound, this mutual friction reduces to the form given in equation (6.22). In the presence of second sound $(\mathbf{v}_s - \mathbf{v}_n)$ must be split into two perpendicular components. The component parallel to the heat current is the same as before and the heat current is not disturbed by the second sound, as is found experimentally. The second sound produces a component perpendicular to the heat current, and it is the collisions between this component and the vortex lines which produce the extra attenuation.

In postulating that the density of vortex lines is proportional to $\overline{(\mathbf{v}_s - \mathbf{v}_n)^2}$, we have assumed that the component of $(\mathbf{v}_s - \mathbf{v}_n)$ due to the second sound is oscillating so rapidly that it does not have time to produce vorticity. The experiments described in the next section confirm that there is an appreciable delay in the setting up of vorticity. Moreover, a mutual friction of the form

$$F_{sn} = A'\rho_s \rho_n \,|\, \mathbf{v}_s - \mathbf{v}_n \,|^2 (\mathbf{v}_s - \mathbf{v}_n) \qquad (6.41)$$

is not consistent with the experimental observation that the extra attenuation in a heat current was independent of the second sound amplitude. The component of $(\mathbf{v}_s - \mathbf{v}_n)$ due to the second sound was large compared with the component due to the heat current and, therefore, if the density of vortex lines had adjusted itself rapidly to the second sound amplitude, the attenuation would have varied approximately as the square of this amplitude. Another difference between equations (6.39) and (6.41) is that (6.39) reduces to zero in the absence of a heat current, whereas (6.41) does not and implies a large attenuation of second sound caused by mutual friction between the two components oscillating in opposite directions. No such large attenuation has been observed at frequencies above 10 c./s. (Peshkov, 1949), from which we conclude that turbulence takes at least 0·1 sec. to be established.

$W_0$ may be a type of critical heat current, but a different, and more sharply defined, critical heat current $W_{\text{crit.}}$ revealed itself near 1·4° K. by an abrupt fall of the excess attenuation to zero, as shown in fig. 64. It is not obvious whether this is caused by a critical value of $v_s$, $v_n$ or $v_s - v_n$. In the smaller tube (case 1) at 1·414° K. it would correspond to a critical $v_s$ of 0·0514 cm. sec.$^{-1}$ or

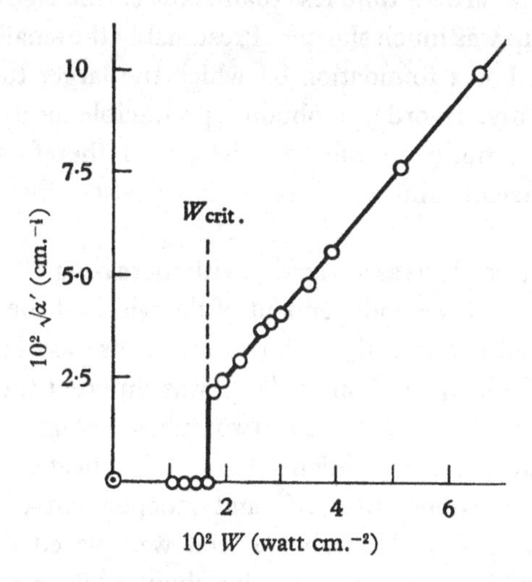


Fig. 64. The excess attenuation $\alpha'$ of second sound propagated perpendicularly to a heat current $W$ (after Vinen, 1957 *a*).

a critical $v_s - v_n$ of 0·644 cm. sec.$^{-1}$. In the larger tube (case 2) at 1·398° K., the corresponding critical velocities are rather smaller; 0·0328 cm. sec.$^{-1}$ for $v_s$ or 0·443 cm. sec.$^{-1}$ for $v_s - v_n$.

### 6.4.2. Transient effects

In a normal liquid turbulence takes a finite time to build up and similar effects were observed in liquid helium. When the heat current was first switched on, the temperature difference was too small to be observed for a time of the order of 1 sec. and then gradually rose to its equilibrium value in a further time of the order of 1 sec. Simultaneously, the extra attenuation of second sound was also gradually building up in a very similar way. The characteristic build-up time $\tau$ was defined as the time taken for the resonant second sound amplitude to fall half-way to its final value. When the heat was switched off, the excess attenuation decayed with a relaxation time of the same magnitude.

It is difficult to explain these phenomena in any way other than as the growth and decay of turbulence. Moreover, there was good evidence that a small amount of turbulence remained as long as 200 sec. after the heat was switched off, because, if the heat was switched on again after a time less than 200 sec., the characteristic time for build-up was much shorter. Presumably the small residual turbulence acted as a foundation on which the larger turbulence could grow rapidly. In order to obtain reproducible measurements of the maximum build-up time $\tau_f$, the liquid therefore had to be given sufficient time to settle down after the previous experiment.

For $W \gg W_{crit.}$, $\tau_f$ decreased rapidly with increasing $W$, approximately as $1/W^{\frac{3}{2}}$, and was independent of the width of the tube. As $W$ was decreased towards $W_{crit.}$, however, $\tau_f$ rose asymptotically to infinity at $W = W_{crit.}$, and since $W_{crit.}$ was different for the two tubes, the build-up times for the two tubes became different. Above 1·5° K. there was no evidence for a critical heat current and $\tau_f$ was then proportional to $1/W^{\frac{3}{2}}$ and independent of channel width for all values of $W$. When a coil was placed down the centre of the tube, $\tau_f$ was reduced by about 30%, presumably because the wires of the coil were able to aid the development of turbulence.

The existence of a critical heat current below 1·5° K. could also be demonstrated in the following way. The heat current was switched off and then, after waiting a sufficiently long time to enable the liquid to settle down, a small heat current $W_1$ was switched on. After another long interval of time, the heat current was suddenly increased to a large value $W_2 > W_{crit.}$ and the growth of the second sound attenuation was observed. With $W_1 > W_{crit.}$ the build-up time was appreciably less than $\tau_f$, presumably because $W_1$ was accompanied by a small turbulence on which the larger turbulence could build. However, with $W < W_{crit.}$ no initial turbulence (or very little turbulence) was present and the build-up time was approximately equal to $\tau_f$. It will be seen from fig. 65 that the transition from one case to the other was very sharp. The critical heat currents measured in this way agreed satisfactorily with those derived from the onset of the extra attenuation (fig. 64). However, the well-defined discontinuity of fig. 65 was present only at the lower temperatures. Upon raising the temperature a drop in $\tau$ appeared below the critical heat current, and above 1·55° K. had become so prominent that the discontinuity was completely smeared out (Vinen, 1958). In a wider channel these effects appeared at lower temperatures. It seems, therefore, that a small amount of turbulence is present even below $W_{\text{crit.}}$ and that this

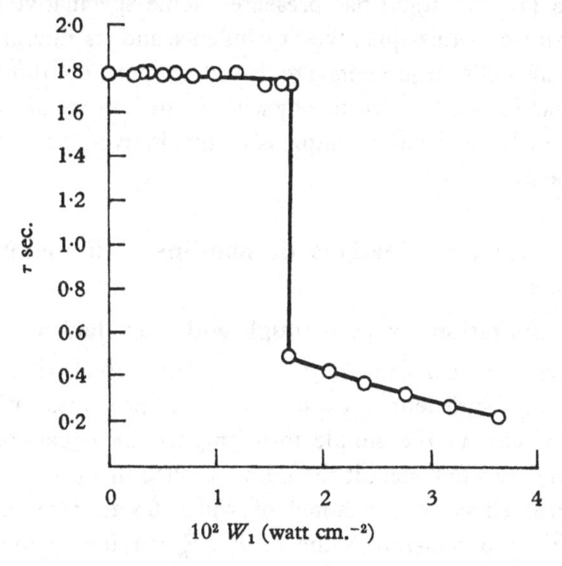


Fig. 65. The build-up time $\tau$ as a function of the heat current $W_1$ initially present. $W_2 = 7{\cdot}6 \times 10^{-2}$ watt cm.$^{-2}$; temperature $= 1{\cdot}41$° K. (after Vinen, 1957*b*).

'subcritical turbulence' increases with increasing temperature or increasing channel width.

In the procedure just described, when $W_1$ was slightly greater than $W_{\text{crit.}}$ the attenuation was found to build up in two separate steps. Vinen interprets this as meaning that $W_1$ produced turbulence in only one part of the tube, so that the build-up was less rapid in the other part. As the superfluid enters the tube one might expect it to flow for a time $\sim \tau_f$ before becoming fully turbulent and, unless turbulence can be propagated backwards against the direction of flow, this would result in an inlet length $L_s \sim v_s \tau_f$ comparatively free from turbulence. This inlet length is much smaller than the length of the tube, except just above $W_{\text{crit.}}$ where $\tau_f$ tends to infinity. If these arguments are valid, the turbulence producing the mutual friction must occur in the superfluid rather than the normal component, because the normal component had a much higher velocity at 1·4° K. and therefore would have had an inlet length $L_n \sim v_n \tau_f$ comparable with the length of the tube, and this would undoubtedly have had a detectable effect on some of the experimental results.

Vinen (1957*c*, 1958) has presented some speculative theories of the nature of this quantized turbulence and its rate of growth and decay. His arguments are based in part on dimensional analysis and in part on an analogy with the turbulence of a classical liquid, but he achieves an impressive qualitative agreement with experiment.

## 6.5. Further investigations of non-linear dissipative processes

### 6.5.1. Gravitational flow through wide capillaries

We shall now consider the evidence that mutual friction alone is not always sufficient to explain the non-linear dissipative processes, at least in the simple form implied by equations (6.20) and (6.21). As a first step, it should be recalled that (*a*) the thermomechanical effect in a channel of width $6 \times 10^{-4}$ cm. provided evidence for a non-zero value of $F_s + F_n$ varying as the square of some velocity, (*b*) thermal conduction experiments in narrow channels also suggested the existence of such a force, even below the critical velocity and (*c*) in the experiments on thermal flow through narrow channels, the overshoot method seems to reveal a frictional force varying linearly with velocity below the critical velocity. Of course, none of these forces is yet established with certainty.

For steady isothermal flow under a pressure gradient, the Gorter-Mellink theory of mutual friction puts the thermohydrodynamical equations in the form

$$\frac{\rho_s}{\rho} \operatorname{grad} p = A \rho_n \rho_s (\mathbf{v}_n - \mathbf{v}_s)^3, \qquad (6.42)$$

$$\frac{\rho_n}{\rho} \operatorname{grad} p = -A \rho_n \rho_s (\mathbf{v}_n - \mathbf{v}_s)^3 + \eta_n \nabla^2 \mathbf{v}_n. \qquad (6.43)$$

In the case of flow through a capillary of radius $r$, these are readily solved to give for the mean velocity of flow

$$\bar{v}=\frac{\rho_s}{\rho}\bar{v}_s+\frac{\rho_n}{\rho}\bar{v}_n$$
$$=\frac{\rho_s}{\rho}\left(-\frac{\operatorname{grad}p}{A\rho\rho_n}\right)^{\frac{1}{3}}+\frac{r^2\operatorname{grad}p}{8\eta_n}. \qquad (6.44)$$

This velocity is the sum of a Gorter-Mellink term independent of $r$ and a viscosity term varying as $r^2$. Atkins (1951) has attempted to separate these two terms by studying gravitational flow through several capillaries whose radii ranged from $10^{-3}$ to $2\times10^{-2}$ cm. A correction had to be applied for end effects, which were found to result in a pressure drop of approximately $\frac{1}{2}\rho\bar{v}^2$ at each end of the capillary. A typical set of corrected results is shown in fig. 66.

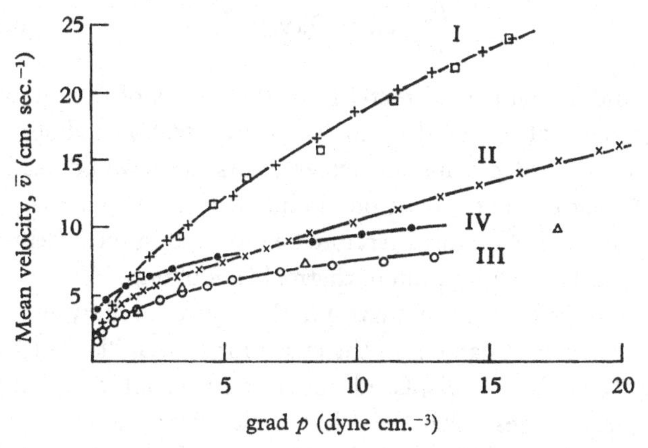


Fig. 66. Flow through wide capillaries. Temperature = 1·22° K.; internal diameters: I, $4{\cdot}40\times10^{-2}$ cm.; II, $2{\cdot}03\times10^{-2}$ cm.; III, $0{\cdot}815\times10^{-2}$ cm.; IV, $0{\cdot}262\times10^{-2}$ cm.

The critical velocity, which plays such an important role in narrow channels, is now certainly not greater than 1 cm. sec.$^{-1}$, except perhaps in the finest capillary, and most of the flow is subject to frictional forces varying with a high power of the velocity. Equation (6.44) is partially successful in explaining these results. For capillary III of bore $0{\cdot}815\times10^{-2}$ cm., the viscosity term $r^2\operatorname{grad}p/8\eta_n$ is small and $\bar{v}$ is, in fact, approximately represented by

$$\frac{\rho_s}{\rho}\left(\frac{\operatorname{grad}p}{A\rho\rho_n}\right)^{\frac{1}{3}}$$

if $A$ is put equal to 40 cm. sec.$^{-1}$, which is in order of magnitude agreement with the value deduced from thermal conduction. The variation of $\bar{v}$ with $r$ cannot, however, be represented by $r^2$ grad $p/8\eta_n$. The extra velocity does not vary as $r^2$ and the values deduced for $\eta_n$ at fixed values of $r$ exceed the accepted values by as much as a factor of four. It seems that the Gorter-Mellink theory does not provide a complete description of the flow under these circumstances.

We can, however, deduce that, if it is sufficient to describe the breakdown of superfluidity in terms of the two equations (6.20) and (6.21), then the mutual friction $F_{sn}$ must be included. If $F_s$ were the only frictional force (apart from $\eta_n \nabla^2 \mathbf{v}_n$) then steady isothermal flow would be described by the equation

$$\frac{\rho_s}{\rho} \text{grad}\, p = F_s(\mathbf{v}_s) \tag{6.45}$$

and $F_s$ could be deduced directly from the curves of fig. 66. The resulting values of $F_s$ would be too small by a factor of about five to explain thermal conduction experiments, and would lead to predicted heat currents about one hundred times larger than are actually observed. $F_{sn}$ must therefore be included in the scheme to give a satisfactory explanation of thermal conduction.

It is interesting that the curve for the finest capillary of bore $2{\cdot}62 \times 10^{-3}$ cm. runs parallel to the curve for the capillary of bore $8{\cdot}15 \times 10^{-3}$ cm., but is displaced above it by about 2 cm. sec.$^{-1}$. This strongly suggests that the critical velocities in the two cases differ by 2 cm. sec.$^{-1}$. Extrapolating the curves back to zero pressure head, we can tentatively assign critical velocities of 1 cm. sec.$^{-1}$ to capillary III and 3 cm. sec.$^{-1}$ to capillary IV. Unfortunately, there is some uncertainty in the extrapolation and also capillary IV was so fine that it was difficult to work with and to clean, so it may not have been entirely free from obstructions.

### 6.5.2. The rotating cylinder viscometer

Hallett (1951, 1953) has made experiments with a conventional rotating cylinder viscometer and has obtained results which cannot be explained solely in terms of the Gorter-Mellink mutual friction. The liquid was contained in an annular gap of width 0·106 cm.

between two co-axial cylinders, the outer of which rotated with velocities up to 3 cm. sec.$^{-1}$, while the inner was suspended by a torsion fibre. The couple transmitted through the liquid to the inner cylinder was measured by observing the angular deflexion of this cylinder in the usual way. This type of arrangement has several advantages for the present purpose. It is a steady state experiment in which $D\mathbf{v}_s/Dt$, $D\mathbf{v}_n/Dt$, grad $p$, grad $T$ and grad div $v_n$ can all be ignored, so that the hydrodynamical equations (6.20) and (6.21) take the particularly simple form

$$F_{sn} + F_s = 0, \tag{6.46}$$

$$F_{sn} - F_n + \eta_n \nabla^2 \mathbf{v}_n = 0. \tag{6.47}$$

Moreover, Taylor (1923) has shown that, in this type of arrangement, the flow of a normal liquid cannot become turbulent at any velocity. The work of Chandrasekhar and Donnelly (1957) suggests that a similar conclusion is valid for liquid helium II: instability is possible only if the *inner* cylinder rotates. Finally, this experiment makes it possible to decide unambiguously whether the Gorter-Mellink theory is alone sufficient to describe the flow phenomena, for equations (6.42) and (6.43) now reduce to

$$A\rho_n \rho_s (\mathbf{v}_s - \mathbf{v}_n)^3 = 0, \tag{6.48}$$

$$A\rho_n \rho_s (\mathbf{v}_s - \mathbf{v}_n)^3 + \eta_n \nabla^2 \mathbf{v}_n = 0 \tag{6.49}$$

from which we deduce immediately that

$$\mathbf{v}_s = \mathbf{v}_n, \tag{6.50}$$

$$\eta_n \nabla^2 \mathbf{v}_n = 0. \tag{6.51}$$

The mutual friction should therefore not influence the flow and the method should measure merely the viscosity of the normal component.

In actual fact, however, the couple on the inner cylinder was not linearly proportional to the velocity of the outer cylinder, as it would have been for a normal viscous liquid. In fig. 67 the dashed curve, representing the torque arising from $\eta_n$ alone, was estimated from the measurements of Heikkila and Hallett (1955) with outer cylinder velocities less than 0·1 cm. sec.$^{-1}$. At velocities greater than about 0·1 cm. sec.$^{-1}$ the torque rose non-linearly above the viscous value. This additional torque obviously cannot be explained by the Gorter-Mellink theory and neither is it reasonable that it could arise from turbulence of the normal component. It might still be related to vorticity in the superfluid component, and might even include a type of mutual friction if there is any possibility that the vortex lines in the superfluid move relative to the normal component.

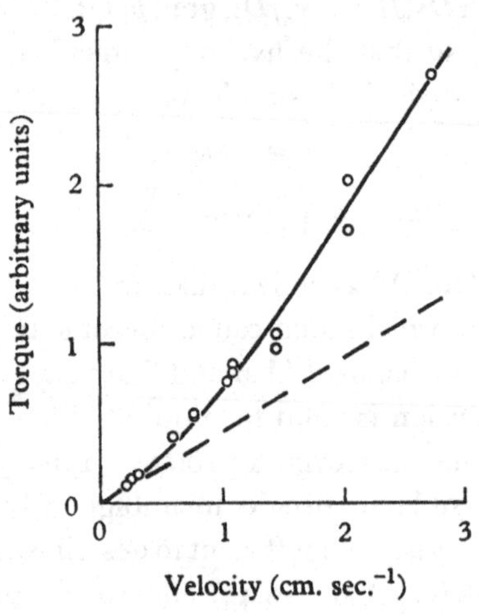


Fig. 67. The torque as a function of velocity in the rotating cylinder viscometer experiment of Hallett (1953). Temperature = 1·84° K.

### 6.5.3. Accelerating systems

The original two-fluid theory supposed that the damping of the torsional motion of an oscillating disk was due entirely to the viscosity of the normal component. We now see that this is likely to be true only for sufficiently small peripheral velocities of the disk, and that at higher velocities the non-linear frictional forces will begin to make their contribution towards the damping. This effect was discovered by Hallett (1950, 1951, 1952) with a single disk. The results of fig. 68 were obtained later by Benson and Hallett (1956), using an oscillating sphere in order to avoid edge effects and extending the measurements up to initial amplitudes as large as 10 radians. In helium I the damping at small amplitudes was explicable solely in terms of the known viscosity of the liquid and the logarithmic decrement was constant up to a critical amplitude $\phi_t$, after which it increased linearly with amplitude.

$\phi_t$ probably indicates the onset of turbulence and corresponds to a Reynolds number

$$R_e = \frac{v_{\text{max.}}\rho\lambda}{\eta}$$

$$= 2R\phi_t\left(\frac{\pi\rho}{\eta\tau}\right)^{\frac{1}{2}}. \tag{6.52}$$

$R$ is the radius of the sphere, $\tau$ is the period, $v_{\text{max.}} = (2\pi/\tau)R\phi_t$ is the maximum velocity at the equator of the sphere and the charac-

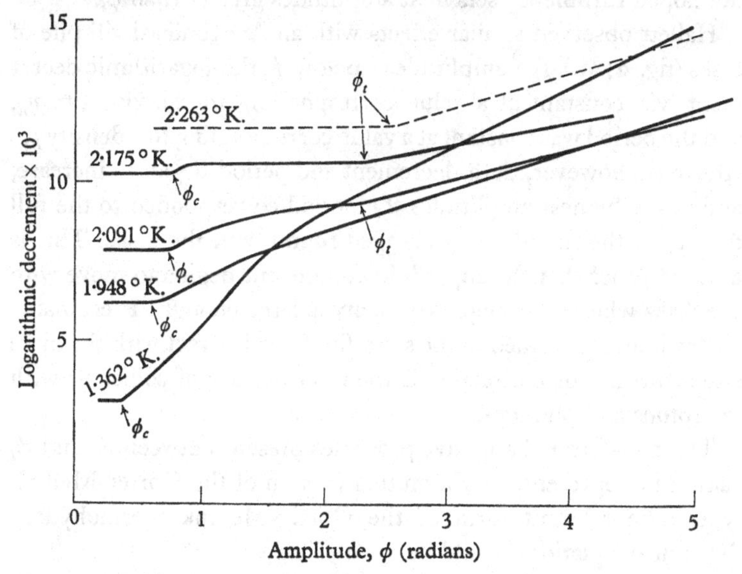


Fig. 68. The logarithmic decrement for the damped motion of a sphere performing torsional oscillations in liquid helium, as a function of the amplitude of oscillation. ——, period = 25·0 sec.; ----, period = 18·7 sec. (after Benson and Hallett, 1956).

teristic length $\lambda = (\eta\tau/\pi\rho)^{\frac{1}{2}}$ is the penetration depth of the viscous motion into the liquid. This Reynolds number was approximately 220, both in liquid helium I and in helium gas at 4·23° K.

In liquid helium II above 2° K. the logarithmic decrement was constant at very small amplitudes and corresponded to the viscosity $\eta_n$ of the normal component, but at a critical amplitude $\phi_c$ ($\sim$ 0·1 radian) the decrement began to increase rapidly and then levelled off at a higher constant value. Finally, at a second critical amplitude, $\phi_t$, the decrement began to increase again linearly in a manner very reminiscent of the onset of turbulence in liquid helium I. Below 2° K. the region of constant decrement just below $\phi_t$ was not observed and so $\phi_t$ itself was not sharply defined. If, in equation (6.52), $\rho$ is taken to be the total density $\rho = \rho_s + \rho_n$ and $\eta$ is calculated from the region of constant decrement preceding $\phi_t$, the Reynolds number $R_e$ corresponding to $\phi_t$ is always approximately 220, as for helium I and helium gas. As this would imply, $\phi_t$ is proportional to $\tau^{\frac{1}{2}}$. It therefore appears to be reasonable to assume that fully developed turbulence sets in at amplitudes greater than $\phi_t$.

Hallett observed similar effects with an Andronikashvili pile of disks (fig. 2) at large amplitudes. Below $\phi_c$ the logarithmic decrement was constant at a value corresponding to the viscosity $\eta_n$, and the period was constant at a value corresponding to a density $\rho_n$. Above $\phi_c$, however, both decrement and period began to increase, and at the highest amplitudes the period corresponded to the full density of the liquid being dragged round with the disks. This is a direct proof that the superfluid component begins to move with the disks when their angular velocity is large enough. Presumably vortex lines are formed in the superfluid and interact with the disks either directly or indirectly via the intermediary of collisions with the rotons and phonons.

The non-linear dissipative processes present between $\phi_c$ and $\phi_t$ cannot be represented by a mutual friction of the Gorter-Mellink type. The relevant form of the Gorter-Mellink thermohydrodynamical equations would be

$$\rho_s \frac{D\mathbf{v}_s}{Dt} = -A\rho_n \rho_s(\mathbf{v}_s - \mathbf{v}_n)^3, \tag{6.53}$$

$$\rho_n \frac{D\mathbf{v}_n}{Dt} = +A\rho_n \rho_s(\mathbf{v}_s - \mathbf{v}_n)^3 + \eta_n \nabla^2 \mathbf{v}_n. \tag{6.54}$$

On the assumption that the logarithmic decrement is only slightly larger than its value below $\phi_c$, these equations have been solved by Zwanniken (1951) for a single disk and by Hallett (1951, 1952) for a pile of disks. Their theoretical predictions are compared with experiment in fig. 69, which gives the increase in logarithmic decrement at an amplitude of 0·215 radian above its value at zero amplitude. The Gorter-Mellink constant $A$ was taken to be 50 cm. sec. g.$^{-1}$ at all temperatures, but the temperature dependent values of fig. 63 would not improve the agreement between experiment and theory. It will be seen that the theoretical values at 1·2° K. are too small by a factor of about ten. Hallett has shown that no possible functional form for $F_{sn}$ can explain the experimental results for both the single disk and the pile of disks.

The critical amplitude $\phi_c$ was approximately proportional to the square root of the period $\tau$, and $\phi_c/\tau^{\frac{1}{2}}$ increased slightly with

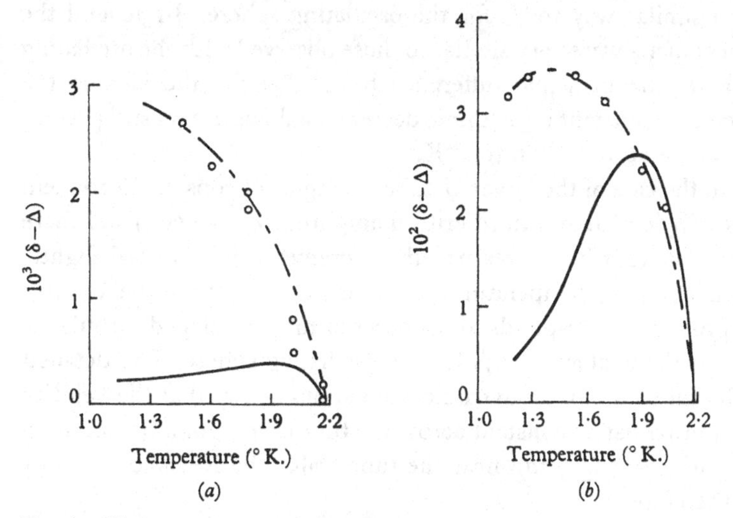


Fig. 69. Temperature variation of the excess decrement at an amplitude of 0·215 radians for (*a*) a single disk, ○, experiment (Hallett, 1952), ——, Gorter-Mellink theory (Zwanniken, 1951); (*b*) a pile of disks, ○, experiment (Hallett, 1952), ——, Gorter-Mellink theory (Hallett, 1951).

increasing temperature, its value at 2·15° K. being about twice its value at 1° K.

Donnelly and Penrose (1956) have studied the damping of the oscillations of liquid in a U-tube. The period $\tau$ was approximately 1 sec. and two U-tubes were used with radii of 0·505 and 0·754 cm. At small amplitudes of oscillation, less than a critical value $h_c$ ($\sim$ 0·1 cm.), the damping could be explained in terms of the separate behaviour of the two components, with the viscosity of the normal component taking its accepted values. Between amplitudes $h_c$ and $h_n$ ($\sim$ 0·3 cm.) the logarithmic decrement steadily increased, but at amplitudes greater than $h_n$ the logarithmic decrement was again constant and could be explained by assuming that the fluid moved as a whole ($v_s = v_n$) and that the effective viscosity $\eta_{\text{eff.}}$ had a value comparable with that deduced from the upper region of constant logarithmic decrement in the experiments on oscillating spheres and disks (fig. 68). The detailed behaviour appeared to depend upon the sharpness of the bend at the bottom of the U-tube, but the following trends were noted. $h_n$ was roughly independent of temperature, but $h_c$ increased slowly with increasing temperature in a similar way to $\phi_c$ for the oscillating sphere. In general the phenomena were very similar to those observed with the oscillating sphere, the principal difference being that for the U-tube the region of constant logarithmic decrement above $h_n$ was still present at temperatures down to 1° K.

In the case of the larger U-tube the region of constant decrement above $h_n$ ended at a third critical amplitude $h_t$ ($\sim$ 1 cm.) and there was then a further increase in the decrement. $h_t$ decreased slightly with increasing temperature and was continuous across the $\lambda$-point. It probably corresponds to the onset of fully developed turbulence and is the analogue of $\phi_t$ for the oscillating sphere. The detailed solution of the hydrodynamical equations shows that the velocity is approximately constant across the tube except for a 'penetration depth' $\lambda \sim (\eta_{\text{eff.}}\tau/\pi\rho)^{\frac{1}{2}}$ near the tube wall. The Reynolds number is therefore

$$R_e = \frac{v_{\text{max.}}\rho\lambda}{\eta_{\text{eff.}}}$$

$$= 2h_t\left(\frac{\pi\rho}{\eta_{\text{eff.}}\tau}\right)^{\frac{1}{2}} \tag{6.55}$$

which is $\sim$255, in excellent agreement with the value 220 for the oscillating sphere.

Hall (1955, 1957) has performed the following experiment with a pile of mica disks clamped inside an aluminium can. The can was filled with liquid helium II at 1·27° K., suspended in the vapour and brought into steady rotation. It was then suddenly brought to rest and the torque exerted on it and the disks by the subsequent motion of the liquid was measured as a function of time. The time integral of this torque gave the total angular momentum imparted to the can by the fluid. For disk spacings of 0·0185, 0·046 and 0·093 cm. and initial angular velocities in the range 0·1 to 1·1 radian sec.$^{-1}$, this total angular momentum corresponded to the total density $\rho$ of the liquid being initially in motion with the can. However, with a disk spacing of $7 \cdot 7 \times 10^{-3}$ cm. and an initial angular velocity of $0 \cdot 2$ radian sec.$^{-1}$, the time integral of the torque was only about 30 % of the maximum possible angular momentum. This result is not entirely unambiguous, but it suggests that either (*a*) the superfluid component was not initially in rotation with the full angular velocity of the can or (*b*) upon stopping the can the vortex lines were swept out of the liquid, leaving behind an irrotational persistent current of superfluid with an angular momentum 70 % of the initial value. This current appeared to have suffered no detectable change after 300 sec.

From the detailed form of the torque versus time curve, it was possible to deduce, for any instant during the motion, the average velocity of the superfluid relative to the disks and also the average force per unit volume exerted by the superfluid on the disks, either directly or through the intermediary of the normal component. This average force per unit volume was proportional to the square of the velocity and approximately independent of the distance apart of the disks, *d*. It could be qualitatively explained as the combined effect of turbulence, which took about 1 sec. to develop, plus a mutual friction

$$\mathbf{F}_{sn} = \frac{B\rho_s \rho_n}{\rho} \,|\,\omega\,|\, (\mathbf{v}_s - \mathbf{v}_n) \tag{6.56}$$

of the type revealed by the attenuation of second sound in rotating liquid (§ 6.1). The dependence of this mutual friction on the angular velocity of the superfluid, $\omega$, was indicated in two ways. First, the frictional forces were too large to be correlated with the results on isothermal flow through capillaries of diameter comparable with the distance between the disks (§ 6.5.1). Secondly, evidence came from a study of the acceleration of the fluid when the can and disks were first set into rotation. The rotation was started and then, after a time $t_1$ too short for the liquid to have achieved a stable rotation, the can was stopped and the total angular momentum of the fluid was measured by the procedure described above. The growth of the rotation was not the inverse of its decay and, in particular, was much more rapid in its final stages, indicating that $\omega$ is relevant as well as the relative velocity of disks and superfluid. The explanation is probably that the frictional forces are proportional to the density of vortex lines.

When the disks were roughened the frictional forces increased by a factor of the order of ten, demonstrating the importance of boundary effects, possibly because a rough surface can generate vorticity more efficiently. However, there was a critical velocity $v_{s,c} \sim 2 \times 10^{-3}/d$ cm. sec.$^{-1}$, below which the frictional forces dropped rapidly to values characteristic of a smooth surface ($d$ is the distance apart of the disks).

Donnelly, Chester, Walmsley and Lane (1956) set a cylindrical vessel into rotation at angular velocities $\sim 100$ radian sec.$^{-1}$ and observed the time taken for the surface of the liquid to assume its final parabolic form. Except in a few cases, which were marked by the formation of a sharp dip at the vertex of the parabola, the growth of the parabolic surface was exponential with a characteristic time of 30 sec. at 1·1° K. and 42 sec. at 2·1° K. With the assumption that the frictional forces can be entirely represented by an effective viscosity $\eta_{\text{eff.}}$, the solution of the Navier-Stokes equation gives a characteristic time

$$\tau = \frac{R^2 \rho}{14 \cdot 6 \eta_{\text{eff.}}}, \tag{6.57}$$

in which $R$ is the radius of the vessel and the effective density of the liquid has been taken to be its total density $\rho$. $\eta_{\text{eff.}}$ ranged from 29 micropoise at 1·1° K. to 21 micropoise at 2·1° K. However, it is quite possible that turbulence occurs during the formation of the meniscus in this type of experiment.

## 6.6. Critical velocities

There are still many uncertainties involved in a consideration of the critical velocities and so only a tentative interpretation can be attempted. The oscillating sphere experiments show that there is one critical amplitude $\phi_c$ related to the breakdown of superfluidity and another $\phi_t$ related to the onset of fully developed turbulence. The oscillations in a U-tube confirm this. Even when considering the lower critical velocity, corresponding to the breakdown of superfluidity, we should not ignore the possibility that there are several different types of process which can produce this breakdown, depending upon the exact circumstances. The process occurring in the thin films, for example, may be very different from that occurring in a channel several millimetres wide. Moreover, the details of the geometry may enter in a complicated way. For a pile of disks, it is not clear whether the critical velocity is identical with that in a parallel-sided channel of width equal to the separation of the disks, or whether the radius of the disks and the angular nature of the motion are also relevant.

TABLE V. *Critical velocities at* 1·4° K.

| Type of experiment | Channel width or characteristic length, $d$ (cm.) | Critical velocity, $v_{s,c}$ (cm. sec.$^{-1}$) | $v_{s,c}d$ (cm.$^3$ sec.$^{-1}$ cm.$^{-1}$) |
|---|---|---|---|
| Unsaturated films | $5 \times 10^{-8}$ | 46 | $2{\cdot}3 \times 10^{-6}$ |
| | $10^{-6}$ | 46 | $4{\cdot}6 \times 10^{-5}$ |
| Saturated films | $3 \times 10^{-6}$ | 25 | $7{\cdot}5 \times 10^{-5}$ |
| Narrow channels (overshoot procedure) | $4 \times 10^{-5}$ | 12 | $4{\cdot}8 \times 10^{-4}$ |
| | $3 \times 10^{-4}$ | 8 | $2{\cdot}4 \times 10^{-3}$ |
| Glass capillaries | $2{\cdot}6 \times 10^{-3}$ | 3 | $7{\cdot}8 \times 10^{-3}$ |
| | $8{\cdot}1 \times 10^{-3}$ | 1 | $8{\cdot}1 \times 10^{-3}$ |
| Oscillations in a U-tube | $2{\cdot}1 \times 10^{-2}$ | 0·62 | $13 \times 10^{-3}$ |
| Oscillating disk | $4{\cdot}2 \times 10^{-2}$ | 0·21 | $8{\cdot}8 \times 10^{-3}$ |
| | $7{\cdot}1 \times 10^{-2}$ | 0·14 | $10 \times 10^{-3}$ |
| Oscillating sphere | $5{\cdot}5 \times 10^{-2}$ | 0·30 | $16{\cdot}5 \times 10^{-3}$ |
| | $6{\cdot}9 \times 10^{-2}$ | 0·26 | $17{\cdot}9 \times 10^{-3}$ |
| | $10{\cdot}7 \times 10^{-2}$ | 0·15 | $16{\cdot}1 \times 10^{-3}$ |
| Rotating cylinder viscometer | $10{\cdot}6 \times 10^{-2}$ | 0·07 | $7{\cdot}4 \times 10^{-3}$ |
| Second sound in a heat current | $24{\cdot}0 \times 10^{-2}$ | 0·051 | $12{\cdot}5 \times 10^{-3}$ |
| | $40{\cdot}0 \times 10^{-2}$ | 0·033 | $13{\cdot}2 \times 10^{-3}$ |

In Table V the critical velocities at 1·4° K. obtained in a large variety of different experiments have been collected together. It has been assumed that it is always the superfluid velocity which is critical at a value $v_{s,c}$. In the case of unsaturated films (§ 7.6) we have accepted the result of Long and Meyer (1955) that the critical velocity is independent of film thickness. The thickness of the saturated film has been taken as $3 \times 10^{-6}$ cm. (§ 7.3) and its rate of flow as $7{\cdot}5 \times 10^{-5}$ cm.$^3$ sec.$^{-1}$ cm.$^{-1}$ (§ 7.5). For narrow channels obtained by pressing together two optically polished glass surfaces, the results quoted are those of Winkel, Delsing and Poll (1955) obtained by the overshoot procedure. The values for glass capil-

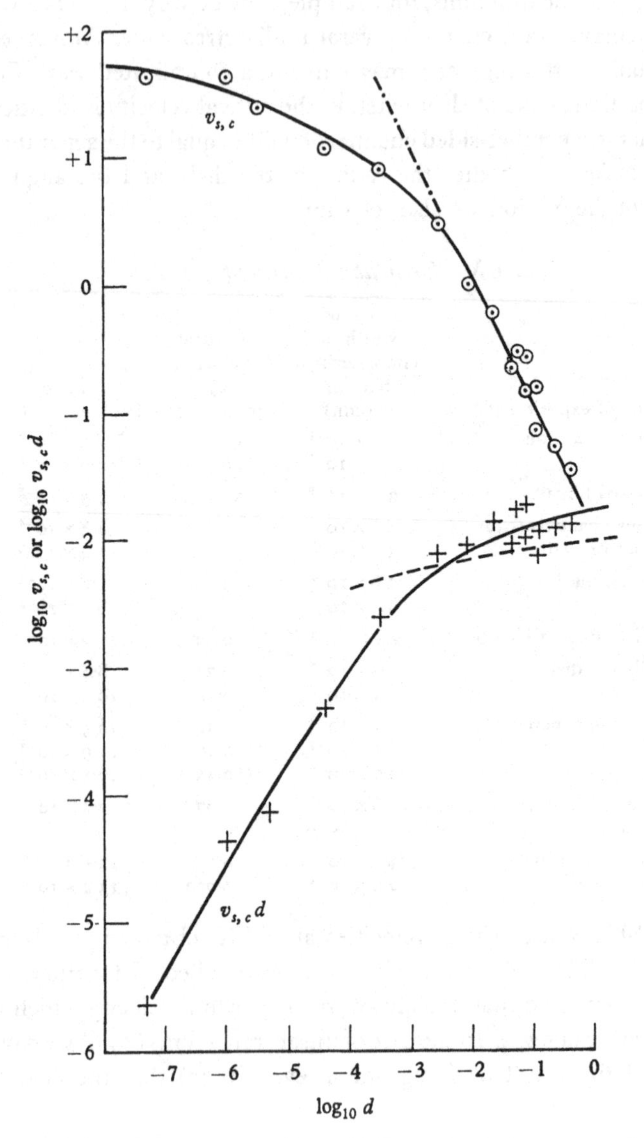


Fig. 70. The critical velocity $v_{s,c}$ and the critical transfer rate $v_{s,c}d$ as functions of the characteristic length $d$. Temperature = 1·4° K.; —·—·—·—, $v_{s,c} \propto 1/d$; ----, $v_{s,c}d = \frac{4\hbar}{m} \ln \frac{d}{4a}$.

laries of circular bore were obtained by extrapolating the results of Atkins (1951) to zero pressure head. For the oscillating disk and sphere and the oscillations in a U-tube, the characteristic length $d$ has been taken to be the penetration depth $(\eta_n \tau/\pi\rho_n)^{\frac{1}{2}}$ of the viscous motion of the normal component.

Fig. 70 is a log-log plot of $v_{s,c}$ against $d$ and of $v_{s,c}d$ against $d$. For the wider channels ($d > 10^{-3}$ cm.) the critical velocity is in good order of magnitude agreement with the equation

$$v_{s,c} = \frac{4\hbar}{md} \ln \frac{d}{4a} \tag{6.58}$$

with $a \sim 2 \times 10^{-8}$ cm. Referring to the discussion of §4.6.3, this equation might imply that the breakdown of superfluidity is due to the formation of vortex rings of diameter equal to one-half the width of the channel. As fig. 70 shows, $v_{s,c}$ is very nearly proportional to $1/d$ for the wider channels, although there is a slight suggestion that $v_{s,c}d$ slowly increases with $d$. Note that, for the oscillating disk and sphere, the penetration depth $\lambda$ is $(\eta_n \tau/\pi\rho_n)^{\frac{1}{2}}$ and the critical velocity is $2\pi R\phi_c/\tau$, so $v_{s,c}\lambda \propto (\phi_c/\tau^{\frac{1}{2}})$, which Benson and Hallett (1956) found to be approximately independent of $\tau$, again implying that $v_{s,c}d$ is almost constant.

For channel widths less than $10^{-3}$ cm., $v_{s,c}$ falls below the value given by equation (6.58) and may become constant for $d < 10^{-6}$ cm. This suggests that the breakdown of superfluidity has a somewhat different character in the narrower channels.

## 6.7. The Kapitza boundary effect

This section is devoted to an important boundary effect: the temperature discontinuity at the surface of a solid when heat flows from the solid into the liquid. Kapitza (1941) investigated heat flow from a hot body freely suspended in the liquid. The body was a rectangular parallelepiped of copper with a heater and a phosphor-bronze thermometer embedded inside it. If we consider the free flow of heat through the bulk liquid in terms of the equations already discussed in this chapter, it is easy to see that, in the region of sub-critical velocities, the temperature gradient is zero, and that, even in the super-critical region, the frictional forces are not large enough to produce a measurable temperature gradient for attainable values of the heat current. Nevertheless, Kapitza found that the excess temperature, $\Delta T$, inside the solid body was large and linearly proportional to $W$, the outward heat flow per unit area of the surface. However, when two such copper parallelepipeds were suspended almost touching one another and only one was heated, the temperature rise in the unheated one was less than 10 % of that in the heated one, indicating that the temperature jump was confined to a small region less than $10^{-3}$ cm. thick in the immediate vicinity of the hot surface. A layer of gold plating on the copper surface had no effect, but a thin film of oil increased the temperature jump by a factor of the order of 2. The effect of the oil was probably to smooth out the irregularities on the rough surface and to decrease its effective area, for when the copper was carefully cleaned and polished the temperature jump again increased and a thin film of oil then had very little further effect. This dependence on the roughness of the surface is additional proof that the temperature drop occurs within an extremely thin layer. For a clean, polished copper surface the order of magnitude of the temperature jump was given by $\Delta T/W \sim 10/T^3$ watt$^{-1}$ cm.$^2$ deg.

White, Gonzales and Johnston (1953) found that $\Delta T/W$ had a minimum near 2·1° K. and increased rapidly just below the $\lambda$-point. Below 2·1° K. $\Delta T/W$ was approximately proportional to $1/T^{2\cdot 6}$. Andronikashvili and Mirskaia (1955) found a similar minimum and sharp increase just below the $\lambda$-point. Between 1·6 and 2·1° K. their values of $\Delta T/W$ were approximately proportional to $1/T^4$, but between 1·4 and 1·6° K. the variation was more nearly as $1/T^3$. Fairbank and Wilks (1955) have made measurements between 0·3 and 1° K. Their results can be summarized by the equation

$$\frac{\Delta T}{W} = \frac{45}{T^2} \text{ watt}^{-1} \text{ cm.}^2 \text{ deg.}$$

Theoretical explanations have been attempted by Gorter (1951), Kronig (1951) and Khalatnikov (1952*a*). In the theories of Gorter and Kronig the temperature drops off exponentially near the solid surface:

$$T - T_0 = W\frac{\delta}{\chi}e^{-z/\delta}. \tag{6.59}$$

This is equivalent to a layer of liquid of thickness $\delta$ with a normal thermal conductivity $\chi$, and the total temperature jump is

$\Delta T = W\delta/\chi$. In Kronig's theory, equation (6.59) is one possible solution of the hydrodynamical equations when the terms involving the viscosity and the thermal conductivity $\chi$ are included. Then

$$\delta_K = \left(\frac{\eta_n \chi}{\rho^2 S^2 T}\right)^{\frac{1}{2}}. \tag{6.60}$$

This predicts too small an effect above 1° K., but the mechanism may be important at lower temperatures. Gorter's treatment takes account of the fact that the superfluid component moves towards the hot surface while the normal component moves away from it, so that there must be a conversion of superfluid into normal fluid in the vicinity of the surface. It is assumed that the rate of conversion is proportional to the excess temperature and the 'continuity' equation for the normal component then becomes

$$\operatorname{div}(\rho_n \mathbf{v}_n) + \alpha\rho(T - T_0) = 0. \tag{6.61}$$

The entropy equation is a modified form of equation (5.97):

$$\rho S \operatorname{div} \mathbf{v}_n = \frac{\chi}{T}\nabla^2 T. \tag{6.62}$$

These two equations give equation (6.59) with

$$\delta_G = \left(\frac{\rho_n \chi}{\alpha\rho^2 S T}\right)^{\frac{1}{2}}. \tag{6.63}$$

Unfortunately the constant $\alpha$ determining the rate of interconversion of the two fluids is not known.

In Khalatnikov's theory the temperature jump does not occur in the liquid, but directly at the solid boundary, and is a consequence of the processes which enable the solid surface to radiate phonons into the liquid. He showed that the total heat current corresponding to all the phonons emitted by a solid surface at a temperature $T$ is

$$W_{\text{tot.}} = \alpha \frac{4\pi^5}{15} \frac{\rho}{\rho_{\text{sol.}}} c \frac{(kT)^4}{(2\pi h c_t)^3}. \tag{6.64}$$

$\rho$ and $\rho_{\text{sol.}}$ are the densities of the liquid and the solid, $c$ is the velocity of phonons in the liquid, $c_t$ is the velocity of transverse phonons in the solid and $\alpha$ is a constant ($\sim 2$) determined by the ratio of the velocities of transverse and longitudinal sound in the solid. The variation of $W_{\text{tot.}}$ with $T^4$ is related to the fact that the energy density of phonons in the solid varies as $T^4$. When the liquid is at the same temperature as the solid, this heat current is exactly compensated by the phonons entering the solid from the liquid but, when the liquid is at a lower temperature $(T-\Delta T)$, the net heat current is obviously

$$W=\alpha\frac{16\pi^5}{15}\frac{\rho}{\rho_{\text{sol.}}}c\frac{k^4T^3}{(2\pi hc_l)^3}\Delta T. \tag{6.65}$$

This agrees with the experiments above 1° K. in order of magnitude and also in the $T^3$ variation. Since the origin of $\Delta T$ is essentially the acoustic mismatch for phonons travelling from the solid to the liquid, the situation may be influenced by the presence of a thin film of solid helium on the surface of the solid body (§7.7). Khalatnikov has suggested that the following mechanism may also become important just below the $\lambda$-point. A phonon or roton striking the solid surface from the liquid is reflected with a change of energy, the energy balance being made up by the emission or absorption of a phonon by the solid.

The Kapitza boundary effect has also been observed in liquid $He^3$ (Fairbank and Lee, 1957). Its magnitude is two or three times larger than in liquid $He^4$ and its variation with temperature is very similar. This suggests that the effect in liquid $He^4$ cannot be primarily a consequence of superfluidity as it is in the theories of Kronig and Gorter. However, Khalatnikov's theory is far from satisfactory. It predicts values of $\Delta T/W$ which are rather too large, and it cannot explain the $1/T^2$ variation below 1° K. Moreover, it predicts a marked decrease of $\Delta T/W$ with increasing pressure, because of the increase in the factor $\rho c$ in equation (6.65), whereas Dransfeld and Wilks (1957) have found that the boundary resistance is almost independent of pressure up to 25 atmospheres. They also found that there was only a factor of 2 difference between a copper surface and a lead surface, whereas equation (6.65) predicts a factor of 14 because of the large difference in $c_l$.

CHAPTER 7

# HELIUM FILMS

## 7.1. Introduction

Some of the fundamental properties of the films have already been discussed, particularly in §§ 1.4 and 4.2.2. This chapter contains a more thorough treatment and includes discussion of several matters which are still controversial.

Helium films present a clear cut case of physical adsorption. The helium atom is chemically inert and there can be no question of chemical bonding between the atoms of the film and the atoms of the wall. Offsetting this initial simplification, however, are the complications associated with the existence of the films at low temperatures: the importance of a large zero-point energy and the quantum effects which give rise to the $\lambda$-transition in the liquid. The superfluidity of the liquid also gives the films unique properties and the flow of the films draws attention to their existence in a dramatic fashion. Thick films of helium are particularly easy to study, because, whereas a thick film of an ordinary substance is readily evaporated by a small amount of stray heat, a helium film under the same circumstances can be replenished by superfluid flow. Multimolecular adsorption is therefore unusually prominent in the case of helium.

The thick film, which is responsible for transfer of liquid into or out of an open vessel, has been called the saturated film, because it exists in equilibrium with the saturated vapour. It is also possible to perform a more conventional type of adsorption experiment in which the film is in equilibrium with gas at a pressure $p$ less than the vapour pressure $p_0$, and films of this type are called unsaturated films. There is a possible connexion between the two cases. Imagine the ideal case of a static saturated film formed on a perfectly smooth, clean, vertical surface dipping into liquid in an accurately isothermal enclosure. The thickness $d$ of the film is a function of the height $H$ above the free surface of the bulk liquid. However, at this height the film is in equilibrium with gas whose pressure $p$ at large distances from the wall is not $p_0$ but

$$p = p_0 e^{-MgH/RT}. \tag{7.1}$$

$M$ is the atomic weight of helium. This seems to imply that an unsaturated film being studied in an adsorption experiment at a relative pressure $p/p_0$ has identical properties with a saturated film in equilibrium at a height $H$ above the surface of the bulk liquid given by

$$H \equiv -\frac{RT}{Mg} \ln (p/p_0). \tag{7.2}$$

## 7.2. Experimental difficulties

It is extremely doubtful whether the ideal case just discussed can be realized, or even approached, in practice. Atkins (1948) and, independently, van den Berg and de Haas (1949) observed that, occasionally, the rate of transfer through the saturated helium film is many times greater than normal. Bowers and Mendelssohn (1950*b*) were able to show that this is a consequence of contamination of the surface by a deposit of some solidified gas, such as solid air, solid hydrogen or ice. Perhaps the deposit is granular and either increases the effective perimeter of the beaker on a microscale or liquid is held in the spaces between the grains, so that the effective thickness of the film is increased. Possibly, even a thin deposit of some impurity is able to modify the forces between the wall and the helium in such a way that the film thickness is increased. The views of McCrum and Eisenstein (1955) on this point will be discussed later. From a practical point of view, it is extremely difficult to avoid accidental contamination of the surface and even the most careful of experiments can be suspect in this connexion.

No surface is perfectly smooth, and even the surface of glass probably contains the so-called Griffith's cracks which are important in determining its strength. For every helium film experiment, therefore, it is relevant to ask whether the results are entirely determined by the film or whether they are significantly influenced by liquid which is held in the cracks by surface tension effects. Recent experiments of Dyba, Lane and Blakewood (1954) and

Brewer and Mendelssohn (1953), confirm that surface tension effects can fill such cracks with liquid, particularly below the $\lambda$-point where any liquid evaporated out of the crack by stray heat can be quickly replenished by film flow.

Consider an adsorption experiment on an unsaturated helium film with a gas pressure $p$ and with the surrounding liquid helium bath at a temperature $T$, where the vapour pressure is $p_0(T)$. A small amount of heat falling on the surface holding the film will raise the temperature in the vicinity to $T+dT$, where the vapour pressure is $p_0(T+dT)$. The experimenter will therefore relate the thickness of the film to the relative pressure $p/p_0(T)$, whereas he should use $p/p_0(T+dT)$. If $dT$ is not too large, this does not matter when $p/p_0$ is appreciably less than 1, but when $p$ is close to $p_0$ the thickness of the film is very sensitive to $p_0-p$ (see fig. 72) and a film which might be expected to be thick because $p_0(T)-p$ is small remains thin because $p_0(T+dT)-p$ is large. For this reason, thick films were not observed above the $\lambda$-point until care was taken to reduce extraneous radiation. Below the $\lambda$-point the situation is somewhat different. The superfluid film is able to flow towards the source of heat and prevent the formation of large temperature differences, but there is then a dynamic type of equilibrium in which the film arriving at the hot region evaporates, returns as vapour to the cold region, condenses as bulk liquid or film and flows back to the hot region. It is relevant to inquire whether a film behaving in this way has the same properties as a static film. Meyer (1955) has suggested, for example, that when the superfluid component in the film is flowing with a velocity $v_s$, the kinetic energy of the film must be included in its free energy and the thermodynamic equilibrium is thereby disturbed in such a way that the film thickness is increased and has a different variation with height. Kontorovich (1956) has also considered this point, but arrives at the opposite conclusion—that a moving film is thinner than a stationary film.

For saturated films of liquid helium II there is another disturbing possibility if a vertical temperature gradient exists. In accordance with the thermomechanical effect, if two volumes of liquid communicate with one another through a narrow channel and there is a temperature difference $\Delta T$ between them, the level in the warmer vessel is raised above the other level by an amount

$$\Delta h = \frac{S}{g} \Delta T. \tag{7.3}$$

If a vertical surface dips into liquid helium II and the temperature increases upwards, one might wonder if the effective height $H$ determining the formation of the film ought to be reduced by $\Delta h$. At 2·0° K. a temperature gradient of only $10^{-4}$ deg. cm.$^{-1}$ would be sufficient to counteract the effect of gravity.

This is a formidable array of possible difficulties and it is clear that the perfect film experiment would be difficult to perform. It is, therefore, not surprising that many of the experiments which have so far been performed have been mutually contradictory. Moreover, no single experiment can be said to be absolutely free from suspicion on one or more of the counts presented above.

## 7.3. Measurements of the thickness of the saturated film

Jackson and his co-workers (Burge and Jackson, 1951; Jackson and Henshaw, 1953; Ham and Jackson, 1954, 1957) have measured the thickness of the saturated film by an optical method, depending upon the fact that, if plane polarized light is reflected from a polished

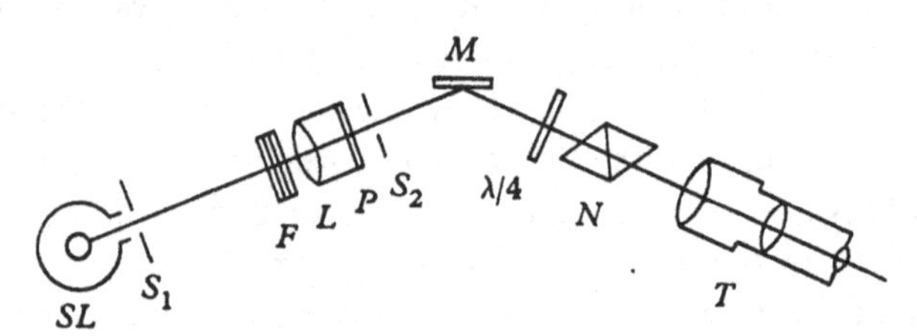


Fig. 71. The optical method of measuring film thickness. *SL*, sodium lamp; $S_1$, $S_2$, slits; *F*, filter of heat absorbing glass; *L*, lens; *P*, polaroid; *M*, mirror; $\lambda/4$, mica compensating plate; *N*, Nicol prism; *T*, telescope (after Burge and Jackson, 1951).

metal surface covered by a film, then the reflected light is elliptically polarized and the angle between the initial plane of polarization and the major axis of the ellipse can be related to the thickness of the film. Referring to fig. 71, light from a sodium lamp (*SL*) was collimated by the slits $S_1$ and $S_2$, passed through a heat absorbing filter *F* and was plane polarized at 45° to the vertical by the polaroid slab $P$. After reflexion from the surface $M$, the elliptically polarized light was converted back into plane polarized light by the compensating plate, $\lambda/4$, and the plane of polarization was determined by rotating the Nicol prism $N$. The upper part of the polished stainless steel mirror $M$ was coated with a monomolecular layer of barium stearate ($2{\cdot}4 \times 10^{-7}$ cm. thick), whereas the lower part had a coating three molecules thick. Different settings of the Nicol prism $N$ were therefore needed to make the two regions appear dark, but an intermediate setting could be found at which the upper and lower sides of the 1-3 boundary appeared equally bright. When the helium film formed on top of the barium stearate layers, this half-shade setting was destroyed, but could be restored by rotating the Nicol prism through a small angle of a few degrees, measurable to within $\pm 8$ min. of arc, and from this angle the film thickness at the 1-3 boundary could be deduced. The lower part of the mirror $M$ dipped into a liquid helium bath and the film thickness measured therefore corresponded to the height $H$ of the 1-3 boundary above the surface of the bath.

As the technique has been developed and some of the difficulties mentioned in §7.2 appreciated, the results have been progressively modified, so although we shall discuss only the latest measurements (Ham and Jackson, 1954, 1957), these may still not be final. Below the $\lambda$-point the film thickness $d$ could be expressed in the form

$$d = k/H^n \text{ cm.} \tag{7.4}$$

$k$ varied from $2{\cdot}96 \times 10^{-6}$ at $1{\cdot}32^\circ$ K. to $3{\cdot}15 \times 10^{-6}$ at $2{\cdot}05^\circ$ K. and $n$ was about $0{\cdot}44$ over this temperature range. At a height of 1 cm. the film was therefore about 80 layers thick. Above the $\lambda$-point, the early experiments gave a value of $k$ of only $4 \times 10^{-7}$ cm., but when special care was taken to avoid extraneous heat, the measured value of $k$ was only slightly smaller than below the $\lambda$-point. To investigate thermal effects below the $\lambda$-point, a heating coil was attached to the top of the mirror and a small heat input of $7{\cdot}75\ \mu$W. was applied; the film thickness was then found to increase by about 20 % and varied less rapidly with height. At $300\ \mu$W. the flow mechanism probably saturated, because then the film was observed to have a sharp upper boundary which retreated down the mirror as the power was increased. At about 30 mW. a very thick film, similar to a surface tension meniscus, was formed and rose to a height of 8 mm. above the bulk liquid. In another experiment the mirror $M$ formed part of the outer wall of a beaker and it was found that the film thickness did not change noticeably when there was film flow out of the beaker.

Bowers (1953*a*) has used a microbalance to weigh the film formed on a roll of aluminium foil. A wire tail hanging from the foil dipped into bulk liquid and the variation in the weight of the film was studied as the distance between the liquid surface and the bottom of the foil was varied. Interpreting the results in terms of equation (7.4), $k$ was $11{\cdot}8 \times 10^{-6}$ and $n$ was $0{\cdot}5 \pm 0{\cdot}07$, so that the height dependence was not inconsistent with the optical method, but the absolute thickness of the film was about three times greater. The thickness was almost independent of temperature below the $\lambda$-point, but within a few thousandths of a degree of the $\lambda$-point suddenly fell to about $4 \times 10^{-7}$ cm. The weight of the film was very sensitive to radiation influx, and when the wire tail did not dip into the bulk liquid, so that the heat contact of the creeping film was lost, the thickness fell to about $2{\cdot}2 \times 10^{-6}$ cm. and was independent of the height of the foil above the liquid surface and almost independent of temperature.

Atkins (1950*a*) has used an oscillation method. As a beaker empties through the film, the rate of film transfer is almost independent of the level difference and so, when the inner level reaches the outer level, the film is still moving rapidly and its momentum causes the level to overshoot and then oscillate about its equilibrium position. This oscillation involves a periodic interchange of energy between the kinetic energy of the film and the potential energy of that liquid inside the vessel which is raised above the outer level, and the period therefore depends primarily on the thickness of the film. To interpret the results it is necessary to make certain assumptions about the nature of the flow, such as that the velocity in the film at a fixed height is independent of distance from the wall, and that the thickness of the film is independent of its velocity. The period of an oscillation can then be shown to be

$$\tau = 2\pi \left[ \frac{\rho}{\rho_s} \frac{r}{2g} \left(1 + \frac{r}{R}\right) \int_0^l \frac{dH}{d} \right]^{\frac{1}{2}}, \tag{7.5}$$

where $r$ and $R$ are the inner and outer radii of the beaker and $l$ is the height of the rim of the beaker above the bath level. Measurements of the period $\tau$ over a range of values of $l$ therefore enabled $d$ to be deduced as a function of $H$.

Over a range of heights from 1 to 5 cm., the results could be represented by equation (7.4), with $k$ varying from $1 \cdot 5 \times 10^{-6}$ at $1 \cdot 1°$ K. to $2 \cdot 2 \times 10^{-6}$ at $2 \cdot 0°$ K., more in agreement with the optical method than the weighing method. The value of $n$ was 0·14, which is much less than the other two methods gave, but there was some evidence that the effective value of $n$ was greater at heights less than 1 cm. However, when the experiment was repeated with a slightly different beaker, the film appeared to be about 30% thicker, so there were obviously some unexplained effects.

## 7.4. Unsaturated films

The basic experiment on unsaturated films is the plotting of an adsorption isotherm, showing how the mass of gas held on a surface varies with the relative pressure $p/p_0$ at constant temperature. This has been done by the usual volumetric method of introducing measured volumes of gas into a system containing a powder of large surface area, such as carbon (Schaeffer, Smith and Wendell, 1949), jeweller's rouge (Frederikse and Gorter, 1950; Long and Meyer, 1949; Strauss, 1952) or rutile (Mastrangelo and Aston, 1951). Bowers (1953*b*) has used the microbalance technique described in the previous section, weighing the film formed on a roll of aluminium foil in an atmosphere of helium gas at a pressure $p$ less than the vapour pressure $p_0$. The optical method of Jackson has been adapted to the unsaturated film by McCrum and Mendelssohn (1954).

Fig. 72 is a schematic representation of the shape of a typical adsorption isotherm. At low pressures there are two plateaus corresponding to the formation of the first two layers (Keesom and Schweers, 1941; Meyer, 1956). The density of the first layer is found to be about twice that of the bulk liquid, and this layer is undoubtedly solid. Its density decreases slightly with increasing temperature.

At larger values of $p/p_0$, the thickness $d$ of the film can be expressed in the form (Bowers, 1953*b*)

$$d = \kappa/[\ln(p/p_0)]^{\frac{1}{3}}. \tag{7.6}$$

Using equation (7.2) to express this as a variation of film thickness with height, it is equivalent to

$$d = k/H^{0.333} \tag{7.7}$$

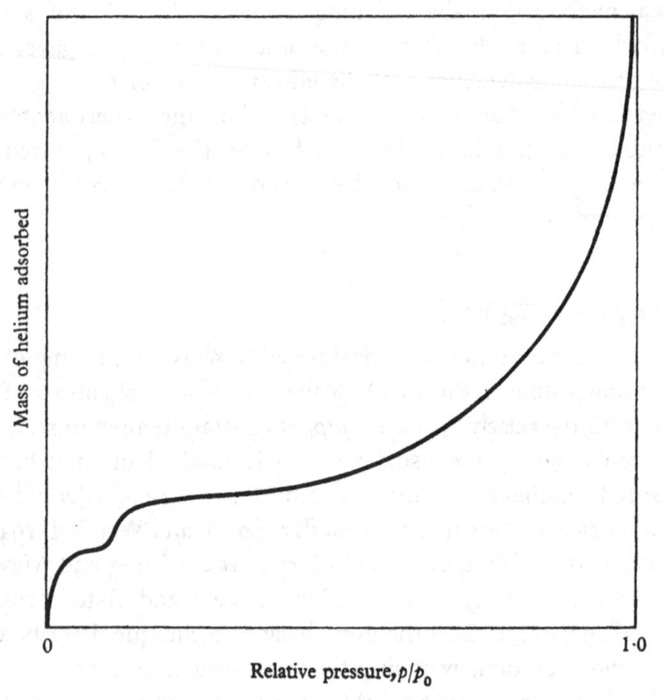


Fig. 72. An adsorption isotherm (schematic).

with values of $k$ similar in order of magnitude to those quoted in § 7.3 but varying more markedly with temperature. It is interesting to note that the thickest film studied in an adsorption experiment had about 20 layers, and the corresponding value of $p/p_0$ was equivalent to a height of about 25 cm. in a saturated film experiment.

From a set of adsorption isotherms at different temperatures, standard thermodynamic arguments can be used to derive the heat of adsorption and the differential entropy of the film. The heat of adsorption is high for the first one or two layers, but soon falls to a value near the latent heat of vaporization of the bulk liquid. Similarly, the differential entropy of the film, $S_f$, is very close to the entropy of the bulk liquid, $S_l$, for films more than one or two layers thick. Both these facts suggest that, apart from a few layers near the wall, the liquid in the film is very similar to the bulk liquid.

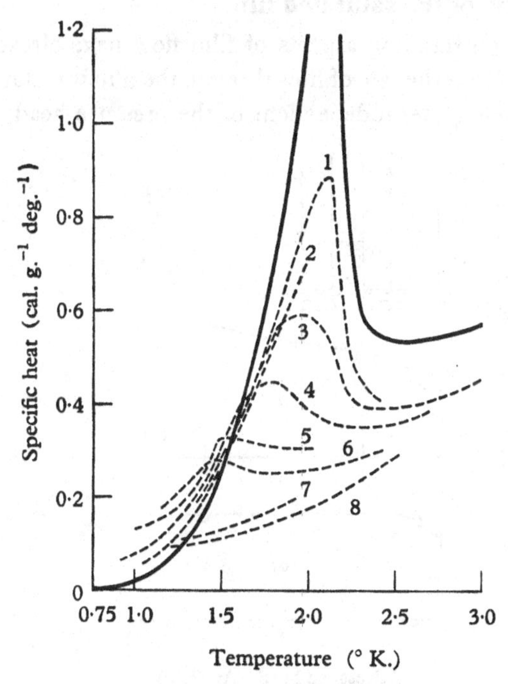


Fig. 73. Average specific heat of helium adsorbed on jeweller's rouge in cal. g.$^{-1}$ deg.$^{-1}$ as a function of temperature for different amounts adsorbed. The full curve is the specific heat of the bulk liquid (after Frederikse, 1949).

| Curve ... | 1 | 2 | 3 | 4 | 5 | 6 | 7 | 8 |
|---|---|---|---|---|---|---|---|---|
| $p/p_0$ | 0·82 | 0·76 | 0·70 | 0·48 | 0·40 | 0·28 | 0·13 | 0·11 |
| Approximate number of layers, $n$ | 9 | 7 | 6 | 4 | 3 | 2·5 | 2 | 2 |

It is interesting to note that $S_f - S_l$ is negative above the $\lambda$-point and positive below the $\lambda$-point (but with a suggestion of negative values for intermediate coverages). The specific heat of films of various thicknesses on jeweller's rouge has been measured by Frederikse (1949). The results (fig. 73) suggest that some sort of smeared out $\lambda$-transition occurs at temperatures lower than the bulk $\lambda$-point. Band (1949) has pointed out that the specific heat of the first layer follows roughly a $T^2$ law, as for a two-dimensional solid with a Debye temperature of 18° K. There is order of magnitude agreement between the specific heat curves and the entropies deduced from the adsorption isotherms.

## 7.5. The flow of the saturated film

The more outstanding aspects of film flow have already been presented. In § 1.4 the rate of flow through the film was stated to be almost (but not quite) independent of the pressure head, and we

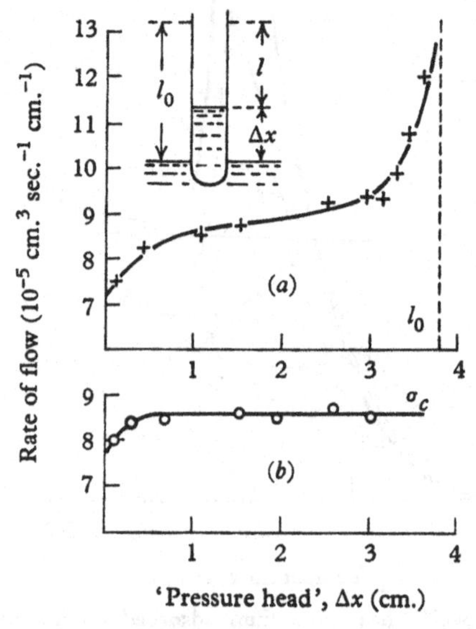


Fig. 74. Dependence of the rate of flow of the film on the level difference. (a) +, emptying, $l_0 = 3{\cdot}78$ cm.; (b) ○, filling, $l_0 = 0{\cdot}82$ cm. (Atkins, 1950b).

have seen how this fact first suggested the idea of a critical velocity. In § 4.2.2 we considered the experimental evidence that flow of the film experiences no detectable friction below the critical velocity. The controversial issue of the small decrease in transfer rate at small pressure heads was also discussed. Some of the finer details of film flow will now be considered.

The transfer rate varies in a complicated way with the level difference and the height of the film (Daunt and Mendelssohn, 1939b; Atkins, 1950b; Eselson and Lasarew, 1952; Smith and Boorse, 1955). During a filling the transfer rate is practically constant (fig. 74*b*), except at very small pressure heads, but the constant rate thus determined depends markedly upon the height of the rim of the beaker above the outer level (fig. 75). When the beaker is emptying the transfer rate does vary with the level difference, but the various observations can be correlated if we assume that the important factor is the height of the rim of the beaker above the *higher* of the two liquid levels. Atkins (1950*b*) has pointed out that, since the film thickness varies with height and since the transfer rate through a channel is known to vary with its width (fig. 70),

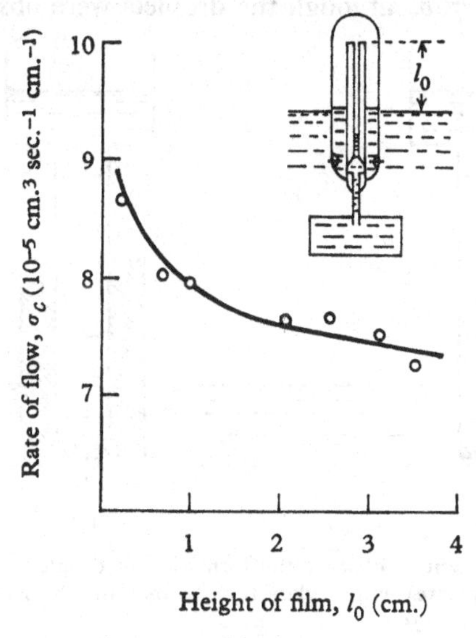


Fig. 75. Variation of the rate of filling with the height of the film. Temperature = 1·47° K. (Atkins, 1950*b*).

one might expect the rate of transfer to depend upon the thickness of the film at the rim of the beaker. Although such an effect may exist, it is also possible that the observed variations are mainly caused by secondary factors, such as capillary flow in cracks (Chandrasekhar, 1952).

When the inner and outer levels are at very different heights, it is pertinent to ask whether the thickness of the film is determined by its height above the inner level, the outer level, or neither. Jackson and Henshaw (1953) and Ham and Jackson (1953) used the optical method (fig. 71) to observe the thickness of the film on the outer surface of the stainless steel beaker shown in fig. 76*a*. They found that this thickness was uniquely determined by the height above the outer level and was quite independent of the position of the inner level. When the inner level was in the wide part of the beaker, *A*, but the outer level was in the region of the narrow part, *B*, the transfer rate was still determined by the perimeter of *A*, but on the outside of *B* droplets were observed to form and run down the wall, carrying away the excess liquid which could not be transferred by the film in this constricted region. However, in the apparatus of fig. 76*b*, although the droplets were observed in the

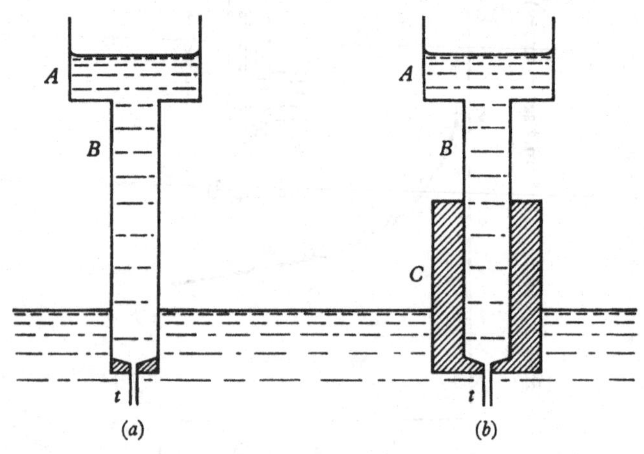


Fig. 76. An experiment to observe the formation of droplets out of the film. The tube *t* communicates with a glass tube to monitor the inside level (after Jackson and Henshaw, 1953).

waist region, *B*, they merged into the film again and were not observed on the outside of *C* where the transfer was again supported wholly by the film. Using the complicated beaker shown in fig. 77, Chandrasekhar and Mendelssohn (1955) observed that the formation of drops occurred in a manner which can be understood by studying the figure. From these experiments we may conclude: first, that the transfer rate is determined by the smallest perimeter lying above the higher of the two liquid levels; secondly, that, if there is a still smaller perimeter below this level, then drops form there to carry away the excess flow; thirdly, that no drops are formed below the higher level if the perimeter is sufficient to enable the film to carry all the flow. Ham and Jackson (1953) estimated the size and velocity of the drops and found them to be in agreement with a theory advanced by Frenkel (1948).

When the surface is contaminated by a solid deposit, the transfer rate is increased and its dependence on the height of the rim above the liquid level is accentuated (Atkins, 1948; van den Berg and de Haas, 1949; Bowers and Mendelssohn, 1950*b*). In such cases, the oscillation method of measuring film thickness gives a period

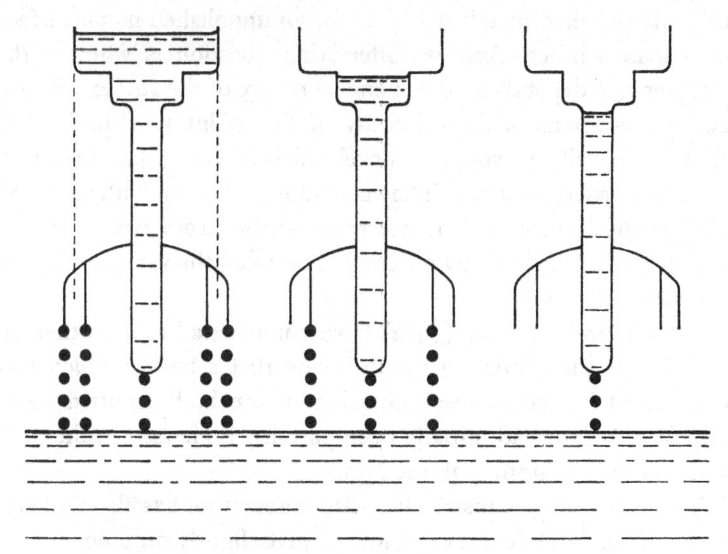

Fig. 77. The formation of droplets out of the film (after Chandrasekhar and Mendelssohn, 1955).

linearly proportional to the height $l_0$ of the film, and, if this is interpreted in terms of equation (7.5), it implies an equivalent film thickness of the form

$$d = k/H, \tag{7.8}$$

with $k$ several times greater than for a clean surface. The detailed explanation of this is still obscure, but might well reward further study. The transfer rate is also increased when the surface is deliberately roughened (Mendelssohn and White, 1950; Dash and Boorse, 1951; Chandrasekhar, 1952; Chandrasekhar and Mendelssohn, 1952; Smith and Boorse, 1955), but again it is not clear whether this should be ascribed to an increase in the micro-perimeter or to bulk liquid held in grooves. At this stage, one might reasonably ask whether all secondary effects, such as the variation of the rate of transfer with the height of the rim, are caused by contamination or surface irregularities, and the experimental difficulties are such that it is not yet possible to give a dogmatic answer. However, these effects have always been present in the most carefully conducted experiments, and Smith and Boorse (1955) found that the height dependence was just as marked for a highly polished metal surface as for an unpolished metal surface or a glass surface. Another interesting question is whether the transfer rate depends on the chemical nature of the surface. Smith and Boorse made a careful study of this point with beakers of aluminium, silver, copper, nickel, nickel-silver, stainless steel, Pyrex, quartz and Lucite. They conclude that, even with the same surface the transfer rate is not reproducible from run to run to better than 10 %, and any differences between the various surfaces were less than this.

Eselson and Lasarew (1952) have complicated still further an already complex situation by observing that a beaker which had been filled by complete submersion in the bath emptied more rapidly than if it had initially been partially immersed in the bath and allowed to fill through the film.

The variation of transfer rate with temperature has already been given in fig. 6. Different experiments give slightly different curves and the maximum near 1·5° K. is not always reproduced. The transfer rate per unit width of perimeter is, of course,

$$\sigma_c = \frac{\rho_s}{\rho} v_{s,c} d, \tag{7.9}$$

where $v_{s,c}$ is the velocity of the superfluid component. The variation between 1° K. and the λ-point is caused mainly by the factor $\rho_s/\rho$, although the weight of the evidence suggests that $v_{s,c}d$ also varies with temperature. Ambler and Kurti (1952) have measured transfer rates below 1° K. and find an increase of about 30 % as the temperature is lowered from 1 to 0·15° K. The experiments of Lesensky and Boorse (1952), Waring (1955) and Hebert, Chopra and Brown (1957) support this conclusion. In order of magnitude

$$v_{s,c} d \sim \frac{h}{4\pi m}, \tag{7.10}$$

where $m$ is the mass of the helium atom, but there is no clear evidence that $v_{s,c}d$ is constant and it probably varies with both temperature and the height of the film.

## 7.6. The flow of unsaturated films

The discovery by Long and Meyer (1950, 1952) of superfluid flow through unsaturated films opened up the possibility of extending our knowledge of superfluidity to much narrower channels and also gave valuable information about the nature of these very thin films. In their excellent review article, Long and Meyer (1953) state: 'The flow measurements and the heat transport experiments show that only $\sim \frac{1}{2}$ statistical layer on top of the densely packed first monolayer is sufficient to produce He II phenomena,' (that is, film flow). The apparatus used in their initial investigations consisted of an adsorption chamber $A$ connected to a measuring chamber $B$ through a superleak, which was either a platinum wire sealed into a Pyrex capillary or two optically ground stainless steel plates pressed together. They adopted two different procedures, which they call Method I and Method II, and which gave apparently contradictory results. In Method I, chamber $A$ contained gas at a pressure $p/p_0$ and there was therefore a thin unsaturated film on the walls of this chamber and on the surface of an adsorbent ($Fe_2O_3$) which was sometimes present to act as a reservoir of film. Chamber $B$ was initially evacuated, then closed off, and the initial rate of rise of pressure in $B$ was taken to represent the rate of flow of the unsaturated film through the superleak. The results are shown in fig. 78, which gives the rate of flow as a function of temperature at the constant relative pressure $p/p_0$ indicated by the figure adjacent to each curve. The range of values of $p/p_0$ covered in these experiments corresponds to film thicknesses from 4 up to 20 atomic layers. The striking feature of these results is that superfluidity sets in sharply at a temperature well below the normal $\lambda$-point, and the thinner the film, the lower the onset temperature.

In Method II the pressure in chamber $B$ was only slightly less than in chamber $A$ so that the flow took place under a small pressure differential. Under these conditions, for all thicknesses of film, flow occurred at all temperatures up to the normal $\lambda$-point. A type of thermomechanical effect was also observed. When there was a temperature difference between the two chambers, the film flowed into the chamber at the higher temperature, even though the pressure there was higher. The discrepancy between Methods I and II has not yet been resolved, but the heat transport experiments described below favour the conclusions of Method I.

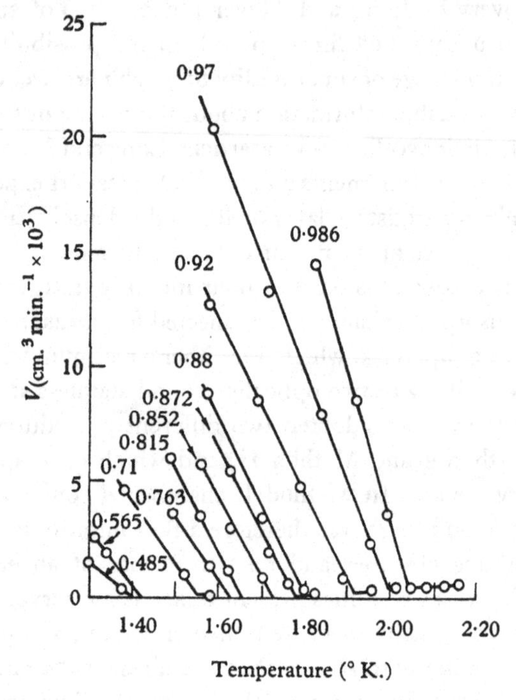


Fig. 78. Flow rate of the unsaturated film as a function of temperature for different relative pressures (Long and Meyer, 1950, Method I).

The experiment performed by Bowers, Brewer and Mendelssohn (1951) is illustrated in fig. 79. The lower part of the chamber was in contact with a helium II bath and the upper part was thermally isolated. The chamber contained helium gas at a relative pressure $p/p_0$. Heat at a rate $\mathring{Q}$ was supplied to a heater near the top of the chamber and a thermometer registered the temperature in the vicinity of the heater. When the heat input was small, no temperature rise was observed, probably because the heat was dissipated by flow of the unsaturated film up the walls of the chamber, evaporation of the film near the heater and flow of the gas down to the bottom of the chamber where it condensed again. However, at a critical heat input, $\mathring{Q}_c$, a large rise in temperature suddenly occurred and $\mathring{Q}_c$ was therefore taken to be a measure of the critical rate of transfer of the unsaturated film. The total rate of transfer over the whole perimeter in grams per sec. must in fact have been

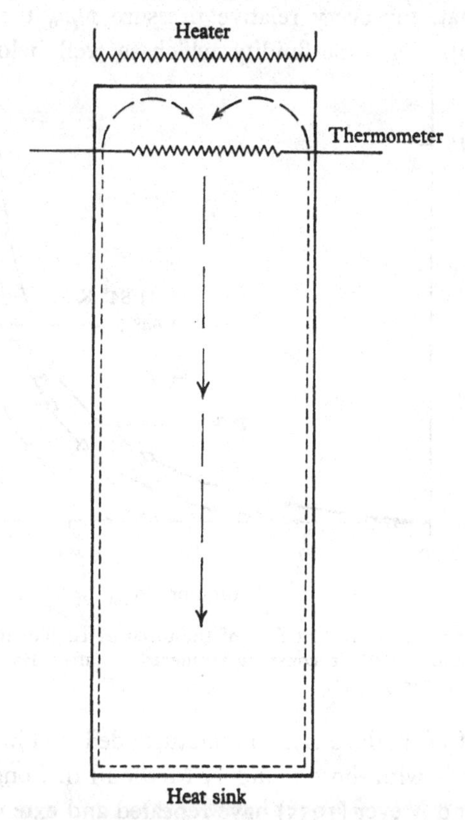


Fig. 79. Apparatus to investigate heat transport in the unsaturated film (after Bowers, Brewer and Mendelssohn, 1951).

$\mathring{Q}_c/(L_B+TS)$, where $L_B$ is the latent heat of the film, which is known to be approximately the same as the latent heat of the bulk liquid. When the pressure in the chamber was raised to the saturated vapour pressure $p_0$, the saturated film was studied and the critical heat input $\mathring{Q}_s$ was found to vary with temperature in the same way as the transfer rate of the saturated film (fig. 6). The quantity $r_c/r_s=\mathring{Q}_c/\mathring{Q}_s$ plotted against $p/p_0$ at various temperatures in fig. 80

is the ratio of the critical transfer rates in the unsaturated and saturated films. The rate of flow of the unsaturated film is seen to be only a small fraction of that of the saturated film. This is important, because it establishes without any doubt that $\sigma_c$ must vary with $d$ in this region of film thicknesses. Examination of fig. 80 reveals that, for every relative pressure $p/p_0$, there is an onset temperature for superfluidity which is well below the normal

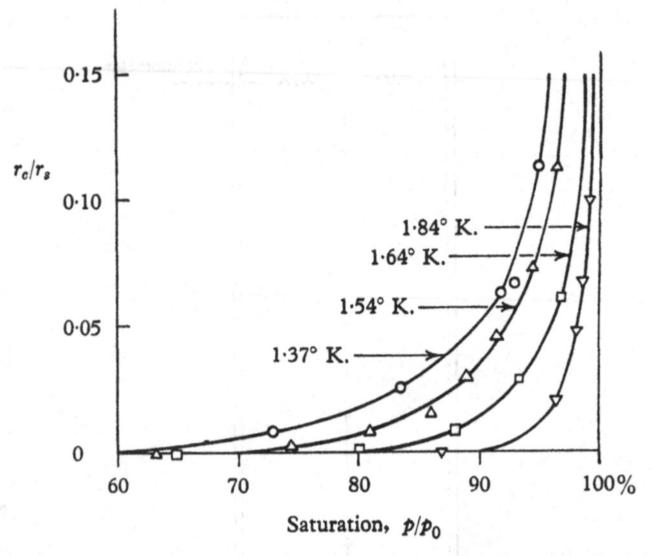


Fig. 80. The relative rate of flow of the unsaturated film plotted against the relative pressure $p/p_0$ at constant temperature (after Bowers, Brewer and Mendelssohn, 1951).

$\lambda$-point. In fact, the onset temperatures deduced in this way agree satisfactorily with those found by Method I of Long and Meyer.

Long and Meyer (1955) have repeated and extended these heat transport measurements. They noted, however, that even when the heat input was less than the critical value $\dot{Q}_c$, there was a small temperature gradient along the film, varying approximately as the cube of the heat input. Using the adsorption isotherm data of Strauss (1952) they were able to express their measurements in terms of the film thickness in atomic layers. They conclude that the critical velocity of the unsaturated film is approximately independent of thickness and varies from 21 cm. sec.$^{-1}$ at 2·0° K. to 48 cm. sec.$^{-1}$ at 1·3° K. Fig. 81 gives their data for the variation of onset temperature $T_0$ with film thickness, expressed as a number $n$ of atomic layers. They interpret fig. 81 to mean that, whatever the thickness of the film at a temperature $T_0$, the first $n$ layers near the wall are immobile and the outer layers are mobile. Another possible interpretation, of course, is that a film $n$ layers thick is completely mobile below $T_0$ and completely immobile above $T_0$ (ignoring the solid layers very close to the wall).

Atkins and Seki (1956) have confirmed that the onset temperature of superfluid flow is less than the bulk $\lambda$-temperature in

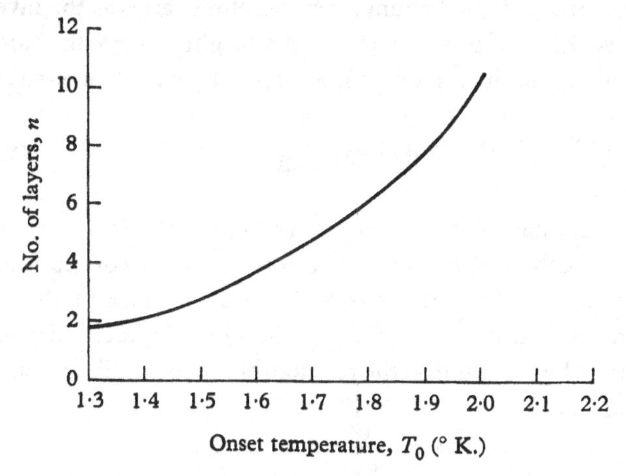


Fig. 81. The onset temperature of superfluidity as a function of film thickness (after Long and Meyer, 1955).

extremely narrow channels. They studied flow through porous Vycor glass, which has a sponge-like porous structure with an average pore diameter of the order of $30 \times 10^{-8}$ cm., corresponding to only a few atomic layers of liquid. The onset temperature of the flow was unmistakably lower than the bulk $\lambda$-point and decreased monotonically throughout a range of specimens of decreasing pore size, being only 1·36° K. for the finest pores. Moreover, when a particular specimen was exposed to the atmosphere for several days so that it adsorbed organic matter, the resulting decrease in pore size was accompanied by a lowering of the onset temperature. Subsequent cleaning of the glass raised the onset temperature to its original value.

## 7.7. Theories of the formation of the film

Schiff (1941) and Frenkel (1940*a*) have discussed the formation of the film solely in terms of the forces of attraction between the helium atoms and the wall. At large distances the force between a helium atom and an atom of the wall is of the van der Waals type and varies as the inverse seventh power of the distance. Integrating over all the atoms of the wall, the force on a helium atom varies as the inverse fourth power of its distance $z$ from the wall, and the potential energy of the helium atom therefore varies as the inverse third power of $z$. If the atom is also at a height $y$ above the surface of the bulk liquid, its total potential energy may be expressed as

$$\phi = mgy - \frac{m\alpha}{z^3}, \tag{7.11}$$

where $m$ is the mass of the helium atom and $\alpha$ is a constant depending on the strength of the interatomic forces. If, in equilibrium, a small quantity of fluid is removed from the surface of the bulk liquid and placed on the surface of the film, the total change in energy must be zero, and so the relationship between film thickness $d$ and height $H$ is

$$mgH - \frac{m\alpha}{d^3} = 0,$$

$$d = \left(\frac{\alpha}{gH}\right)^{\frac{1}{3}}$$

$$= \frac{k}{H^{0\cdot33}}. \tag{7.12}$$

Schiff has calculated the magnitude of the van der Waals forces for several surface materials and the resulting values of $k$ are given in Table VI. Comparing these theoretical predictions with the somewhat uncertain experimental results described in §7.3, it will be seen that the order of magnitude of the film thickness is correctly predicted, but that there is some experimental evidence that the exponent of $H$ is actually slightly larger than 0·33. An important prediction of the Schiff-Frenkel theory is that the film thickness should be independent of temperature and should be the same above the $\lambda$-point as below, the peculiarities of He II being quite irrelevant to the theory. The latest experimental results are not inconsistent with this (Ham and Jackson, 1957). Hill (1949) has made a detailed analysis of multilayer physical adsorption and his results are essentially in agreement with equation (7.12) for thick films.

TABLE VI. *The constant k in equation* (7.12) *as calculated by Schiff*

| Material of wall | Copper | Silver | Glass | Rocksalt |
|---|---|---|---|---|
| $k \times 10^6$ | 4·3 | 4·7 | 4 | 2·2 |

Atkins (1954) has extended these ideas by treating the film as a continuous fluid distributed over the potential field given by equation (7.11). It is a well-known classical result that the hydrostatic pressure of a fluid varies from point to point in such a field, and in this particular instance the pressure is given approximately by

$$p = p_0 + \rho\left(\frac{\alpha}{z^3} - gy\right). \tag{7.13}$$

The pressure therefore increases as one goes inwards from the surface of the film towards the wall, for exactly the same reason that the hydrostatic pressure increases as one goes downwards in any fluid in the earth's gravitational field. The analysis is permissible as long as the pressure does not change appreciably over a distance equal to the interatomic separation, since the assumption of a continuum is then a reasonable approximation. The argument therefore breaks down near the wall where the atomistic nature of the film must be taken into account, but even here the continuum approach might be expected to give a qualitative understanding of some of the important characteristics of the film. The first important consequence of equation (7.13) is that, if we put $p$ equal to the vapour pressure $p_0$, we obtain equation (7.12) for the contour of the surface of the film. The pressure inside the film may then be rewritten as

$$p = p_0 + \rho\alpha\left(\frac{1}{z^3} - \frac{1}{d^3}\right). \tag{7.14}$$

At a distance of about $7 \times 10^{-8}$ cm. from the wall, $p$ becomes equal to the solidification pressure of 25 atmospheres. As the interatomic distance in the solid is about 3·3 Å., this suggests that there may be two solid layers near the wall. Even outside these solid layers the increasing pressure implies that the density of the liquid film increases towards the wall. Just below the $\lambda$-point there is a point in the film where the pressure rises above the $\lambda$-curve, and the film therefore divides up into three regions, an outer region of He II, an intermediate region of He I and the solid region near the wall.

Bijl, de Boer and Michels (1941) adopted an entirely different approach. At the time it was popular to treat liquid helium as an ideal Bose-Einstein gas, and they pointed out that, for such a model, the ground-state wave function has a node at the wall and at the surface of the film, with a wavelength $\lambda = 2d$. The zero-point energy per atom would then be $h^2/8md^2$ and the total energy per cm.$^2$ of film would be

$$U = nd\,mgH + nd\frac{h^2}{8md^2}, \tag{7.15}$$

$n$ being the number of atoms per cm.$^3$ of film. Varying the film thickness $d$ in order to minimize this energy, one readily obtains

$$n\,mgH - n\frac{h^2}{8md^2} = 0,$$

$$d = \frac{h}{2m}\left(\frac{1}{2gH}\right)^{\frac{1}{2}}$$

$$= \frac{11 \cdot 2 \times 10^{-6}}{H^{0 \cdot 50}}. \tag{7.16}$$

This is consistent with the experiments of Bowers (1953 *a*) but with none of the other measurements of film thickness. In this treatment the forces of attraction to the wall have been ignored, but this is readily seen to be untenable, since the excess energy given by equation (7.15) is always positive, so that the film would be metastable. Moreover, Mott (1949) has pointed out that the ground-state wave function assumed in the theory implies a peculiar variation of density from a maximum at the centre to zero at the wall and at the surface.

Actually, it is now well established that the zero-point energy of liquid helium is not $h^2/8md^2$ but the much larger quantity $h^2/8m\delta^2$, where $\delta$ has the order of magnitude of the interatomic distance. Atkins (1954, 1955 *b*) has based a theory upon the fact that the properties of the film would be seriously influenced if this large zero-point energy varied with film thickness by only 1 part in $10^6$.

To illustrate this, he considers the longitudinal Debye modes which are known to exist in liquid helium and shows that the zero-point energy of these modes for a slab of thickness $d$ has the form

$$z = z_{\infty}\left[1 - \frac{1}{6}\frac{\lambda_{c,\infty}}{d} + \frac{1}{32}\left(\frac{\lambda_{c,\infty}}{d}\right)^2 + \ldots\right], \tag{7.17}$$

where $z_{\infty}$ is the zero-point energy of the bulk liquid and $\lambda_{c,\infty}$ is the cut-off wavelength of these normal modes in the bulk liquid. The next step is to write down the total energy of the film, including gravitational energy, van der Waals' energy and zero-point energy, and then to minimize this by varying both the density and thickness of the film. The mean density is found to be slightly greater than that of the bulk liquid:

$$\frac{\rho(d) - \rho(\infty)}{\rho(\infty)} = \tfrac{1}{6} z_{\infty} K\rho\left(1 + \frac{\rho}{c}\frac{\partial c}{\partial \rho}\right)\frac{\lambda_{c,\infty}}{d}, \tag{7.18}$$

where $K$ is the compressibility and $c$ is the velocity of first sound. The equilibrium thickness of the film is given by

$$gH = \frac{\alpha}{d^3} + \left(\frac{\lambda_{c,\infty}}{d}\right)^2\left[\tfrac{1}{32} z_{\infty} - \tfrac{1}{72} K\rho z_{\infty}^2\left(1 + \frac{\rho}{c}\frac{\partial c}{\partial \rho}\right)^2\right]. \tag{7.19}$$

The results of Jackson (§7.3) can be expressed in this form:

$$H = \left(\frac{a}{d}\right)^3 + \left(\frac{b}{d}\right)^2, \tag{7.20}$$

with $a = 1{\cdot}89 \times 10^{-6}$ and $b$ varying from $2{\cdot}54 \times 10^{-6}$ at $1{\cdot}32^{\circ}$ K. to $2{\cdot}78 \times 10^{-6}$ at $2{\cdot}05^{\circ}$ K. The order of magnitude of the term in $1/d^2$ is correctly predicted by equation (7.19).

The increased density suggested by equation (7.18) may be very important in the case of unsaturated films. The effect, which varies as $1/d$, becomes quite large for thin films and is further augmented by the large increase in pressure very close to the wall (equation (7.14)). It provides a ready explanation of the differential entropy of the film. For He I the entropy of the liquid is known to decrease with increasing density, whereas for He II the entropy increases with increasing density, and this is mirrored in the behaviour of the differential entropy of the film, which is less than the entropy of the bulk liquid above the $\lambda$-point but greater than the entropy of the bulk liquid below the $\lambda$-point. The increase in density may also explain the low temperatures for the onset of superfluidity observed in thin films (see fig. 81). The negative slope of the $\lambda$-curve implies that the $\lambda$-point of the bulk liquid decreases with increasing density and it is therefore very plausible that the $\lambda$-point of a thin film should be lower if its density is higher. One might object that the $\lambda$-curve meets the solidification curve at 1·75° K., whereas onset temperatures have been observed down to 1·3° K., but it is not necessarily permissible to apply the phase diagram of the bulk liquid to extremely thin films. A more serious objection is that, even for thick films, the increase in density given by equation (7.18) seems to be too small by a factor of about ten. This may be because the spectrum of normal modes has been oversimplified, or because only the zero-point energy has been considered and the thermal excitations have been ignored. It is also possible that the $\lambda$-transition involves long-range effects (Feynman, 1955) which are modified in a thin film. Franchetti (1955) has advanced a theory based upon the fact that the thickness of the film is comparable with the wavelength of the excitations, but full details are not yet available.

McCrum and Eisenstein (1955) have drawn attention to the possibility that the surface might be contaminated by small quantities of a polar substance and that the dipoles might become aligned perpendicular to the surface, particularly in the case of a metal surface. Even a small non-uniformity of the dipole density would produce strong electric fields near the wall and would thereby play an important role in determining the film thickness. Full details of this treatment have not yet been published.

There is obviously much still to be done to straighten out the theory of helium films. From what has already been said, it can be seen that the following factors should all be given serious consideration:

(1) The basic factor is probably the force between the wall and a helium atom in the film. This may be a simple van der Waals type of force or something more complicated, as in the theory of McCrum and Eisenstein.

(2) The zero-point energy per gram of the film may vary with its

thickness. More generally, one should consider, not only the zero-point energy, but the total free energy.

(3) It is possible that the density of the film is not the same as that of the bulk liquid, even at points remote from the wall.

(4) The density of the film probably increases towards the wall. This also suggests that, in some circumstances, the outer region of the film may be superfluid while the region nearer the wall is not. In flow phenomena, the fraction of superfluid $\rho_s/\rho$ may have to be taken to be a function of distance from the wall.

(5) Looming in the background of any theory of a perfect film is the possibility that the properties of an experimentally realizable film may be determined primarily by surface imperfections, surface roughness, surface contamination, temperature inhomogeneity or flow effects.

CHAPTER 8

# HELIUM THREE

## 8.1. Introduction

$He^3$ is present in well helium to the extent of about 1·4 parts in $10^7$ and in atmospheric helium to the extent of about 12 parts in $10^7$. Using the processes which will be described in a later section, this concentration could be increased to almost 100 % $He^3$, but the complete operation would be very laborious. Fortunately, almost pure $He^3$ can be produced in a nuclear reactor by bombarding lithium with neutrons to form tritium, which $\beta$-decays with a lifetime of about 12·5 years to give $He^3$:

$$Li_3^6 + n_0^1 \rightarrow H_1^3 + He_2^4, \tag{8.1}$$

$$H_1^3 \xrightarrow{12\cdot 5 \text{ yr.}} He_2^3 + e^-. \tag{8.2}$$

The tritium can readily be separated from the helium by allowing it to diffuse through a heated palladium thimble, and so the $He^3$ produced by this method contains less than 0·1 % of $He^4$.

The $He^3$ atom contains an odd number of fundamental particles and therefore requires antisymmetric wave functions and Fermi-Dirac statistics. The absence of a $\lambda$-transition in liquid $He^3$ therefore suggests that the transition in $He^4$ is a consequence of Bose-Einstein statistics or of symmetric wave functions. This may perhaps be related to the mathematical peculiarities inherent in the statistical mechanical treatment of an ideal Bose-Einstein gas, but not in the corresponding treatment of a Fermi-Dirac gas. Alternatively, the symmetry of the wave functions may enter in a more subtle way, as in Feynman's theory of the elementary excitations. Apart from this important question of statistics, we would still expect the two isotopes to behave very differently, because the difference in mass is large enough to have a profound effect on the thermodynamic properties of the liquid and solid phases, and also the $He^3$ nucleus has a spin of $\frac{1}{2}\hbar$ with an accompanying magnetic moment, which introduces the possibility of magnetic interactions and orientation effects. The electronic wave functions are almost identical for the two isotopes and the interatomic force between two $He^3$ atoms is therefore almost identical with that between two $He^4$ atoms. However, there is a small difference of about 1 part in $10^4$ due mainly to the difference in the 'reduced mass' of the electrons.

Review articles on solid and liquid $He^3$ and $He^3$-$He^4$ mixtures have been written by Daunt (1952), Abraham, Osborne and Weinstock (1953), Hammel (1955) and Beenakker and Taconis (1955).

## 8.2. Some basic properties

The density of liquid $He^3$ in equilibrium with its vapour was first measured by Grilly, Hammel and Sydoriak (1949) and then, with increased accuracy, by Kerr (1954). Kerr's method was to measure out the $He^3$ in a gas pipette at room temperature and then to transfer successive shots from this pipette into a glass bulb in the cryostat until the liquid $He^3$ level was near a fiducial mark on the narrow neck of the bulb. The same apparatus was used to measure the density of the saturated vapour. The results are given in Table VII. Note that the density at the critical point is very nearly one-half of the density at 0° K. and the rectilinear diameter is therefore parallel to the temperature axis, which is an unusual situation.

TABLE VII

| Temp. (° K.) | Density of liquid (g. cm.$^{-3}$) | Density of saturated vapour (g. cm.$^{-3}$) | Latent heat (cal. mole$^{-1}$) |
|---|---|---|---|
| 0·0 | 0·08235 | 0 | 5·05 ± 0·25 |
| 0·5 | — | — | 7·15 |
| 1·0 | 0·08185 | 0·00058 | 9·10 |
| 1·2 | 0·08147 | 0·00098 | 9·76 |
| 1·4 | 0·08093 | 0·00154 | 10·32 |
| 1·6 | 0·08020 | 0·00228 | 10·71 |
| 1·8 | 0·07924 | 0·00325 | 11·02 |
| 2·0 | 0·07801 | 0·00450 | 11·15 |
| 2·2 | 0·07645 | 0·00608 | 11·05 |
| 2·4 | 0·07448 | 0·00806 | 10·71 |
| 2·6 | 0·07200 | 0·01056 | 10·11 |
| 2·8 | 0·06882 | 0·01376 | 9·19 |
| 3·0 | 0·06462 | 0·01798 | 7·88 |
| 3·1 | 0·06193 | 0·02067 | 7·01 |
| 3·2 | 0·05861 | 0·02400 | 5·91 |
| 3·3 | 0·05416 | 0·02845 | 3·36 |
| 3·34 | 0·04131 | 0·04131 | 0 |

Since the interatomic forces are almost the same for $He^3$ as for $He^4$, the potential energy is practically the same for identical configurations. If this were the only consideration, we would expect the two liquids to have the same configuration at 0° K. and the density of liquid $He^3$ would be three-quarters of the density of $He^4$, or 0·109 g. cm.$^{-3}$, whereas an extrapolation of Kerr's data gives 0·0824 g. cm.$^{-3}$. The explanation, of course, is that the zero-point energy discussed in § 2.3 varies inversely as the atomic mass and its 'blowing up' effect is greater in liquid $He^3$.

The variation of density with pressure has been measured by Walters and Fairbank (1956*a*). They worked with a constant volume of liquid and assumed that the size of their nuclear resonance absorption signal (§ 8.4) was proportional to the number of nuclei present. Sherman and Edeskuty (1957) adopted the more direct approach of measuring the volume of gas at S.T.P. needed to fill a known volume with liquid to a measured pressure. Their results are summarized in Table VIII, which gives the molar volume in cm.$^3$ as a function of pressure at three temperatures.

TABLE VIII. *Molar volume of* $He^3$ *as a function of pressure*

| Pressure (atmospheres) ... | 1·2° K. | 1·4° K. | 1·6° K. |
|---|---|---|---|
| Vapour pressure ... | 37·11 | 37·40 | 37·77 |
| 1 | 35·95 | 36·13 | 36·44 |
| 2 | 34·91 | 35·06 | 35·30 |
| 5 | 32·62 | 32·69 | 32·79 |
| 10 | 30·20 | 30·25 | 30·31 |
| 15 | 28·73 | 28·76 | 28·80 |
| 20 | 27·60 | 27·62 | 27·65 |
| 25 | 26·68 | 26·69 | 26·71 |
| 30 | 25·91 | 25·92 | 25·93 |
| 35 | 25·28 | 25·28 | 25·28 |
| 40 | 24·72 | 24·72 | 24·72 |
| 50 | — | 23·78 | 23·78 |

The velocity of sound at the vapour pressure has been measured at 5 Mc./s. by Laquer, Sydoriak and Roberts (1957) and at 14 Mc./s. by Flicker and Atkins (1957). It varies from 115 m. sec.$^{-1}$ at 3° K. to 180 ± 5 m. sec.$^{-1}$ at 0° K. The compressibility at 0° K. is therefore 0·038 ± 0·001 atmospheres$^{-1}$.

The vapour pressure curve has been determined by Sydoriak, Grilly and Hammel (1949), Abraham, Osborne and Weinstock (1950) and Roberts and Sydoriak (1957*a*). Table IX contains the smoothed data quoted by Roberts and Sydoriak. The pressure is in millimetres of mercury at 0° C. and the 55*E* temperature scale has been used. The boiling point is

$$T_B = 3{\cdot}19^\circ \text{ K.} \tag{8.3}$$

The critical point is

$$T_c = 3{\cdot}33 \pm 0{\cdot}02^\circ \text{ K.,} \tag{8.4}$$

$$p_c = 875 \pm 20 \text{ mm. Hg,} \tag{8.5}$$

$$V_c = 73{\cdot}23 \text{ cm.}^3 \text{ mole}^{-1}. \tag{8.6}$$

The vapour pressure of liquid $He^3$ is considerably larger than that of liquid $He^4$. By pumping on a bath of liquid $He^3$ it should be possible to reach temperatures in the vicinity of 0·3° K.

TABLE IX. *The vapour pressure of liquid* $He^3$

| Temp. (° K.) (55*E* scale) | Vapour pressure (mm. Hg at 0° C.) | Temp. (° K.) (55*E* scale) | Vapour pressure (mm. Hg at 0° C.) |
|---|---|---|---|
| 0·3 | 0·00150 | 1·9 | 124·69 |
| 0·4 | 0·02405 | 2·0 | 150·55 |
| 0·5 | 0·1418 | 2·1 | 179·68 |
| 0·6 | 0·4985 | 2·2 | 212·28 |
| 0·7 | 1·291 | 2·3 | 248·52 |
| 0·8 | 2·744 | 2·4 | 288·60 |
| 0·9 | 5·092 | 2·5 | 332·71 |
| 1·0 | 8·564 | 2·6 | 381·02 |
| 1·1 | 13·384 | 2·7 | 433·73 |
| 1·2 | 19·765 | 2·8 | 491·00 |
| 1·3 | 27·913 | 2·9 | 553·01 |
| 1·4 | 38·03 | 3·0 | 619·92 |
| 1·5 | 50·30 | 3·1 | 691·88 |
| 1·6 | 64·91 | 3·2 | 769·04 |
| 1·7 | 82·06 | 3·3 | 851·50 |
| 1·8 | 101·93 | | |

The latent heat of vaporization, $L$, may be derived from the Clausius-Clapeyron equation

$$\left(\frac{dp}{dT}\right)_{\text{sat.}} = \frac{L}{T(V_v - V_l)}. \tag{8.7}$$

The results are given in Table VII. The value extrapolated to 0° K. is $5{\cdot}05 \pm 0{\cdot}25$ cal. mole$^{-1}$ and this, of course, is the internal energy of the liquid at 0° K. (see § 2.2).

The surface tension has been measured by Lovejoy (1955) and Zinoveva (1954, 1955), both using the capillary rise method. The results are shown in fig. 82. The variation with temperature is in order of magnitude agreement with the theory of Atkins (1953*b*) based on the importance of surface vibrations (see § 2.5.5).

The adsorption of $He^3$ on activated charcoal has been studied by Hoffman, Edeskuty and Hammel (1956). The adsorption isotherm was of the Langmuir type, with no multilayer adsorption. The volume of gas needed to form a monolayer was almost the same for $He^3$ and $He^4$, implying the same interatomic spacing in

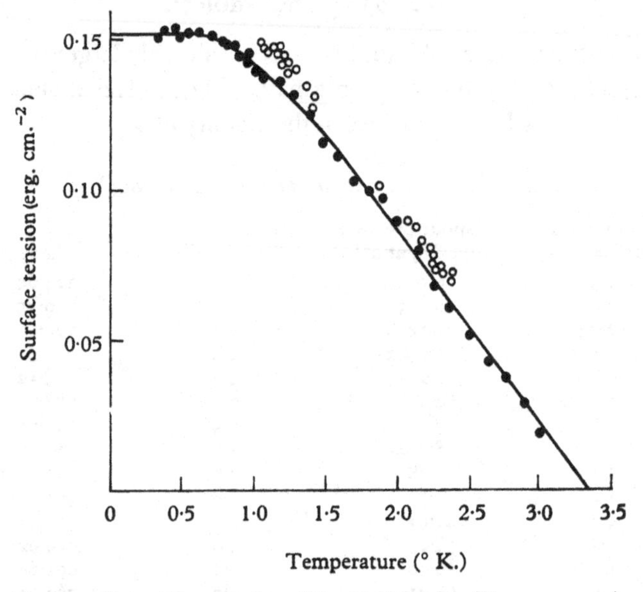


Fig. 82. The surface tension of liquid $He^3$ (after Zinoveva, 1955). ○, Lovejoy (1955); ●, Zinoveva (1955).

the first layer for the two isotopes. The explanation may be that the first layer is so tightly packed that the situation is entirely dominated by the strong repulsive forces for close approach of two atoms, and these forces are almost the same for the two isotopes.

## 8.3. The entropy and specific heat

Fig. 83 summarizes the measurements of the specific heat under the saturated vapour pressure (de Vries and Daunt, 1953, 1954; Roberts and Sydoriak, 1955; Abraham, Osborne and Weinstock, 1955; Brewer, Sreedhar, Kramers and Daunt, 1958). Since the specific heat is not known below 0·1° K. and is still large there, the entropy cannot be derived by direct integration of the specific heat. Entropy differences can be calculated from the equation

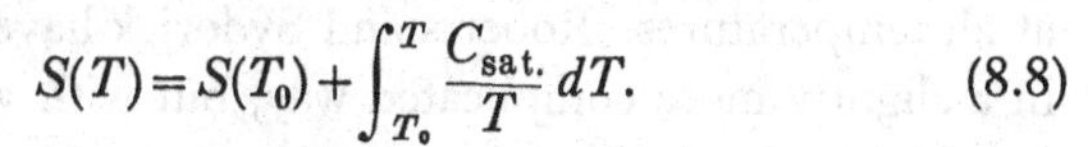

$$S(T) = S(T_0) + \int_{T_0}^{T} \frac{C_{\text{sat.}}}{T}\,dT. \qquad (8.8)$$

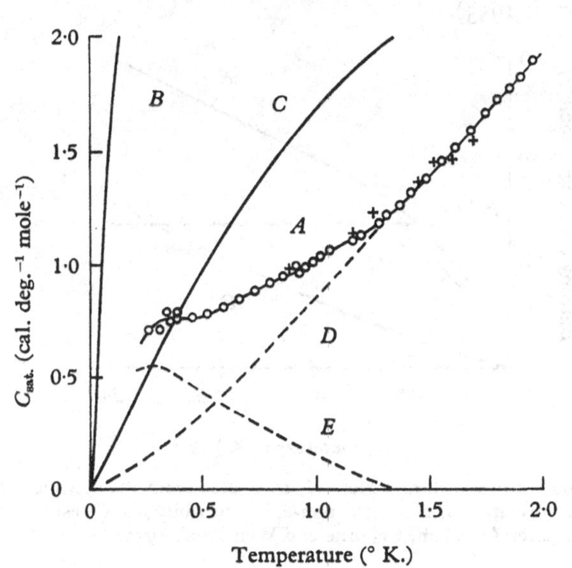


Fig. 83. The specific heat of liquid $He^3$. *A*, experimental: ○, Abraham, Osborne and Weinstock (1955); +, Roberts and Sydoriak (1955). *B*, ideal Fermi-Dirac gas with a degeneracy temperature of 0·45° K. *C*, ideal Fermi-Dirac gas with a degeneracy temperature of 4·98° K. *D*, estimated non-spin contribution. *E*, estimated spin contribution (after Abraham, Osborne and Weinstock, 1955).

The entropy at some convenient temperature $T_0$ can then be derived from the slope of the vapour pressure curve, using the Clausius-Clapeyron equation

$$S_l(T_0) = S_v(T_0) - (V_v - V_l)\left(\frac{dp}{dT}\right)_{\text{sat.}}. \qquad (8.9)$$

The virial coefficients of the vapour are accurately known (Keller, 1955) and so the entropy of the vapour, $S_v$, can be calculated precisely. A term $R \ln 2$ must be included to represent the two possible orientations of the nuclear spin of a $He^3$ atom. At 1·5° K., where the slope of the vapour pressure curve is known with reasonable accuracy, this procedure gives $2{\cdot}52 \pm 0{\cdot}17$ cal. deg.$^{-1}$ mole$^{-1}$ for the entropy of the liquid. The entropy curve then has the form shown in fig. 84, with a possible consistent error of $\pm 0{\cdot}17$ cal. deg.$^{-1}$ mole$^{-1}$ at all temperatures. Roberts and Sydoriak have analysed the data in a slightly more complicated way, but with very similar results (see Hammel, 1955).

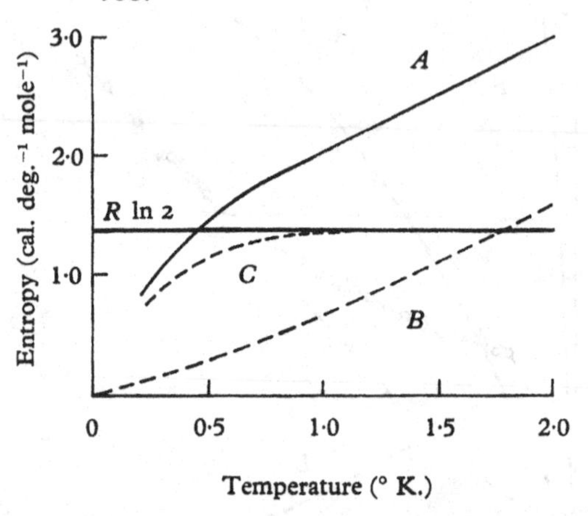


Fig. 84. The entropy of liquid $He^3$. *A*, calculated from the specific heat and vapour pressure data. *B*, estimated non-spin contribution. *C*, estimated spin contribution (after Abraham, Osborne and Weinstock, 1955).

If each $He^3$ nucleus in the liquid were free to orient its spin in two directions with equal probability, the entropy would be at least $R \ln 2$, or $1{\cdot}38$ cal. deg.$^{-1}$ mole$^{-1}$. The measured entropy at $0{\cdot}23°$ K. is only $0{\cdot}86 \pm 0{\cdot}17$ cal. deg.$^{-1}$ mole$^{-1}$. It follows that, at this temperature, there is some degree of alignment amongst the nuclei: that is, the orientation of a nuclear spin is influenced by the orientation of its neighbours, as in a ferromagnetic or antiferromagnetic material. Figs. 83 and 84 show how the specific heat and entropy might be divided into two contributions, spin and non-spin. However, it is by no means certain that such a division is permissible. We shall return to this point in §§ 8.4 and 8.6.4.

An interesting question (Daunt, 1952; Mikura, 1955) is whether the specific heat includes a phonon contribution analogous to that found in liquid $He^4$. Experimentally, sound waves of frequency $\sim 10^7$ sec.$^{-1}$ have been propagated through the liquid above $0{\cdot}4°$ K. and it is clear that phonons of this frequency must therefore exist, but the phonons excited thermally at 1° K. would have a frequency $\sim 10^{10}$ sec.$^{-1}$ and it is conceivable that sound waves with this higher frequency might be strongly attenuated. Ignoring this last possibility, Flicker and Atkins (1957) have used their measurements of the velocity of sound to calculate the phonon contribution to the specific heat and have shown that it would be negligible below 1° K., but would seriously influence the shape of the specific heat versus temperature curve above 1° K., being about 30 % of the total specific heat at 2° K.

## 8.4. Magnetic properties

The diamagnetic susceptibility of the extra-nuclear electrons in liquid $He^3$ can be calculated to be about $-5\times 10^{-8}$. At temperatures sufficiently high for the nuclear spins to behave independently, the nuclear paramagnetic susceptibility would be

$$\chi = \frac{n\mu^2}{kT}$$

$$\simeq +\frac{1.33\times 10^{-8}}{T}, \qquad (8.10)$$

where $n$ is the number of nuclei per cm.$^3$ and $\mu$ is the nuclear magnetic moment. Neither contribution to the susceptibility is large enough to measure directly. At a sufficiently low temperature, coupling between the nuclear spins probably produces an ordered antiferromagnetic or ferromagnetic alignment. The absence of ferromagnetism down to 0·9° K. was demonstrated experimentally by Hammel, Laquer, Sydoriak and McGee (1952).

The nuclear paramagnetic susceptibility has been measured indirectly by a nuclear magnetic resonance technique (Fairbank, Ard, Dehmelt, Gordy and Williams, 1953; Fairbank, Ard and Walters, 1954; Walters and Fairbank, 1956*a*). In the presence of a steady magnetic field $H$ ($\sim 10^4$ gauss) there are $n_p$ nuclear spins per cm.$^3$ aligned parallel to $H$ and a smaller number $n_a$ per cm.$^3$ aligned antiparallel to $H$. The paramagnetic susceptibility is then

$$\chi = \frac{(n_p - n_a)\,\mu}{H}. \qquad (8.11)$$

If an oscillating magnetic field (30 Mc./s.) is then applied perpendicularly to the steady field, the power absorbed from it is proportional to $(n_p - n_a)$ if saturation effects have been avoided (Bloembergen, Purcell and Pound, 1948). The height of the nuclear magnetic resonance absorption signal is therefore directly proportional to the nuclear paramagnetic susceptibility. Fairbank and his co-workers calibrated their apparatus with $He^3$ gas at 4·2° K. and 900 mm. of mercury pressure, under which conditions an ideal Fermi-Dirac gas would depart from Curie's Law by only 4 %. In fig. 85, $\chi T/C$ for the liquid is plotted against temperature, $C$ being

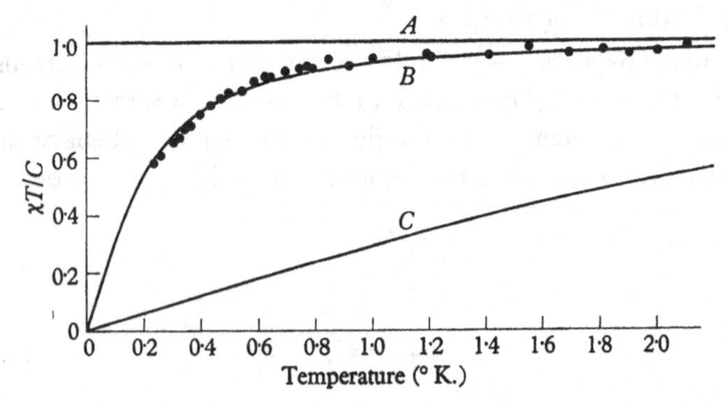


Fig. 85. The nuclear magnetic susceptibility. *A*, Curie's Law. *B*, ideal Fermi-Dirac gas with a degeneracy temperature of 0·45° K. *C*, ideal Fermi-Dirac gas with a degeneracy temperature of 4·98° K., which corresponds to the density and atomic mass of liquid $He^3$ (after Fairbank, Ard and Walters, 1954).

the Curie constant for independent nuclear spins. Above 1° K. departures from Curie's Law are seen to be small, but below 1° K. the susceptibility falls well below the value to be expected from this law. We conclude that, above 1° K., each spin is effectively free to assume any orientation independently of the other spins, but that, below 1° K., there is already some ordering of the antiferromagnetic type. This is consistent with the behaviour of the entropy (fig. 84). Preliminary results at higher pressures (Walters and Fairbank, 1956*a*) suggest that the higher the pressure the lower the temperature at which departures from Curie's Law become apparent.

Goldstein (1954) has suggested that the susceptibility data may be used to estimate that part of the entropy associated with disordering of the spins. His basic assumption is that $\nu(T)$ spins per cm.$^3$ are free to assume either of two orientations and hence to respond to the application of a magnetic field, whereas the remaining $n-\nu(T)$ spins are firmly bound to a fixed scheme of orientations. It follows simply that the ratio of the actual susceptibility to the susceptibility expected from Curie's Law is

$$\frac{\chi(T)}{\chi_0(T)}=\frac{\nu(T)}{n}. \tag{8.12}$$

The contribution of the spins to the entropy is

$$S_\sigma=\frac{\nu(T)}{n}R\ln 2 \tag{8.13}$$

$$=\frac{\chi(T)}{\chi_0(T)}R\ln 2, \tag{8.14}$$

so $S_\sigma$ may be derived directly from the $\chi(T)/\chi_0(T)$ curve. This principle was used in fig. 84 to divide the entropy into spin and non-spin contributions. In a later paper, Goldstein (1956) has extended the argument to obtain the contributions of the spins to all the thermodynamic functions and, in particular, has suggested that the coefficient of expansion may be large below 1° K. The basic assumptions underlying Goldstein's treatment are valid for simple models, such as the ideal Fermi-Dirac gas, but it is questionable whether they are still true if the ordering of the spins is a highly co-operative phenomenon (see § 8.6.4).

## 8.5. Transport properties

The various properties of liquid He$^3$ discussed above provide no evidence for a $\lambda$-transition down to 0·1° K. A direct search for superfluidity was made by Osborne, Weinstock and Abraham (1949). They investigated the flow of the liquid through a superleak made by sealing a platinum wire into a Pyrex capillary. When this arrangement was cooled down, the difference in the coefficients of expansion of the platinum and glass resulted in a narrow annular gap $7\times10^{-5}$ cm. wide. Liquid helium was condensed on one side of this leak and the other side was evacuated by a Toepler pump. The rate of flow of liquid through the leak was then deduced from the rate of rise of the pressure on the evacuated side. In a trial run with $He^4$ the rate of flow through the leak increased sharply at the $\lambda$-point in an unmistakable manner. With liquid $He^3$, however, the rate of flow fell monotonically as the temperature was lowered right down to 1·05° K., which was the lowest temperature investigated. The absence of superfluidity was therefore demonstrated down to a temperature 0·31 times the critical temperature, whereas the $\lambda$-point of $He^4$ is 0·42 times its critical temperature.

From the actual values of the rate of flow through the superleak they deduced the viscosity of liquid $He^3$ after making approximate corrections for the fact that the liquid vaporizes at some point within the leak. The viscosity was found to be 30·4 micropoise at 1·04° K. and 22 micropoise at 2·79° K. Later oscillating disk measurements by Dash and Taylor (1957*b*) are in approximate agreement with these earlier results. Zinoveva (1958) has extended the measurements to 0·35° K. using the method of Poiseuille flow through a capillary. She found a steady increase from 16 micropoise at 3·2° K. to 50 micropoise at 0·35° K. Between 0·5 and 1·1 °K. the viscosity varied approximately as $1/T^{\frac{1}{2}}$, and increased still more rapidly below 0·5° K. This increase in viscosity with decreasing temperature is characteristic of a normal liquid, but is the opposite of the behaviour of all gases and of liquid $He^4$ above the $\lambda$-point. Since the density of liquid $He^3$ is considerably less than that of liquid $He^4$, it might be expected to behave even more like a gas. Early theories of the viscosity of liquid $He^3$ therefore treated the liquid as a degenerate ideal Fermi-Dirac gas (Tomonaga, 1938; Singwi and Kothari, 1949; ter Haar and Wergeland, 1949; Pomeranchuk, 1950; Buckingham and Temperley, 1950). During the collision of two $He^3$ atoms, the Pauli exclusion principle prevents the atoms from being scattered into states which are already occupied, and the probability of scattering is thereby reduced. The more degenerate the gas, the smaller the effective cross-section of the atoms becomes. The effective mean free path, and hence the viscosity, therefore increase rapidly with decreasing temperature. At temperatures well below the degeneracy temperature the viscosity should vary at $1/T^2$ and the thermal conductivity as $1/T$. However, we shall see in the next section that, if liquid $He^3$ bears any resemblance to an ideal Fermi-Dirac gas, the degeneracy temperature is certainly not very large compared with 1° K.

The thermal conductivity has been measured by Fairbank and Lee (1957) and by Challis and Wilks (1957). It decreases from about $1.9 \times 10^{-4}$ watt cm.$^{-1}$ deg.$^{-1}$ at 3° K. to about $0.6 \times 10^{-4}$ watt cm.$^{-1}$ deg.$^{-1}$ at 0.2° K. There is therefore no indication of a $1/T$ variation over this temperature range. If $K$ represents the thermal conductivity, $\eta$ the viscosity and $C$ the specific heat, then, between 1 and 2° K.,

$$\frac{K}{\eta C} \simeq \frac{5}{2}. \tag{8.15}$$

This equation is well known to be true for an ideal classical gas.

The self-diffusion of $He^3$ atoms in pure liquid $He^3$ has been investigated by Garwin and Reich (1958) using a spin echo technique. The diffusion coefficient is about $2 \times 10^{-5}$ cm.$^2$ sec.$^{-1}$ at 1° K. and increases only slightly with increasing temperature up to 3° K.

## 8.6. Theories of liquid $He^3$

### 8.6.1. de Boer's modified law of corresponding states

Before any of the fundamental thermodynamic properties of liquid $He^3$ had been measured, de Boer had made what turned out to be a very successful prediction of their values, based upon a modified form of the reduced equation of state (de Boer, 1948; de Boer and Lunbeck, 1948). The basic assumption is that the intermolecular potential between two simple molecules has the form

$$V_{ij} = \epsilon f\left(\frac{r_{ij}}{\sigma}\right). \tag{8.16}$$

The energy $\epsilon$ and the length $\sigma$ are characteristic of the substance considered, but $f$ is a universal function for all simple molecules. If we express all lengths in units of $\sigma$ and all energies in units of $\epsilon$, the usual statistical mechanical treatment of an assembly of $N$ interacting particles gives a partition function in the form

$$Z = Z(V^*, T^*, \Lambda^*), \tag{8.17}$$

$$V^* = \frac{V}{N\sigma^3}, \tag{8.18}$$

$$T^* = \frac{kT}{\epsilon}, \tag{8.19}$$

$$\Lambda^* = \frac{h}{\sigma\sqrt{(m\epsilon)}}. \tag{8.20}$$

In addition to the reduced volume $V^*$ and the reduced temperature $T^*$, the factor $\Lambda^*$ introduces the quantum effects which are normally absent in the classical treatment of corresponding states. Since all the thermodynamic quantities can be derived from the partition function, $Z$, it follows that any thermodynamic quantity can also be expressed as a function of $V^*$, $T^*$ and $\Lambda^*$ only, and these functions, of course, are universally applicable to all simple substances. In particular, the reduced pressure is

$$p^* = \frac{p}{\sigma^3/\epsilon}, \tag{8.21}$$

which can readily be derived from the definition:

$$p^* = \frac{T^*}{N}\frac{\partial}{\partial V^*}\ln Z(V^*, T^*, \Lambda^*), \tag{8.22}$$

which is the equation of state in the form

$$F(p^*, V^*, T^*, \Lambda^*) = 0, \tag{8.23}$$

$F$ being a universal function for all simple substances.

Readily condensable substances with high critical temperatures have strongly interacting molecules, a high value of $\epsilon$ and a value of $\Lambda^*$ very close to zero. The variation of $\Lambda^*$ amongst these substances therefore has a negligible effect on the function $F$, and the equation of state may be simplified to

$$F'(p^*, V^*, T^*) = 0. \tag{8.24}$$

The critical point $p_c^*$, $V_c^*$, $T_c^*$ is the same for all 'classical' substances and so the critical pressure, volume and temperature for a particular substance are:

$$p_c = \frac{p_c^*\sigma^3}{\epsilon}, \quad V_c = V_c^* N\sigma^3, \quad T_c = \frac{T_c^*\epsilon}{k}, \tag{8.25}$$

and the reduced equation of state could equally well have been written

$$F''\left(\frac{p}{p_c}, \frac{V}{V_c}, \frac{T}{T_c}\right) = 0, \tag{8.26}$$

which leads to the classical law of corresponding states (see, for example, Guggenheim, 1950).

For less easily condensable substances, however, $\Lambda^*$ is not small and $p_c^*$, $V_c^*$ and $T_c^*$ are then not constants, but functions of $\Lambda^*$. de Boer and Lunbeck therefore proceed as follows. From measurements of the second virial coefficients of the gases, the constants

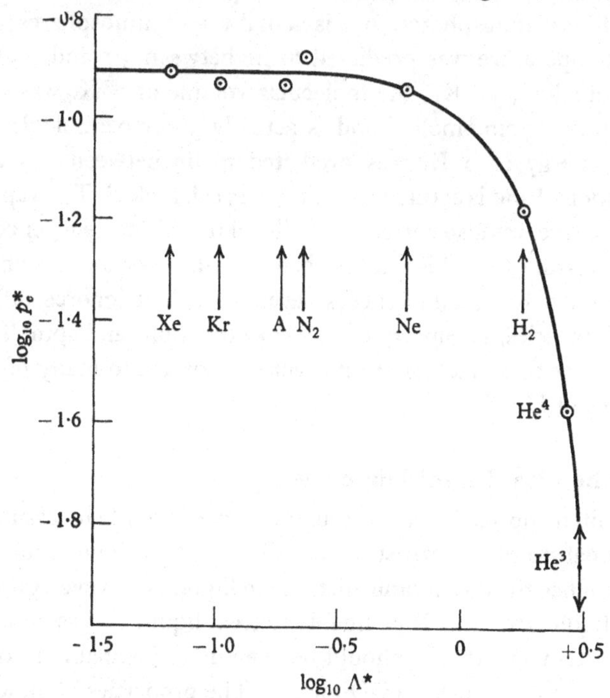


Fig. 86. $\log_{10} p_c^*$ plotted against $\log_{10} \Lambda^*$ to illustrate de Boer's modified law of corresponding states (after de Boer and Lunbeck, 1948).

$\sigma$ and $\epsilon$ can be deduced for each substance, the intermolecular forces being assumed to have the form

$$V_{ij} = 4\epsilon\left[\left(\frac{\sigma}{r_{ij}}\right)^{12} - \left(\frac{\sigma}{r_{ij}}\right)^{6}\right]. \tag{8.27}$$

The critical constants of Xe, Kr, A, $N_2$, Ne, $H_2$ and $He^4$ have been measured, so it is possible to plot a graph of $p_c^*$ against $\Lambda^*$, as in fig. 86, and then to extrapolate to the value of $\Lambda^*$ applicable to $He^3$ (which is $\sqrt{\frac{4}{3}}$ of the value for $He^4$, since the intermolecular forces and hence $\epsilon$ and $\sigma$ are the same for the two isotopes (see equation (8.20)). A similar procedure gives the reduced value of any thermodynamic function. Fig. 86 shows that the classical law of corresponding states is approximately obeyed for Xe, Kr, A, $N_2$ and Ne ($p_c^* =$ constant), but that there are strong quantum effects for $H_2$, $He^4$ and $He^3$.

In this way the critical pressure was predicted to lie between 0·93 and 1·35 atmospheres, and is actually 1·15 atmospheres; the critical temperature was predicted to lie between 3·1 and 3·5° K. and is actually 3·33° K.; the molecular volume at 0° K. was predicted to be 33 cm.$^3$ mole$^{-1}$ and is actually 36·5 cm.$^3$ mole$^{-1}$; the internal energy at 0° K. was predicted to lie between $-3$ and $-5$ cal. mole$^{-1}$ and is actually $-5{\cdot}05 \pm 0{\cdot}25$ cal. mole$^{-1}$. The vapour pressure curve was also correctly predicted to within a few per cent. It is interesting that this simple theory of de Boer and Lunbeck achieves such remarkable success even though it ignores differences in statistics, symmetry of the wave functions and spin. This suggests that these factors are not very important to many of the properties of $He^3$.

### 8.6.2. The ideal Fermi-Dirac gas

The initial approach to a molecular theory of liquid $He^3$ assumed that the nature of the statistics was all-important. The argument was that, since the interatomic distance in liquid $He^3$ is even greater than in liquid $He^4$, if it is justifiable to treat liquid $He^4$ as an ideal Bose-Einstein gas, then it should be even more justifiable to treat liquid $He^3$ as an ideal Fermi-Dirac gas. The properties of an ideal Fermi-Dirac gas are familiar in connexion with the free electron model of a metal. At the density of liquid $He^3$ the degeneracy temperature is

$$T_D = \frac{h^2}{8k}\left(\frac{3\rho}{\pi}\right)^{\frac{2}{3}}\left(\frac{1}{m}\right)^{\frac{5}{3}}$$

$$\sim 5^\circ \text{ K.} \qquad (8.28)$$

The specific heat curve for this degeneracy temperature is shown in fig. 83 and is in reasonable order of magnitude agreement with the experimental results, although quite wrong in its detailed shape. However, the paramagnetic susceptibility of the nuclear spins would not obey Curie's Law at any temperature below 5° K. and would be in violent disagreement with the experimental results (fig. 85). A curve corresponding to a degeneracy temperature of

0·45° K. would fit these nuclear susceptibility results rather well, but the corresponding specific heat curve would be totally inconsistent with fig. 83. There is therefore nothing to be gained by introducing a large effective mass for the $He^3$ atoms and, if the Fermi-Dirac gas model is to be retained, it must be drastically modified by the introduction of interatomic interactions.

Pomeranchuk (1950) adopts an approach analogous to that taken by Landau in the case of liquid $He^4$, but assumes that the elementary excitations are similar in nature to states near a Fermi surface

$$\epsilon = \frac{p_0}{2\mu}(p - p_0), \tag{8.29}$$

and that the exclusion principle applies. From this point on, the statistical mechanics is identical with that for an ideal Fermi-Dirac gas. The effective mass $\mu$ can be chosen to give the best agreement with the specific heat data, but, since the elementary excitations are not necessarily free $He^3$ atoms, the nuclear spins can be considered separately. This is an interesting phenomenological treatment but requires a firmer foundation in molecular theory.

### 8.6.3. The Fermi liquid

In § 3.4 we considered the progress that has been made with the problem of a Bose-Einstein gas of interacting particles. We shall now consider the effect of interactions on the properties of a Fermi-Dirac gas. Many of the basic ideas are due to Landau (1956) or to Brueckner and Gammel (1957, 1958). Illuminating discussions of some of the points involved have been given by Buckingham (1957) and Penrose (1957).

The starting point is an ideal gas with the same density as the liquid. In the ground state let there be $\nu_a^0$ $He^3$ atoms in a single-particle state described by a wave vector $\mathbf{k}_a$ and a spin quantum number $s_a$. In an excited state $\phi$, let there be $\nu_a = \nu_a^0 + \hat{\nu}_a$ atoms in this single-particle state. For a Fermi-Dirac gas $\hat{\nu}_a$ will be $+1$ for a state above the Fermi level into which an atom has been excited, and $\hat{\nu}_a = -1$ for a state below the Fermi level which has lost its atom and has become a 'hole'. Now slowly switch on the interatomic forces and adiabatically transform the state $\phi$ into a stationary state $\psi$ of the liquid. If there are no violent transitions during this process the liquid state $\psi$ has a one-to-one correspondence with the ideal gas state $\phi$, and $\psi$ could be unambiguously labelled by giving all the quantities $\nu_a$. Moreover, each particle of the ideal gas is transformed into a 'quasi-particle', which, in the language of field theory, is a 'bare' particle surrounded by a cloud of virtual excitations. Thus $\hat{\nu}_a = +1$ corresponds to a quasi-particle excited above the Fermi level, while $\hat{\nu}_a = -1$ corresponds to a quasi anti-particle (or hole) below the Fermi level.

The energy of the liquid state $\psi$ can be expanded in the form

$$E = E_0 + \sum_a \hat{\nu}_a \epsilon_a + \tfrac{1}{2} \sum_a \sum_b \hat{\nu}_a \hat{\nu}_b f_{ab} + \dots. \tag{8.30}$$

$E_0$ is the ground state energy; $\epsilon_a$ is the energy of the quasi-particle corresponding to the state $a$; and $f_{ab}$ is related to the interaction between quasi-particles in states $a$ and $b$. The quasi-particles, like the original ideal gas particles, must obey Fermi-Dirac statistics, but the dependence of $\epsilon$ upon $p$ is considerably modified by the interactions and, moreover, as Landau has emphasized, $\epsilon$ is also a functional of the distribution function of all the other quasi particles amongst their states and is therefore temperature dependent. Similarly, the density of states is not the same as for the ideal gas and is temperature dependent. However, it can be shown that, at very low temperatures, the expectation value of $\nu_a$ has a form very similar to that for the ideal gas:

$$\langle \nu_a \rangle = \frac{1}{e^{(\epsilon_a - \epsilon_F)/kT} + 1}. \tag{8.31}$$

The specific heat at these very low temperatures (probably $< 0{\cdot}1^\circ$ K.) can then be related to the density of states $g_0$ near the Fermi level

$$C_v = \tfrac{1}{3}\pi^2 g_0 k^2 T. \tag{8.32}$$

This is linearly proportional to the temperature, as for the ideal gas, and may also be written

$$C_v = \frac{m^*}{m} C_{\text{ideal}}, \tag{8.33}$$

where $C_{\text{ideal}}$ is the specific heat of the ideal gas, $m$ is the mass of a $He^3$ atom and $m^*$ is an effective mass equal to $p/(\partial\epsilon/\partial p)$ evaluated at the Fermi level. A reasonable extrapolation of the specific heat measurements to temperatures below 1° K. (Brewer, Sreedhar, Kramers and Daunt, 1958) suggests that $C_v=3{\cdot}75T$ cal. deg.$^{-1}$ mole$^{-1}$, from which $g_0$ can be estimated and $m^*/m$ shown to be about 1·8.

If the interatomic forces were independent of the orientations of the nuclear spins the nuclear magnetic susceptibility could be calculated in the same way as the paramagnetism of the electrons in a metal and, at very low temperatures, would approach the constant value

$$\chi_{\text{Ind.}}=\frac{g_0\mu^2}{V}, \tag{8.34}$$

where $\mu$ is the nuclear magnetic moment, $V$ is the volume and $g_0$ is known from the specific heat measurements. The susceptibility measurements shown in fig. 85 indicate that $\chi$ is approaching a constant value near 0° K., but it turns out to be several times larger than the value given by equation (8.34). To explain this we must assume that the mutual potential energy of two $He^3$ atoms includes an exchange energy which is negative when the nuclear spins are parallel and positive when they are antiparallel. The magnetic susceptibility can then be shown to be

$$\chi=\frac{g_0\mu^2}{V(1-\alpha)}. \tag{8.35}$$

$\alpha$ is related to the exchange energy. As the temperature is lowered there are two opposing effects. The exchange effects favour a parallel alignment of the spins, which would lead to ferromagnetism, but the Pauli exclusion principle requires that the atoms fall into pairs of states with opposite spins, and this latter effect predominates to give a type of antiferromagnetism. The exchange energy seems to be comparable in magnitude with the Fermi energy, so that if the exchange effects were just a little bit stronger, liquid $He^3$ might be ferromagnetic.

Brueckner and Gammel (1957, 1958) have applied the Brueckner treatment of the quantum mechanical many body problem, which was discussed in § 3.4 in connexion with liquid $He^4$. In the case of liquid $He^3$ it was possible to give a more complete treatment, using the best available expression for the interatomic force with both attractive and repulsive parts. The results are still restricted to temperatures just above 0° K., but the Brueckner method enables the various parameters to be calculated once the form of the interatomic force is known. The energy of a quasi-particle is first expressed approximately as

$$\epsilon = \frac{p^2}{2m} + V(p), \qquad (8.36)$$

where $V(p)$ is a sort of self-consistent potential which depends, however, on the momentum of the particle. The interaction between quasi-particles is then taken into account with due regard to statistics and correlations in spin and relative momenta. Finally, corrections are applied for the fact that $V(p)$ really depends upon the distribution function of all the quasi-particles, which means that certain 'multiple excitations' have to be considered.

The average energy per particle at 0° K. was calculated as a function of liquid density and was a minimum at a density of 0·067 g. cm.$^{-3}$, as compared with an actual density of 0·082 g. cm.$^{-3}$. The energy at the minimum was −1·92 cal. mole$^{-1}$, whereas the experimental value is −5·05 cal. mole$^{-1}$. The calculated compressibility was 0·053 atmospheres$^{-1}$, whereas the experimental value is 0·038 atmospheres$^{-1}$. All these quantities are very sensitive to the exact form of the interatomic potential. In the vicinity of the Fermi level equation (8.36) may be written

$$\epsilon = \frac{p^2}{2m^*} + V(p_F), \qquad (8.37)$$

where $p_F$ is the Fermi momentum and $m^*$ is the effective mass near the Fermi level. $m^*/m$ was found to be 1·84 at the vapour pressure, which is in very good agreement with an extrapolation of the specific heat data to temperatures below 0·1° K. $m^*/m$ was found to increase with increasing density, implying that the specific heat and entropy should both increase with increasing pressure below 0·1° K. Since $(\partial V/\partial T)_p = -(\partial S/\partial p)_T$, the coefficient of expansion would then be negative.

### 8.6.4. Transport properties and zero sound

Using the Landau formulation of the theory of a Fermi liquid, Abrikosov and Khalatnikov (1957) have calculated the viscosity and thermal conductivity near 0° K. The probability that a quasi-particle will be scattered is very much influenced by the Pauli exclusion principle and the resulting scarcity of available states into which the particle can be scattered. The mean free path varies as $1/T^2$, as in the theory of an ideal Fermi-Dirac gas (§ 8.5) and the corresponding relaxation time $\tau = \beta/T^2$. The uncertainty in the energy of the quasi-particle must therefore be at least $\hbar/\tau = \hbar T^2/\beta$. The distribution function of the quasi-particles is modified over an energy range $\sim kT$ near the Fermi level, and, if this description is to have any significance, we must have $\hbar T^2/\beta \ll kT$. This will always be so below some characteristic temperature

$$T_c = k\beta/\hbar \sim 0{\cdot}1^\circ \text{ K}.$$

Above this temperature the whole of our previous discussion of a Fermi liquid breaks down. At temperatures below 0·1° K. Abrikosov and Khalatnikov estimate that the viscosity $\eta \sim 10^{-6}/T^2$ poise and the thermal conductivity $\kappa \sim 40/T$ erg cm.$^{-1}$ sec.$^{-1}$ deg.$^{-1}$. The measurements described in § 8.5 do not extend to low enough temperatures for this theory of a Fermi liquid to be valid. The measured thermal conductivity increases with increasing temperature above 0·2° K. The viscosity is increasing rapidly with decreasing temperature near 0·35° K., but less rapidly than $1/T^2$.

At 0° K. the mean free path of a quasi-particle becomes infinite. Landau (1957) has argued that, since a compressional wave cannot be propagated in a gas when the wavelength is small compared with the mean free path, first sound does not exist in liquid $He^3$ at 0° K. It is replaced by a new type of wave propagation called 'zero sound', which is a periodic oscillation in the shape of the Fermi surface. In first sound the Fermi surface remains spherical, but its centre oscillates about the origin of momentum space. In zero sound, at a particular instant the Fermi surface is considerably elongated in the forward direction of propagation and slightly shortened in the backward direction (like an egg), but half a cycle later it is slightly elongated in the backward direction and considerably shortened in the forward direction, the amplitude of oscillation being greater at the forward pole than at the backward pole. At a *finite* temperature, first sound is propagated at long wavelengths for which $\omega\tau \ll 1$ and zero sound is propagated at short wavelengths for which $\omega\tau \gg 1$. When $\omega\tau \sim 1$ both types of sound are heavily attenuated.

In a Fermi-Dirac gas with very weak interactions the velocity of zero sound, $u_0$, is equal to the particle velocity, $v_F$, at the Fermi surface. The velocity of first sound, $u_1$, is equal to $v_F/\sqrt{3}$, and so $u_0=\sqrt{3}\,u_1$. In liquid $He^3$ $u_0$ and $u_1$ are probably not very different in magnitude. When $\omega\tau \ll 1$ the attenuation of *first* sound is given by an equation similar to (5.33) and therefore varies as $\omega^2/T^2$. When $\omega\tau \gg 1$ the attenuation of *zero* sound is proportional to $T^2$ and independent of $\omega$ in the 'classical region', where $\hbar\omega \ll kT$, but is proportional to $\omega^2$ and independent of $T$ in the 'quantum region' where $\hbar\omega \gg kT$ (Landau, 1957; Khalatnikov and Abrikosov, 1957).

Landau (1957) and Silin (1957) have considered the possibility of spin-wave propagation utilizing the magnetic moments of the $He^3$ nuclei. They conclude that spin waves probably do not exist in zero applied magnetic field, but may well exist in a finite applied magnetic field.

### 8.6.5. Cell models and other theories

Several theories are modified forms of the cell model of a liquid in which each molecule is treated as an independent unit, but is assumed to move inside a potential well or 'cell' representing the smoothed-out effect of interactions with all the neighbouring molecules. Prigogine and Philippot (1952, 1953) (see also Prigogine, 1954), have developed a theory of liquid $He^4$ along these lines, introducing the possibility of density fluctuations by allowing a cell to contain 0, 1 or 2 atoms. The application of Bose-Einstein statistics to the $He^4$ case gave a maximum in the specific heat near the $\lambda$-point, but the use of Fermi-Dirac statistics for the $He^3$ case shifted this maximum to a temperature above the critical temperature of $He^3$. The introduction of nuclear spin creates difficult conceptual problems, since any tendency for the $He^3$ atoms to divide up into pairs with antiparallel spins might produce a reversion to the Bose-Einstein type of behaviour.

A more promising approach seems to be to consider larger cells containing two atoms. The thermal excitations are then of two types: (*a*) Debye waves involving collective motions of the centres of gravity of all the cells and leading to a phonon specific heat term; (*b*) internal motions of the atoms within a cell, considered to be independent of the motions of all atoms outside the cell. The simplest treatment of these internal motions assumes that the atom pair behaves like a rigid rotating diatomic molecule with rotational energy levels $R(R+1)\hbar^2/md^2$ (Price, 1955; de Boer and Cohen, 1955). Even values of the rotational quantum number $R$ require antiparallel nuclear spins, whereas odd values of $R$ occur only for parallel spins. The specific heat curve corresponding to these energy levels is well known in connexion with ortho- and para-hydrogen, and it has a qualitative similarity to the experimental curve for liquid $He^3$ (fig. 83). In this theory the nuclear para-magnetic susceptibility arises only from those molecules with odd $R$, and at a temperature $T_0 = \hbar^2/md^2k \sim 0{\cdot}5°$ K., the number of such molecules begins to decrease rapidly as the rotators all drop into their ground states with $R=0$. In this way one can obtain qualitative, but not quantitative agreement with the experimental facts.

Rice (1955) obtains better agreement with experiment by postulating that the influence of neighbouring atoms restricts the rotation to a plane with a variable potential, and he therefore applies the theory of a hindered plane rotator. He also ignores all except the first two rotational levels ($R=0$ and 1). Temperley (1955 *a*, *b*) has obtained moderately good agreement with experiment by choosing empirical values for the Debye temperature and the scheme of energy levels for the internal motions within the cell. These empirical energy levels are very similar to those applicable to Rice's hindered plane rotator (Rice, 1956). Buckingham (1955) has argued that the environment would vary considerably from one rotator to another, the degree of hindering would therefore vary and the energy levels would be broadened by an amount comparable with their separation, giving a continuous rather than a discrete spectrum.

de Boer and Cohen (1955) have independently developed a cell model of the above type. In addition to cells containing a rotating pair, they also consider larger cells containing four atoms capable of rotating in a manner similar to a rigid tetrahedron. They show that moderately good agreement with experiment is obtained if the cell is an elongated ellipsoid of revolution, which is a step in the direction of Rice's hindered plane rotator.

Houston and Rorschach (1955) have attempted a novel approach in which the nuclei are imagined to move in a periodic potential field representing the smeared out effect of the electrons. This is the reverse of the Bloch treatment of electrons in a metal.

Mazo and Kirkwood (1955) have presented a theoretical curve describing the radial distribution function of the atoms in liquid $He^3$.

## 8.7. The melting curve

We have seen that the entropy of *liquid* $He^3$ falls below $R \ln 2$ at about 0·5° K. when the nuclear spins begin to take up an ordered arrangement. Pomeranchuk (1950) has suggested that this does not occur in *solid* $He^3$ until much lower temperatures, because the nuclei are then localized near lattice sites and their amplitude of oscillation is smaller than the interatomic distance, so that the electronic wave functions do not overlap sufficiently to give the strong exchange effects needed to align the nuclear spins. In the complete absence of exchange effects, the nuclear alignment would be caused by direct magnetic dipole interaction between the spins and would occur at a temperature

$$T_M \sim \frac{\mu^2}{kb^3} \sim 10^{-7}\,^\circ\text{K.}, \tag{8.38}$$

where $b$ is the interatomic distance. Exchange effects probably take over at a temperature $T_s^*$ which is greater than this, but nevertheless smaller than the corresponding temperature for the liquid. All of this is supported by the preliminary experiments of Walters and Fairbank (1956*a*) which indicate that increasing the density of the liquid shifts departures from Curie's Law to lower temperatures. Nuclear magnetic resonance experiments on the solid (Walters and Fairbank, 1957*a*, *b*) have not provided a decisive answer, since under some circumstances there is evidence for antiparallel alignment of the spins, whereas under other circumstances the evidence favours parallel alignment, although in both cases the alignment occurs at temperatures a little lower than the temperature at which departures from Curie's Law set in for the liquid.

Now let us compare the entropy of the melting solid with that of the freezing liquid (fig. 87). The entropy of the freezing liquid falls below $R \ln 2$ just below 0·5° K. According to Pomeranchuk, however, the entropy of the solid remains above $R \ln 2$ until a much lower temperature. It would follow that the two curves must cross, and at very low temperatures we would get the unusual situation that the entropy of the solid would be greater than that

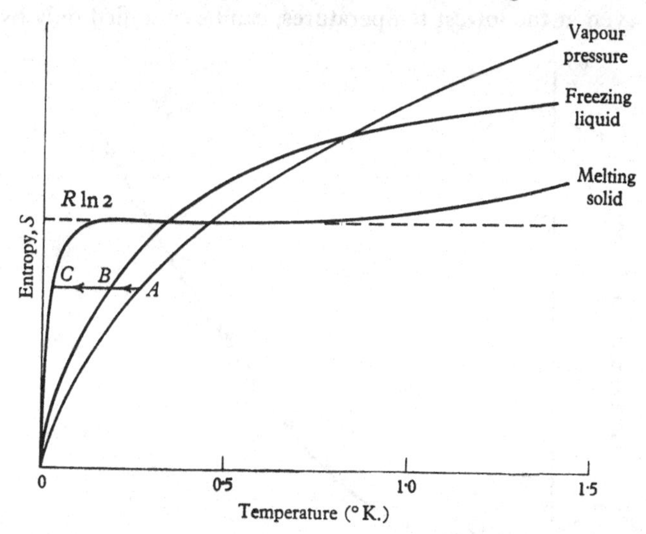


Fig. 87. The entropies of solid and liquid He³ below 1° K. according to Pomeranchuk (1950). The line *ABC* represents the process of cooling by adiabatic compression (schematic).

of the liquid. Moreover, from the Clausius-Clapeyron equation, the slope of the melting curve is

$$\left(\frac{dp}{dT}\right)_{\text{m.p.c.}} = \frac{S_l - S_s}{V_l - V_s}. \tag{8.39}$$

If $S_l - S_s$ were negative at very low temperatures and if $V_l - V_s$ remained positive, the slope of the melting curve would be negative. There would therefore be a minimum in the melting curve just below 0·5° K., and then a point of inflexion at the temperature corresponding to the ordering of the nuclei in the solid, allowing the curve to become horizontal at 0° K. in accordance with the third law of thermodynamics (fig. 88).

Fig. 88 shows the melting pressures obtained experimentally by Osborne, Abraham and Weinstock (1951, 1952) using the blocked capillary technique. The pressure was slowly increased on one side of the liquid $He^3$ in a U-tube capillary and the reading of a Bourdon gauge on the far side was watched. When the solidification pressure was reached, a block of solid $He^3$ formed in the capillary and the gauge on the far side ceased to respond to further increases in pressure. The results indicate that $He^3$, like $He^4$, has no triple point and, even at the lowest temperatures, can be solidified only by the

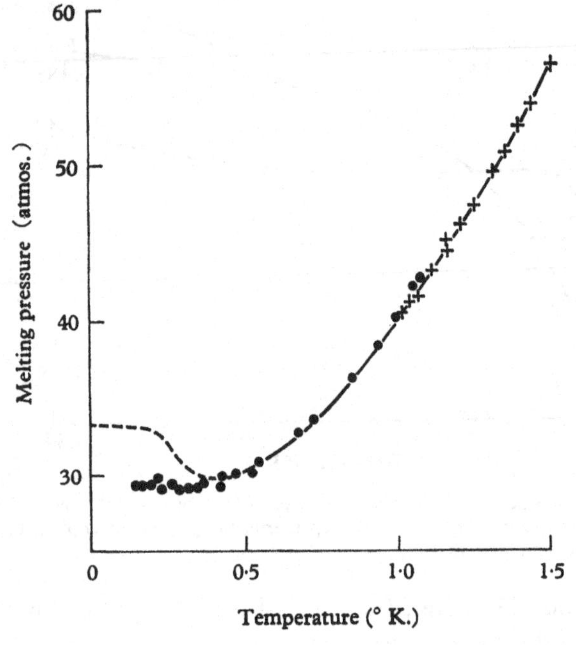


Fig. 88. The melting curve of $He^3$. ●, +, experiment (Osborne, Abraham and Weinstock, 1951, 1952); ----, type of curve to be expected from Pomeranchuk's theory.

application of a high pressure. Below 0·5° K. the experimental melting curve becomes horizontal at about 29·3 atmospheres, but there are two reasons why this result cannot be accepted with complete confidence. To obtain these low temperatures a magnetic cooling technique was employed and the experimenters were not certain that the U-tube capillary was in good thermal contact with the demagnetized salt. Moreover, let us suppose that there really is a minimum just below 0·5° K. Then, even with the bottom of the capillary at 0·2° K., say, there would be a point higher up the capillary containing liquid at the temperature of the minimum, and this would be the point at which solidification would first occur as the pressure was slowly increased from zero upwards. The measured solidification pressure would therefore *appear* to be constant at all temperatures below the minimum.

If Pomeranchuk's views are correct, adiabatic compression and solidification of the liquid, starting at a temperature below 0·5° K., should produce a cooling down to the temperature at which the spins in the solid begin to align themselves. This is illustrated by the line *ABC* in fig. 87. Walters and Fairbank (1957*a*) have observed the reverse effect, a rise in temperature when the solid at 0·2° K. was melted by a rapid decrease of pressure.

The melting-pressure measurements of Osborne, Abraham and Weinstock extended up to 1·5° K. Sherman and Edeskuty (1957) have covered the temperature range from 1·0 to 3·1° K. and have also measured the molar volume of the liquid along the melting curve between 1·2 and 1·6° K. Mills and Grilly (1955, 1957) have measured melting pressures between 2 and 31° K., where the melting pressure has increased to 3500 atmospheres. Over this temperature range they have also obtained data on the molar volume of the liquid along the melting curve and the changes in molar volume and entropy upon melting.

# CHAPTER 9

# $He^3$-$He^4$ MIXTURES

## 9.1. Thermodynamic properties

### 9.1.1. The theory of ideal solutions

The thermodynamic properties of $He^3$-$He^4$ mixtures will first be considered in relation to the classical theory of solutions and the special theories of these mixtures will then be evaluated in the light of what emerges. The subscript $m$ will be used to designate a mean thermodynamic quantity for the liquid solution as a whole, the subscripts 3 and 4 will indicate partial thermodynamic quantities for $He^3$ and $He^4$ respectively, and the subscript 0 will indicate a property of an isotopically pure liquid. If there are $N_3$ atoms of $He^3$ and $N_4$ atoms of $He^4$ in the liquid phase, the mole fraction of $He^3$ is

$$x_3 = N_3/(N_3+N_4), \tag{9.1}$$

and the mole fraction of $He^4$ is

$$\begin{aligned} x_4 &= N_4/(N_3+N_4) \\ &= 1 - x_3. \end{aligned} \tag{9.2}$$

A detailed treatment of the classical theory of solutions will be found in Guggenheim (1950), Hildebrand and Scott (1950) and Rushbrooke (1949). The application of standard thermodynamic reasoning to $He^3$-$He^4$ mixtures has been considered exhaustively by Morrow (1951, 1953*a*, *b*).

The theory of *ideal solutions* is known to be applicable to solutions of isotopes of a heavy element or, for that matter, of two very similar molecules such as chlorobenzene and bromobenzene. Its basic assumption is that the activities may be written as

$$\lambda_3 = x_3 \lambda_{3,0}, \tag{9.3}$$

$$\lambda_4 = (1 - x_3)\,\lambda_{4,0}. \tag{9.4}$$

The partial potentials then become

$$\mu_3 = RT \ln \lambda_3$$

$$= \mu_{3,0} + RT \ln x_3, \tag{9.5}$$

$$\mu_4 = \mu_{4,0} + RT \ln (1 - x_3). \tag{9.6}$$

The Gibbs free energy of the solution as a whole is

$$G_m = x_3 \mu_3 + (1 - x_3) \mu_4$$

$$= x_3 \mu_{3,0} + (1 - x_3) \mu_{4,0} + RT[x_3 \ln x_3 + (1 - x_3) \ln (1 - x_3)]. \tag{9.7}$$

The last equation helps us to understand the fundamental nature of our assumption, which is that the two atoms are so similar that a $He^4$ atom can be replaced by a $He^3$ atom without influencing the energy levels of the remaining $He^4$ atoms, and that the energy levels of the inserted $He^3$ atoms are the same as they would be in the pure liquid. The first two terms therefore represent an admixture of the Gibbs functions of the two pure liquids in the correct proportion, and the third term arises from an entropy of mixing, corresponding to the $(N_3 + N_4)!/N_3!N_4!$ ways of permuting the $N_3$ $He^3$ atoms and $N_4$ $He^4$ atoms.

Considering now the equilibrium between the liquid solution and its vapour, the activity of a component must be the same in the two phases and the activity in the vapour phase is proportional to the partial pressure (assuming the vapour to be an ideal gas). Equations (9.3) and (9.4) may therefore be rewritten in terms of partial pressures:

$$p_3 = x_3 p_{3,0} \qquad \text{Henry's law,} \tag{9.8}$$

$$p_4 = (1 - x_3) p_{4,0} \qquad \text{Raoult's law.} \tag{9.9}$$

$p_{3,0}$ and $p_{4,0}$ are, of course, the vapour pressures of the pure liquids and we recall that $x_3$ is the mole fraction of $He^3$ in the *liquid* phase. Dividing (9.8) by (9.9) and rearranging we get

$$\frac{C_v}{C_l} = \frac{p_{3,0}}{p_{4,0}} \qquad \text{Distribution law.} \tag{9.10}$$

Here $C_v$ and $C_l$ represent the ratio of the number of $He^3$ atoms to the number of $He^4$ atoms $(N_3/N_4)$ in the vapour and liquid phases respectively.

### 9.1.2. The vapour pressure and distribution coefficient

Early experiments attempted to measure the distribution coefficient $C_v/C_l$ directly by withdrawing a sample of the vapour and analysing it mass spectrographically (Fairbank, Lane, Aldrich and Nier, 1947, 1948, 1949; Daunt, Probst and Smith, 1948). These early results are now known to be unreliable below the λ-point, because of two major experimental difficulties. First, the creep of the film up the walls of the containing vessel involves flow of $He^4$ alone (§ 9.4.1) and, when the creeping film evaporates at a region of high temperature, it tends to increase the proportion of $He^4$ in the vapour there. The second effect occurs whenever there is a heat current in the liquid mixture, caused, for example, by the gas evaporating from the film returning to the liquid surface and condensing there, with evolution of its latent heat. The heat current causes a counter flow of the two components and the $He^3$ is known to move with the normal component (§ 9.4.1), and therefore to concentrate away from the source of heat (the 'heat flush' effect). It is the concentration of $He^3$ near the liquid surface which determines the liquid-vapour equilibrium and it may be completely different from the mean concentration of the whole liquid.

A simpler experimental approach is to measure the vapour pressure of the solution relative to pure liquid $He^4$ (i.e. $p_3+p_4-p_{4,0}$), but it is still necessary to avoid the pitfalls just mentioned. We shall disregard the rather inaccurate early measurements (Rollin and Hatton, 1948; Taconis, Beenakker, Aldrich and Nier, 1949; Eselson, Lasarew and Alekseevski, 1950; Weinstock, Osborne and Abraham, 1950; Daunt and Heer, 1952) and confine the discussion to more recent measurements.

Sommers (1952) has made a particularly careful investigation over the temperature range from 1·30 to 2·18° K. with liquid concentrations up to 13 % and vapour phase concentrations up to 80 % of $He^3$. A copper or brass sample chamber was immersed in a liquid helium bath and was connected through a metal capillary filling tube to a manometer system which measured the sample pressure, the bath pressure and the differential pressure between the two. Two experimental procedures were used. The first procedure was a determination of the dew point by slowly adding gaseous mixture to the sample chamber until the saturated vapour pressure was reached, after which further addition of gas produced only a very slight change in pressure. This gave the vapour pressure corresponding to a known composition of the *vapour* phase. The vapour pressure corresponding to a known composition of the *liquid* phase was then determined by filling the chamber completely with liquid, leaving a negligible mass of the mixture in the vapour phase. A careful investigation was made of the possible influence of film effects and heat flush effects. The reader is referred to the original paper for the details of this investigation, but the conclusion was that these effects did not disturb the measurements, except possibly in the case of the most dilute solution which had a liquid concentration of about $\frac{1}{2}$%. Daunt and Heer (1952) have pointed out a very good reason why heat flush effects might not be important at high concentrations. Imagine a heat flow from the surface downwards, producing, if possible, a concentration $x_3$ at the surface and a larger concentration $x_3'$ at a depth $h$. The vapour pressure of the more concentrated solution would be $p(x_3')$, but the actual pressure would be $p(x_3)+\rho gh$, and the maximum permissible difference in concentration would be determined by the equation

$$p(x_3')-p(x_3)=\rho gh,$$

because otherwise bubbles would form and the liquid would be stirred by rapid boiling. The principal uncertainty in Sommer's experiments was revealed when the sample chamber was filled with pure $He^4$ and the measured vapour pressure was found to be slightly higher than the bath pressure, particularly below 1·5° K. The explanation is unknown, but a correction was made on the assumption that the excess pressure to be subtracted from the measured vapour pressure of a mixture was the same as for pure $He^4$.

Wansink, Taconis and Staas (1956) have made similar measurements in the concentration range up to 2%. To avoid the heat flush effect they reduced heat leaks to a minimum and worked with a thin layer of liquid mixture only 0·3 mm. deep. They concentrated their mixture as the experiment proceeded by withdrawing $He^4$ through a superleak. Their results are in satisfactory agreement with those of Sommers if one ignores his measurements at 0·58% concentration and extrapolates to lower concentrations the more reliable results he obtained at concentrations of 2 % and higher. Daunt and Tseng (1955) have measured the vapour pressure of a 4 % solution between 1·5 and 2·5° K.

A very thorough and comprehensive investigation has been made by Eselson and Berezniak (1956). They covered the temperature

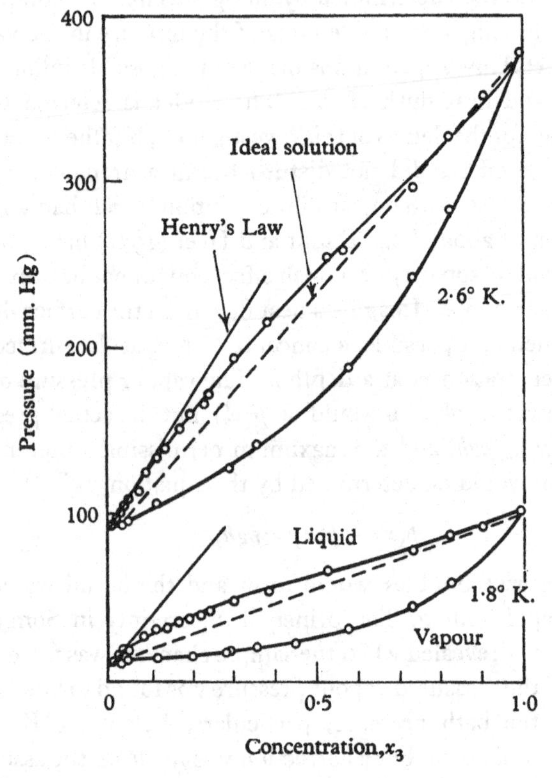


Fig. 89. The phase diagram of $He^3$-$He^4$ mixtures at two temperatures. The dashed curve shows the behaviour to be expected from an ideal solution (after Eselson and Berezniak, 1956).

range from 1·35 to 3·2° K. and made dew-point measurements over a range of vapour concentrations from 1·9 to 82·4 % and vapour pressure measurements over a range of liquid concentrations from 0·4 to 90·8 %. To ensure equilibrium in the liquid they stirred it with a vibrating iron vane magnetically excited. Their results are in good agreement with those of Sommers, but cover a wider range of concentrations.

Fig. 89 shows the vapour pressure as a function of both liquid and vapour concentrations at two temperatures. For any sufficiently dilute solution, ideal or not,

$$p_3 = hx_3 \quad \text{Henry's law,} \tag{9.11}$$

where $h$ is a constant and

$$p_4 = p_{4,0}(1 - x_3) \quad \text{Raoult's law.} \tag{9.12}$$

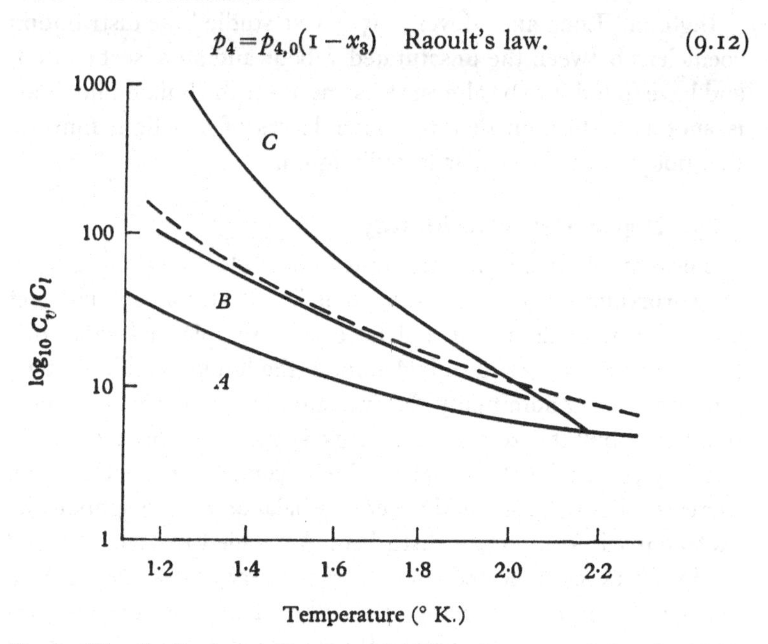


Fig. 90. The distribution coefficient as a function of temperature for very dilute solutions. The dashed curve represents the experimental data and the full curves represent the following theories. *A*, an ideal solution. *B*, the Daunt-Heer theory modified by Mikura (1955). *C*, the Taconis hypothesis in the form developed by de Boer and Gorter (1950).

Thus,

$$\Delta p = p_3 + p_4 - p_{4,0}$$

$$= (h - p_{4,0})\, x_3. \tag{9.13}$$

Experimentally this linear law (Henry's law) is obeyed up to concentrations of about 4 %. However, for an ideal solution the constant $h$ should be equal to $p_{3,0}$ (equation (9.8)), whereas the experimental value of $h$ is larger by a factor of 1·4 at 2·1° K. and by a factor of 2·9 at 1·3° K. The distribution coefficient ($C_v/C_l = h/p_{4,0}$) at low concentrations is plotted against temperature in fig. 90.

It is considerably greater than the value for an ideal solution given by equation (9.10) and the discrepancy becomes worse as the temperature is lowered. The distribution coefficient decreases rapidly with increasing concentration. Notice that the vapour phase is much richer in $He^3$ than the liquid phase, particularly at the lower temperatures.

Inghram, Long and Meyer (1955) have studied the distribution coefficient between the unsaturated vapour and an adsorbed film, and have found it to be almost the same as for the bulk liquid. This is another indication that the outer layers of a helium film are thermodynamically similar to bulk liquid.

### 9.1.3. Departures from ideality

These strong departures from the laws of ideal solutions are not too surprising when we consider that liquid $He^3$ and liquid $He^4$ are really quite dissimilar and have, for example, markedly different molar volumes. If a $He^4$ atom in the liquid is replaced by a $He^3$ atom, then admittedly the interatomic forces remain almost unaltered, but the zero-point energy is increased by a factor of about 4/3. The result is that the 'cell' containing the $He^3$ atom increases slightly in size and the energy balance of the neighbouring $He^4$ atoms is completely disturbed. A detailed treatment would obviously be complicated, but there is every reason to expect a change of volume on mixing and a heat of mixing. For an ideal solution there is no change of volume on mixing and the heat of mixing is identically zero.

Measurements of the densities of the mixtures (Kerr, 1957*b*; Ptukha, 1958) show that there is indeed a sizeable decrease in volume when pure liquid $He^4$ and pure liquid $He^3$ are mixed. The effect increases with increasing temperature, presumably because of the increasing disparity between the molar volumes of liquid $He^4$ and $He^3$. At 3° K. there is an 8·5% decrease in volume on mixing a 50% solution.

The next degree of complication after ideal solutions is the theory of regular solutions. The basic assumption here is that the replacement of a $He^4$ atom by a $He^3$ atom involves a change in volume and a change in energy, but that there is no change in entropy other than that involved in the permutation of the $He^3$ and $He^4$

atoms. If $\epsilon_4$ is the energy of a $He^4$ atom in the pure liquid, $\epsilon_3$ is the energy of a $He^3$ atom in pure liquid $He^3$, $\epsilon_{34}$ is the energy of a $He^3$ atom in a very dilute solution and

$$\omega = N[\epsilon_{34} - \tfrac{1}{2}\epsilon_3 - \tfrac{1}{2}\epsilon_4], \tag{9.14}$$

then (see Rushbrooke, 1949)

$$\lambda_3 = \lambda_{3,0} x_3 \exp\left[\frac{\omega}{RT}(1-x_3)^2\right], \tag{9.15}$$

$$\lambda_4 = \lambda_{4,0}(1-x_3) \exp\left[\frac{\omega}{RT}x_3^2\right]. \tag{9.16}$$

The energy $\omega$ is assumed to be independent of temperature. Assuming the vapour to be an ideal gas

$$p_3 = p_{3,0} x_3 \exp\left[\frac{\omega}{RT}(1-x_3)^2\right] \tag{9.17}$$

$$\rightarrow p_{3,0} x_3 e^{\omega/RT} \quad \text{for small } x_3 \quad \text{Henry's law,} \tag{9.18}$$

$$p_4 = p_{4,0}(1-x_3) \exp\left[\frac{\omega}{RT}x_3^2\right] \tag{9.19}$$

$$\rightarrow p_{4,0}(1-x_3) \quad \text{for small } x_3 \quad \text{Raoult's law.} \tag{9.20}$$

The distribution coefficient becomes

$$\frac{C_v}{C_l} = \frac{p_{3,0}}{p_{4,0}} \exp\left[\frac{\omega}{RT}(1-2x_3)\right]. \tag{9.21}$$

The experimental results can be used to solve this equation for $\omega$, which is plotted against temperature in fig. 91. The variation of $\omega$, particularly near the $\lambda$-point, shows that even the theory of regular solutions is not adequate and that we should consider changes in entropy other than the simple permutation effect. However, $\omega$ is approximately constant near 1·2° K. and the formalism of regular solutions may be applicable in this temperature region.

For a regular solution $\omega$ is the heat of mixing, defined as the heat required to keep the temperature constant when 1 mole of liquid $He^3$ is added to a large excess of liquid $He^4$. Sommers, Keller and Dash (1953) have measured this quantity directly. Two adjacent, thermally isolated chambers containing pure liquid $He^3$ and liquid $He^4$ were mixed together to give an 8·6 % solution by rupturing the partition dividing them. During this procedure the temperature fell from 1·02 to 0·78° K., demonstrating without doubt the existence of a positive heat of mixing. A semi-quantitative measurement was then made by increasing the volume available

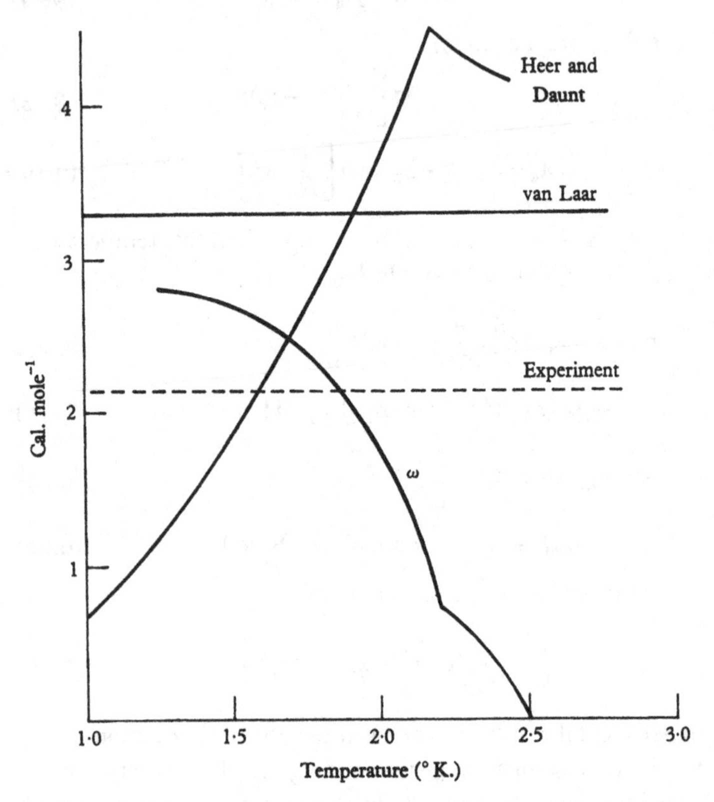


Fig. 91. The heat of mixing according to the theories of Heer and Daunt (see Nanda, 1954) and van Laar (see Sommers, 1952). The dashed curve is the experimental result of Sommers, Keller and Dash (1953). $\omega$ was obtained from the vapour pressure data of Sommers (1952) using equation (9.18). The heat of mixing is $\omega - T d\omega/dT$.

for vapour above the liquid $He^3$ in such a way that the mixing was followed by a condensation of $He^3$ vapour, which liberated sufficient heat to keep the temperature approximately constant. The heat of mixing was then estimated from the amount of latent heat given up, but this undoubtedly gave too low a value, since some of the condensing vapour was warmer than the liquid and brought in additional heat. The lower limit to the heat of mixing obtained in this way was 2·16 cal. mole$^{-1}$, which is in satisfactory agreement with the values derived from the vapour pressure measurements (fig. 91). For non-regular solutions the $\omega$ of equation (9.21) is the change in the Gibbs free energy on mixing, excluding the permutation entropy term $RT[x_3 \ln x_3 + (1-x_3)\ln(1-x_3)]$, whereas the directly measured heat of mixing gives the change in enthalpy. It follows that the contribution of the entropy term $-TS$ cannot be greater than 1 cal. mole$^{-1}$ at 1° K. and may be much smaller.

Starting with the vapour pressure data and using only rigorous thermodynamic arguments, Wansink (1957) has calculated all the thermodynamic functions of the solutions for concentrations up to 7%. He finds that the heat of mixing increases steadily with increasing temperature, but that the increase becomes less pronounced the more concentrated the solution.

A regular solution with a positive heat of mixing can be shown to become unstable below a critical temperature $T_c$ and to separate into two phases of different compositions. The critical temperature is related to the heat of mixing:

$$T_c = \frac{\omega}{2R}$$

$$\sim 0{\cdot}7^\circ \text{ K. for He}^3\text{-He}^4 \text{ mixtures.} \qquad (9.22)$$

Walters and Fairbank (1956*b*) have found that $He^3$-$He^4$ solutions show a phase separation of this type below a critical temperature of 0·83° K. They observed nuclear magnetic resonance lines (see § 8.2.3) from a mixture condensed into a container consisting of three separate chambers vertically above one another and connected through small holes. As there was a magnetic field gradient in the vicinity of this container, three separate resonance lines were observed and their relative amplitudes were a measure of the number of $He^3$ nuclei in each chamber. Below the critical temperature the lowest chamber was found to contain a much lower concentration of $He^3$ than the uppermost chamber, and in this way it was possible to plot the phase diagram of fig. 92 showing the two concentrations which can be in equilibrium at any temperature below $T_c$. The area underneath the hump corresponds to unstable concentrations, which split up into two phases having the two concentrations on the curve at the temperature in question. Peshkov and Zinoveva (1957) have observed this phase separation visually and have photographed the boundary separating the two phases at 0·5° K.

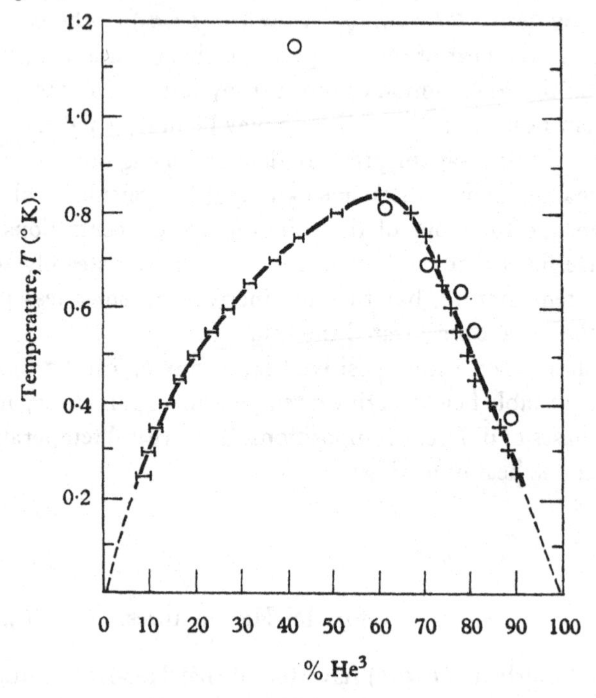


Fig. 92. Phase separation in $He^3$-$He^4$ mixtures. At a temperature $T$° K. a concentration represented by a point ⊢⊣ can coexist with a concentration represented by a point +, but intermediate concentrations are unstable. The open circles, ○, represent the measurements of Daunt and Heer (1950) on the variation of λ-temperature with concentration (after Walters and Fairbank, 1956*b*).

The Leiden workers have measured the specific heats of solutions with concentrations of 1, 2·5 and 7·13 % over the temperature range 1·05 to 2·13° K. (Dokoupil, van Soest, Wansink and Kapadnis, 1954; Kapadnis and Dokoupil, 1955). The specific heat was found to increase steadily with increasing concentration, the fractional change being much greater at lower temperatures. Linhart and Price (1956) have shown how the specific heat measurements can be related thermodynamically to the vapour pressure measurements and that the two sets of data are consistent within their accuracy. At low concentrations the specific heat can be expressed as

$$C_m = (1 - x_3)\, C_{4,0} + x_3 f R. \tag{9.23}$$

Near 1° K. the experimentally determined value of $f$ is very close to 3/2. This is very important, for it indicates that the $He^3$ atoms can be treated as free particles with the specific heat of an ideal monatomic gas, and that they make no other major contributions to the specific heat. Above 1° K., however, $f$ increases rapidly to a value of about thirty near the $\lambda$-point. This must be caused by the same effect which gives a dip in the $\omega$-$T$ curve near the $\lambda$-point (fig. 91).

Eselson, Kaganov and Lifshitz (1957) have considered the thermodynamics applicable to a $\lambda$-transition in a solution and have shown that, if the transition is second order in pure $He^4$ and in the mixture, then there should be discontinuities in the heat of mixing, the heat of vaporization, and the temperature derivatives of the vapour pressure and the distribution coefficient $C_v/C_l$. These discontinuities are related to the discontinuity in specific heat and the variation of the $\lambda$-temperature with concentration, $\partial T_{s,\lambda}/\partial x_3$ (§ 9.3), by equations similar to those derived in § 2.6.1. Discontinuities in the slopes of the vapour pressure curves at the $\lambda$-temperatures have been observed (Eselson, Berezniak and Kaganov, 1956; Sreedhar and Daunt, 1957; Roberts and Sydoriak, 1957*b*).

## 9.2. Theories of $He^3$-$He^4$ mixtures

We have just seen the extent to which the mixtures depart from the classical theory of ideal solutions, and we shall now consider the theoretical attempts to derive the numerical magnitude of these deviations. In later sections these theories will be tested against the variation of the $\lambda$-point with concentration, the hydrodynamics of the mixtures, and the velocity of second sound.

Sommers (1952) has pointed out that an early theory of van Laar (see Hildebrand and Scott, 1950, ch. VII) gives a good value for the heat of mixing near 1° K. The theory is based upon the application of van der Waals' equation of state, the necessary constants being obtained empirically from the critical constants of the pure components. The heat of mixing is calculated to be 3·2 cal. mole$^{-1}$ and is independent of temperature. This is in good agreement with the experimental values near 1° K., but gives no explanation of the observed variation with temperature (fig. 91).

When Taconis, Beenakker, Aldrich and Nier (1949) first measured vapour pressures in excess of the values predicted by the theory of ideal solutions, they suggested that the $He^3$ forms an ideal solution in the normal component of the liquid $He^4$ only and does not dissolve in the superfluid component. We shall refer to this as the 'Taconis hypothesis'. In terms of our present ideas about the nature of the two components, it is difficult to give this hypothesis concrete physical significance, unless we believe with Rice (1954) that the two components are actually separated in configurational space. However, the hypothesis is consistent with the fact that the dissolved $He^3$ always participates in the motion of the normal component (§9.4.1) and, in any case, we can treat it as a mathematical formalism, and compare its consequences with experiment, as was done by de Boer and Gorter (1950). For very dilute solutions the distribution coefficient would be

$$\frac{C_v}{C_l} = \frac{\rho}{\rho_n} \frac{p_{3,0}}{p_{4,0}}. \tag{9.24}$$

Fig. 90 shows that this is in bad disagreement with experiment, particularly at the lowest temperatures. Nanda (1954) has pointed out that the heat of mixing is also predicted incorrectly. Moreover, the theory would give agreement with the laws of ideal solutions above the $\lambda$-point, which is contrary to observation. Nanda (1955) has attempted to remove this defect by assuming that the $He^3$ forms a non-ideal solution with the normal component, but this worsens the situation below the $\lambda$-point.

The theory of Heer and Daunt (1951) is based upon London's treatment of liquid $He^4$ as a degenerate ideal Bose-Einstein gas (§3.1.1). A $He^3$-$He^4$ mixture is considered to be an ideal mixture of two ideal gases, a degenerate Bose-Einstein gas of $He^4$ atoms and a non-degenerate Fermi-Dirac gas of $He^3$ atoms. Pure liquid $He^4$ is taken to be a degenerate ideal Bose-Einstein gas of $He^4$ atoms moving in a smoothed potential well of depth $-\chi_{4,0}$, and pure liquid $He^3$ is taken to be a non-degenerate ideal Fermi-Dirac gas of $He^3$ atoms moving in a smoothed potential well of depth $-\chi_{3,0}$. The depth of the smoothed potential well for the mixture is assumed to be

$$\chi_m = N_3\chi_{3,0} + N_4\chi_{4,0}. \tag{9.25}$$

It is also assumed that there is no change of volume on mixing, so that, if $v_{3,0}$ and $v_{4,0}$ are the mean volumes per atom in pure liquid $He^3$ and $He^4$ respectively, then the total volume of the mixture is

$$V_m = N_3 v_{3,0} + N_4 v_{4,0}. \tag{9.26}$$

The partial potential of each component in the mixture is then calculated as though it were quite independent of the presence of the other component, due regard being taken to the importance of quantum statistics in the case of the degenerate $He^4$ component. The principal effect influencing the partial potentials, therefore, is that the mean density of a component in the mixture is not the same as in the pure liquid. For a very dilute solution below its λ-point ($T_{s,\lambda}$) the distribution coefficient is calculated to be

$$\frac{C_v}{C_l} = \frac{p_{3,0}}{p_{4,0}} \frac{v_{3,0}}{v_{4,0}} \exp\left[1 - 0{\cdot}54\left(\frac{T}{T_{s,\lambda}}\right)^{\frac{3}{2}} \frac{v_{3,0}}{v_{4,0}}\right]. \tag{9.27}$$

This is in reasonably good agreement with experiment (fig. 90) even though there are no arbitrary adjustable parameters. Above the λ-point of the solution

$$\frac{C_v}{C_l} = \frac{p_{3,0}}{p_{4,0}} \frac{v_{3,0}}{v_{4,0}} \exp\left[1 - \frac{v_{3,0}}{v_{4,0}}\left\{1 - 0{\cdot}462\left(\frac{T_{s,\lambda}}{T}\right)^{\frac{3}{2}} - 0{\cdot}022\left(\frac{T_{s,\lambda}}{T}\right)^{3}\right\}\right], \tag{9.28}$$

which gives positive deviations from ideal solution theory, in qualitative agreement with experiment.

It may seem surprising that this theory gives such satisfactory results when the corresponding assumptions are so inadequate for the pure liquids. Part of the reason is that the quantities $p_{3,0}$ and $p_{4,0}$ in equations (9.27) and (9.28) come from the smoothed potentials $\chi_{3,0}$ and $\chi_{4,0}$ and are arbitrarily given the correct values. Thus the theory need only concern itself with departures from the laws of ideal solutions and these departures are explained solely in terms of the details of the quantum statistics. However, we should notice that the heat of mixing calculated from this theory by Nanda (1954) is too small by a factor of two at 1·3° K. (fig. 91).

Heer and Daunt (1951) have also treated the case when pure liquid $He^3$ has to be considered a degenerate gas, but the $He^3$ in solution is non-degenerate, and the further case when the $He^3$ in solution is also degenerate. Since the degeneracy temperatures are not known, these treatments cannot give numerical results, but they lead to the interesting conclusion that $C_v/C_l$ may fall below its value for ideal solutions at very low temperatures.

Mikura (1954*a*, *b*, 1955) has obtained even better agreement with experiment by introducing modifications which recognize that liquid $He^4$ is not really an ideal gas. The energy-momentum relationship of the 'particles' in liquid $He^4$ was assumed to be

$$\epsilon = -\chi_{4,0} + \Delta_4 + \frac{p^2}{2\mu}, \tag{9.29}$$

with an effective mass $\mu$ of about 8·8 atomic masses and an energy gap $\Delta_4$ between the ground state and the first excited state equal to $8{\cdot}6k$ for the pure liquid, but varying with concentration in accordance with the formula

$$\Delta_4 = \Delta_{4,0}\left(\frac{N_4 v_{4,0}}{N_3 v_{3,0} + N_4 v_{4,0}}\right)^{0{\cdot}4}. \tag{9.30}$$

The values of $\mu$ and $\Delta_{4,0}$ were chosen to give the best agreement with the observations on specific heat, $\rho_n/\rho$, and the velocity of second sound in pure liquid $He^4$. The assumed variation of $\Delta_4$ with concentration was necessary to fit the observations on the variation of the $\lambda$-temperature with concentration described in the next section.

The Russian school of thought, based on Landau's treatment of liquid $He^4$, is represented by the theory of Pomeranchuk (1949). He considers only very dilute solutions and can therefore assume that the $He^3$ atoms do not interact with one another and that the assembly of $He^3$ atoms is non-degenerate. A dissolved $He^3$ atom is looked upon as the centre of a small localized region of perturbation which can move freely through the liquid, rather like an electron in a metal. The excess energy associated with the $He^3$ atom is taken to be

$$\epsilon_3 = E_0 + \frac{p^2}{2\mu}, \tag{9.31}$$

where $E_0$ is the energy needed to replace a $He^4$ atom by a $He^3$ atom and perturb the atoms in the immediate vicinity, and $p^2/2\mu$ is the kinetic energy associated with the translational motion of the perturbed region, which has an effective mass $\mu$. (Pomeranchuk also considered an energy-momentum relationship of the type used to describe rotons (equation (3.20)) but this gave poor agreement with the second sound experiments and so we shall simplify the discussion by ignoring it.)

Although equation (9.31) makes a $He^3$ atom resemble a roton, the important difference is that the number of $He^3$ atoms is conserved, whereas the number of rotons is not. The statistical mechanics of the $He^3$ atoms is therefore the same as for an ideal gas and the entropy and specific heat per mole of the solution are found to be

$$S_m = S_{4,0} - N_3 k \ln x_3 + N_3 k \ln \tfrac{3}{2} T^{\frac{3}{2}}, \tag{9.32}$$

$$C_m = C_{4,0} + N_3 \tfrac{3}{2} k. \tag{9.33}$$

The second term in the entropy is the entropy of mixing and the third term is the ideal gas term. Notice that equation (9.33) differs from equation (9.23) inasmuch as $C_{4,0}$ is not multiplied by $(1 - x_3)$. This implies that the $He^3$ atom is not considered to *replace* a $He^4$ atom, but rather to create a small localized disturbance which is ***superimposed on*** normal liquid $He^4$. Near 1° K. $C_{4,0}$ is small compared with $\frac{3}{2}R$ and the difference between (9.33) and (9.23) is unimportant. Equation (9.33) is consistent with the experimental results near 1° K., but not at higher temperatures. This is probably because it is not permissible to assume that $E_0$ is independent of temperature.

The $He^3$ atoms collide freely with the rotons and phonons and move with the normal component. They must, therefore, be incorporated in the value of $\rho_n$ for the solution. If there are $n_3$ $He^3$ atoms per cm.$^3$ and $\rho_{n,0}$ is the density of the normal component for pure liquid $He^4$, then Landau's method of calculating $\rho_n$ leads to

$$\rho_n = \rho_{n,0} + n_3 \mu, \tag{9.34}$$

which has an obvious interpretation in terms of an effective mass $\mu$ per $He^3$ atom. When Pomeranchuk first developed this theory he was particularly interested in the velocity of second sound in $He^3$-$He^4$ mixtures and we shall consider it again in this connexion later.

## 9.3. Variation of the λ-temperature with concentration

Addition of $He^3$ to liquid $He^4$ lowers its λ-temperature. This effect was first discovered by Abraham, Weinstock and Osborne (1949), who determined the λ-point of the solution by the onset of rapid flow through a superleak consisting of a platinum wire sealed into Pyrex glass. They worked with initial concentrations up to 28·2% and obtained λ-temperatures down to 1·56° K. Eselson and Lasarew (1950*b*,*c*) obtained the λ-point of a 1·5 % solution by measuring the rate of film flow and taking the λ-point as the temperature at which the extrapolated flow was zero. Daunt and Heer (1950) extended the data to still lower λ-temperatures by using concentrations in the range from 42 to 89 %. Temperatures below 1° K. were obtained by a magnetic cooling technique, with the vessel containing the solution embedded in the salt and connected to higher temperatures by a fine metal capillary. When the solution was below its λ-temperature, the warm-up rate was large because of film flow up the capillary, but as soon as the λ-temperature was reached there was a marked decrease in warm-up rate. The 89 % solution had a λ-temperature as low as 0·38° K.

More recent, and probably more accurate, measurements have been made by a variety of techniques. King and Fairbank (1953) and Fairbank and Elliott (1957) estimated the λ-temperature by extrapolating the velocity of second sound to zero. Dash and Taylor (1955, 1957*c*) and Berezniak and Eselson (1956) performed an Andronikashvili type experiment (fig. 2) in the solution and took the λ-temperature to be the temperature at which $\rho_n$ became equal to the total density $\rho$. Eselson, Berezniak and Kaganov (1956), Sreedhar and Daunt (1957) and Roberts and Sydoriak (1957*b*) observed a discontinuity in the slope of the vapour pressure curve at the λ-temperature of the mixture. Kerr (1957*b*) and Ptukha (1958) measured the density of the mixture and assumed that its maximum corresponded to the λ-temperature. The Leiden group (Dokoupil, van Soest, Wansink and Kapadnis, 1954; Kapadnis and Dokoupil, 1955) were able to estimate the λ-temperatures from their measurements on the specific heats of mixtures, because the peak at the transition was sharp and well-marked. In all these investigations the depression of the λ-temperature was linearly proportional to concentration up to concentrations of 10% and the slope $\partial T_{s,\lambda}/\partial x_3$ was $-1\cdot5°$ K. per mole fraction.

The $\lambda$-temperature as a function of concentration is shown in fig. 93. In view of the phase separation discovered by Walters and

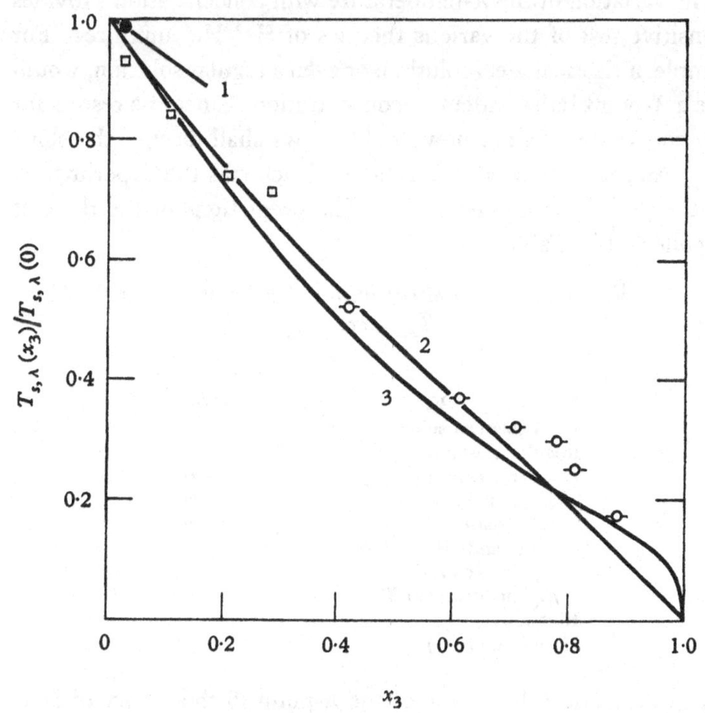


Fig. 93. The $\lambda$-temperature as a function of concentration. Experiments: □, Abraham, Weinstock and Osborne (1949); -○-, Daunt and Heer (1950); ●, Eselson and Lasarew (1950*b*, *c*); curve 1 is the slope near $x_3 = 0$ according to the experiments of King and Fairbank (1953), Dash and Taylor (1955) and Kapadnis and Dokoupil (1955). Theories: curve 2, theory of Daunt and Heer (see Mikura, 1954*a*); curve 3, theory of de Boer and Gorter (1950).

Fairbank (1956*b*), the results of Daunt and Heer below 0·8° K. cannot be accepted with complete confidence. It will be seen that their points fall very close to the phase-separation curve (fig. 92), and this suggests that the appearance of film flow in their experiment was a consequence of the separating out of a very dilute superfluid phase. The $T_{s,\lambda} - x_3$ curve may really intersect the phase separation curve at an angle.

Elliott and Fairbank (1958) have used the second sound method to investigate the variation of the $\lambda$-temperature of a 30·4% mixture with pressure. Up to 11·4 atmospheres the variation was approximately linear with a slope of $-0{\cdot}013°$ K. atmospheres$^{-1}$, as compared with $-0{\cdot}011°$ K. atmospheres$^{-1}$ for pure liquid $He^4$.

The variation of the $\lambda$-temperature with concentration provides a sensitive test of the various theories of $He^3$-$He^4$ mixtures. For example, a classical ideal solution, or even a regular solution, would have a $\lambda$-point independent of concentration. Since the results for high concentrations are now doubtful, we shall discuss the slope $(\partial T_{s,\lambda}/\partial x_3)$ for small concentrations, which has the experimental value $-1{\cdot}5°$ K. per mole fraction. The predictions of the theories are collected in Table X.

TABLE X. *Theoretical initial slope of the curve of $T_{s,\lambda}$ versus $x_3$*

| Theory | $\partial T_{s,\lambda}/\partial x_3$ (° K. per mole fraction) |
|---|---|
| Ideal classical solution | 0 |
| Regular classical solution | 0 |
| Heer and Daunt (1951) | −1·9 |
| Mikura (1954*a*) | −2·7 |
| Stout (1949) | −3·4 |
| de Boer and Gorter (1950) | |
| $\mu_{4,0}$ linear in $T$ | −2·7 |
| $\mu_{4,0}$ quadratic in $T$ | −1·6 |
| Rice (1950) | −3·2 |
| Harasima (1951) | −2·6 |

The theoretical derivation of the $\lambda$-point in the theory of Heer and Daunt is readily understood. This theory treats the $He^4$ component as an ideal gas, quite independent of the presence of the $He^3$, and the transition temperature of this ideal Bose-Einstein gas is

$$T_\lambda = \frac{h^2}{2\pi m_4 k}\left(\frac{N_4}{2{\cdot}612\,V_m}\right)^{\frac{2}{3}}. \tag{9.35}$$

The change in the $\lambda$-temperature is therefore merely a consequence of the change in $V_m/N_4$, which is given by equation (9.26), and so

$$\frac{T_{s,\lambda}(x_3)}{T_{s,\lambda}(0)} = \left[\frac{1-x_3}{1+x_3\left(\dfrac{v_{3,0}}{v_{4,0}}-1\right)}\right]^{\frac{2}{3}}. \tag{9.36}$$

The success of this simple treatment for $He^3$-$He^4$ solutions is somewhat offset by the fact that the $\lambda$-temperature of pure liquid $He^4$ is known to increase with decreasing density. Mikura's modification of the theory overcomes this difficulty by the purely *ad hoc* assumption embodied in equation (9.30).

The other theories in Table X are all based upon Taconis' hypothesis of solubility in the normal component only, and they differ mainly in their assumptions concerning the functional form of the partial potential of the $He^4$. The most successful theory is that of de Boer and Gorter, who take

$$\mu_{4,0} = -E\left[1-\left(\frac{\rho_n}{\rho}\right)^{\frac{7}{6}}\right] - \frac{1}{2}\frac{S_\lambda}{T_\lambda}\left(\frac{\rho_n}{\rho}\right)^{\frac{5}{6}} T^2. \tag{9.37}$$

This equation was chosen empirically to fit the known properties of pure liquid $He^4$.

In the case of Pomeranchuk's theory, we can work backwards and use the experimental value of $\partial T_{s,\lambda}/\partial x_3$ to calculate the effective mass $\mu$ of the $He^3$ atoms near the $\lambda$-point. The $\lambda$-temperature is defined as the temperature at which the density of the normal component becomes equal to the total density

$$\rho_n = \rho_{n,0} + n_3\mu = \rho, \tag{9.38}$$

whence

$$\frac{\partial T_{s,\lambda}}{\partial x_3} = -\frac{\mu\rho}{m_4(\partial\rho_{n,0}/\partial T)_{T_\lambda}}. \tag{9.39}$$

In this way we obtain $\mu \sim 8m_3$. We shall see later that Andronikashvili-type disk experiments and second sound experiments give values of $\mu$ varying from $2{\cdot}0m$ at $0{\cdot}2°$ K. to $3{\cdot}6m_3$ at $1{\cdot}8°$ K. We again conclude that Pomeranchuk's treatment is not very satisfactory near the $\lambda$-point.

## 9.4. The thermohydrodynamics of $He^3$-$He^4$ mixtures

### 9.4.1. Non-participation of $He^3$ in superfluid flow

We shall now examine the evidence that the dissolved $He^3$ forms part of the normal component and does not participate in the flow of the superfluid component. This was first established by Daunt, Probst, Johnston, Aldrich and Nier (1947*a*, *b*) for the case of flow through the film into a small glass Dewar vessel closed at the top by a ground glass plug. The empty vessel was partially immersed in a bath of liquefied atmospheric helium containing a fraction $1 \cdot 2 \times 10^{-6}$ of $He^3$, and the flow was then initiated by supplying heat to the inside of the vessel. A mass spectrographic analysis of the liquid collecting inside the Dewar showed that the abundance ratio of $He^3$ to $He^4$ was less than $5 \times 10^{-8}$. The transfer rate of $He^3$ through the film was therefore not greater than 4 % of the rate for $He^4$ (provided that there was no complication caused by heat flush effects or an unusual distribution of the two isotopes between the liquid, the vapour and the film). Film flow has also been studied by Eselson and Lasarew (1950*b*, *c*). Addition of $He^3$ lowers the $\lambda$-point and also reduces the transfer rate through the film at a fixed temperature, in qualitative agreement with the hypothesis that the $He^3$ joins the normal component, increasing $\rho_n/\rho$ and decreasing $\rho_s/\rho$. In fact, Eselson, Svetz and Bablidze (1958) have shown that, in the immediate vicinity of the $\lambda$-temperature of the solutions, the rate of transfer can always be expressed as $3 \cdot 2 \times 10^{-5} \rho_s/\rho$ cm.$^3$ cm.$^{-1}$ sec.$^{-1}$ for a range of concentrations from 0 to 20%.

The 'heat flush effect' was discovered by Lane, Fairbank, Aldrich and Nier (1948). Atmospheric helium was liquefied inside a glass U-tube, heat was supplied to the liquid in one limb and a sample was withdrawn by evaporation from the opposite limb. The concentration of $He^3$ in this sample was many times greater than in a similar sample withdrawn when the heater was turned off. The suggested explanation was that the $He^3$ had moved with the normal component away from the source of heat.

The non-participation of $He^3$ in superflow must not be assumed to have a direct bearing on the question of whether the unusual properties of liquid $He^4$ are related to Bose-Einstein statistics. First, the solutions are usually so dilute that the dissolved $He^3$ can be regarded as an assembly obeying Maxwell-Boltzman statistics, except at extremely low temperatures. Secondly, Landau and Pomeranchuk (1948) have suggested that any dissolved impurity would form part of the normal component, including $He^6$ which obeys Bose-Einstein statistics. (An investigation by Guttman and Arnold (1953) strongly suggests that the short-lived isotope $He^6$ does not in fact take part in superfluid flow.) Landau and Pomeranchuk's argument is that the impurity atom can readily collide with the phonons and rotons and thereby interact with the normal component. However, the only way for it to interact with the flow of the superfluid component would be by the creation of phonons and rotons, and the laws of conservation of energy and momentum forbid this unless the velocity of the $He^3$ atom relative to the superfluid component is comparable with the velocity of sound.

Some experiments have suggested that $He^3$ does participate to a small extent in the flow of the superfluid component. Abraham, Weinstock and Osborne (1949) found that there was a small amount of $He^3$ in the fluid flowing through a superleak $7 \times 10^{-5}$ cm. wide made by sealing a platinum wire into a glass capillary. Atkins and Lovejoy (1954) obtained similar results with a superleak of the same type. Hammel and Schuch (1952) placed such a platinum-in-glass superleak above the surface of the liquid, with the object of studying flow from the liquid through the film to the superleak and then into an evacuated volume. The pressure in the evacuated volume first rose rapidly to the vapour pressure of pure $He^4$ and then more slowly to the total vapour pressure of the solution. They concluded that $He^3$ flows through the film with a velocity between 3 and 20% of the velocity of the $He^4$ and independent of the pressure difference across the superleak. However, Wansink and Taconis (1955) have made an extensive study with a similar apparatus and conclude that the transfer rate is not determined by the film but by conditions at the superleak. The superleak is probably full of liquid held there by surface tension, and transfer of $He^3$ from the bulk liquid to the superleak may possibly take place by evaporation and condensation. Furthermore, in all these experiments it seems reasonable to explain the flow of $He^3$ through the superleak as a normal viscous flow driven by the very large osmotic pressure which exists when there is an appreciable difference in concentration at opposite ends of the leak. There is therefore no clear evidence for superfluid flow of $He^3$.

Osmotic pressure effects were first detected by Daunt, Probst and Johnston (1948). A vessel containing a $He^3$-$He^4$ solution communicated via the film with a vessel containing almost pure liquid $He^4$. When film flow ceased the level of the solution came to rest a few millimetres above the level of the pure liquid. This did not occur when there was pure liquid $He^4$ in both vessels, and it was therefore obviously an osmotic pressure, with the film acting as a semipermeable membrane preventing the flow of $He^3$. Quantitative measurements of this effect have been made by Taconis, Beenakker and Dokoupil (1950) and by Wansink and Taconis (1955, 1957*a*). Since they used concentrations up to 4 %, they were obliged to measure osmotic pressures corresponding to level differences of several metres. They overcame this difficulty by raising the temperature of the vessel containing the pure liquid $He^4$ above the temperature of the solution until the osmotic pressure was exactly counterbalanced by the thermomechanical effect. Since they were working at concentrations small enough for Raoult's law to be obeyed (fig. 89 and equation (9.12)), the theory of their method is independent of any particular model of the solutions. Imagine a vessel $A$ containing a solution of concentration $x_3$ at a temperature $T_A$ communicating via the film with a vessel $B$ containing pure liquid $He^4$ at a temperature $T_B > T_A$. Since $He^4$ can flow freely through the film, the partial potential of $He^4$ must be the same in both vessels when equilibrium is reached

$$\mu_4(A) = \mu_4(B). \tag{9.40}$$

But, for a sufficiently dilute solution, even if not ideal,

$$\mu_4 = \mu_{4,0} - RTx_3, \tag{9.41}$$

whence

$$\mu_{4,0}(T_A) - RT_A x_3 = \mu_{4,0}(T_B) \tag{9.42}$$

or

$$RT_A x_3 = \int_{T_A}^{T_B} S_{4,0}\, dT. \tag{9.43}$$

The experimental results are in agreement with this equation within the limits of their accuracy, providing further proof that $He^3$ cannot flow through the film, which is therefore a perfect semipermeable membrane. Notice that the thermomechanical pressure corresponding to the temperature difference $T_B - T_A$ is

$$\Delta p = \frac{\rho}{M_4} \int_{T_A}^{T_B} S_{4,0}\, dT. \tag{9.44}$$

So the osmotic pressure which balances this is

$$\Pi = \frac{\rho}{M_4} RT_A x_3, \tag{9.45}$$

which is the well-known Van't Hoff's law. In this derivation we have ignored the thermodynamical consequences of small changes in volume and pressure, but these are not entirely negligible.

### 9.4.2. The thermohydrodynamical equations

The following treatment is based on the work of Pomeranchuk (1949), Mazur and Prigogine (1951), Gorter, Taconis and Beenakker (1951), Khalatnikov (1952c) and Price (1953). We shall assume that, although $\rho_s$ may be different from its value in the pure liquid, the motion of the superfluid component involves the movement of $He^4$ atoms only. We shall also ignore second order terms in the velocities and terms involving irreversible effects such as arise from the coefficients of viscosity and thermal conduction. $\rho_3 = n_3 m_3$ and $\rho_4 = n_4 m_4$ are the partial densities of $He^3$ and $He^4$ respectively. $\rho_{n,4}$ is the contribution of the $He^4$ to the density of the normal component, $\rho_n$, and $\rho_{n,3}$ is the corresponding contribution of the $He^3$, which may be very different from $\rho_3$ if the effective mass of a $He^3$ atom is different from its actual mass:

$$\rho_n = \rho_{n,3} + \rho_{n,4}. \tag{9.46}$$

We must not assume that the bulk velocity, $\mathbf{v}_3$, of the $He^3$ is the same as the velocity, $\mathbf{v}_{n,4}$, of the phonons and rotons which form the contribution of the $He^4$ to the normal component. Consider, for example, a heat current emerging from a hot surface forming one end of a box. Phonons and rotons are created at the surface and stream away from it, but this is not possible for the $He^3$ atoms. We shall see in the next section that the $He^3$ atoms eventually assume a state with $\mathbf{v}_3 = 0$, but with a variation in concentration near the surface.

The equations expressing conservation of mass for the two isotopes separately are

$$\frac{\partial \rho_3}{\partial t} + \operatorname{div} \rho_3 \mathbf{v}_3 = 0, \tag{9.47}$$

$$\frac{\partial \rho_4}{\partial t} + \operatorname{div} [(\rho_4 - \rho_s) \mathbf{v}_{n,4} + \rho_s \mathbf{v}_s] = 0. \tag{9.48}$$

If $s$ is the entropy *per gram*, the equation of conservation of entropy is

$$\frac{\partial}{\partial t}(\rho s_m) + \operatorname{div} (\rho s_m \mathbf{v}_{n,4}) = 0. \tag{9.49}$$

Here we have made the approximation of ignoring the relative velocity between the $He^3$ and the phonons and rotons, but this usually involves only a small correction, particularly for very dilute solutions. The equation of motion of the liquid as a whole is

$$\rho_{n,3}\frac{\partial \mathbf{v}_3}{\partial t}+\rho_{n,4}\frac{\partial \mathbf{v}_{n,4}}{\partial t}+\rho_s\frac{\partial \mathbf{v}_s}{\partial t}+\operatorname{grad} p=0. \tag{9.50}$$

By analogy with equation (4.11), the equation of motion of the superfluid component is taken to be

$$\frac{\partial \mathbf{v}_s}{\partial t}=-\operatorname{grad}\bar{\mu}_4, \tag{9.51}$$

where $\bar{\mu}_4$ is the partial potential of $He^4$ *per g*. In many situations, such as the heat conduction experiment discussed in the next section, it is important to consider the diffusion of the $He^3$ atoms. Since the mean free path determining this diffusion is probably terminated by a collision with a phonon or roton, we write down a diffusion equation for the velocity of the $He^3$ atoms relative to the normal component of the $He^4$:

$$\rho_3(\mathbf{v}_3-\mathbf{v}_{n,4})=-D\operatorname{grad}\rho_3-\rho_3 D_T\operatorname{grad} T-\rho_3 D_p\operatorname{grad} p. \tag{9.52}$$

For dilute solutions the terms involving $D_T$ and $D_p$ can usually be ignored in comparison with the term involving the concentration diffusion coefficient $D$.

For very dilute solutions we can use equation (9.41) in the modified form

$$\bar{\mu}_4=\bar{\mu}_{4,0}-\frac{RT\rho_3}{M_4\,\rho}. \tag{9.53}$$

The equation of motion of the superfluid component then becomes

$$\rho_s\frac{\partial \mathbf{v}_s}{\partial t}=-\frac{\rho_s}{\rho}\operatorname{grad} p+\rho_s s_{4,0}\operatorname{grad} T+\rho_s\operatorname{grad}\frac{RT}{M_4}\frac{\rho_3}{\rho}. \tag{9.54}$$

The last term is the osmotic pressure.

Khalatnikov (1956*c*) has presented a very detailed set of thermo-hydrodynamical equations applicable to high concentrations and incorporating second order terms and dissipative effects.

### 9.4.3. Heat conduction in $He^3$-$He^4$ mixtures

Unlike pure liquid $He^4$, the solutions have a finite heat conductivity, which is related to the heat flush effect. Imagine a solution to be contained in a flat cylindrical box with its axis in the $z$ direction. Let the end $z=0$ be at a temperature $T$ and the end $z=h$ be at a temperature $T+\Delta T$, with a heat current in the negative $z$ direction of magnitude

$$W=\rho s_{4,0} T \mathbf{v}_{n,4}. \tag{9.55}$$

Here we have assumed that the entropy involved in the counterflow of the two components of the $He^4$ is $s_{4,0}$, which is approximately true for very dilute solutions. The flow of the normal component sweeps the $He^3$ towards the cold end, $z=0$, but the increased concentration there produces a back diffusion of $He^3$ atoms and eventually a steady state is reached in which the $He^3$ has assumed a suitable concentration distribution and $\mathbf{v}_3=0$. The diffusion equation (9.52) therefore takes the form

$$\begin{aligned}\rho_3 \mathbf{v}_{n,4} &= D\frac{\partial \rho_3}{\partial z}\\ &= \rho_3 W/\rho s_{4,0} T.\end{aligned} \tag{9.56}$$

The terms in $D_T$ and $D_p$ have been assumed to be negligibly small. The solution of (9.56) gives an exponential distribution of the $He^3$ concentration in the direction of heat flow

$$\rho_3(z)=\rho_3(0)\exp\left[-\frac{Wz}{\rho s_{4,0} TD}\right]. \tag{9.57}$$

In the steady state the time derivatives of all the velocities are zero, and equation (9.50) therefore shows that $\operatorname{grad} p=0$. For very dilute solutions the equation of motion of the superfluid component is a simplified form of equation (9.54):

$$s_{4,0}\frac{\partial T}{\partial z}+\frac{R\rho_3}{M_4\rho}\frac{\partial T}{\partial z}+\frac{RT}{M_4\rho}\frac{\partial \rho_3}{\partial z}=0. \tag{9.58}$$

Ignoring the second term, which is small compared with the first term for small $\rho_3/\rho$, and substituting $\partial\rho_3/\partial z$ from (9.56)

$$s_{4,0}\frac{\partial T}{\partial z}=-\frac{R\rho_3 W}{M_4\rho^2 s_{4,0} D}. \tag{9.59}$$

The effective thermal conductivity $W/(\partial T/\partial z)$ is therefore

$$K_m = \frac{M_4(\rho s_{4,0})^2 D}{R\rho_3}. \tag{9.60}$$

This is inversely proportional to the concentration of $He^3$ and the temperature also varies exponentially in the vicinity of the cold end. However, by integrating the above equations we can readily show that the mean value of the thermal conductivity for the whole box is

$$\overline{K}_m = \frac{Wh}{\Delta T} = \frac{M_4(\rho s_{4,0})^2 D}{R\bar{\rho}_3}, \tag{9.61}$$

where $\bar{\rho}_3$ is the average density of $He^3$ over the whole box.

TABLE XI. *The mean free path and collision cross-section of a* $He^3$ *atom dissolved in* $He^4$

| Temp. (°K.) | Mean free path, $l_3$ (cm.) | Cross-section, $\sigma_{3r}$ (cm.$^2$) |
|---|---|---|
| 1·2 | $7{\cdot}0 \times 10^{-6}$ | $3{\cdot}4 \times 10^{-15}$ |
| 1·4 | $1{\cdot}3 \times 10^{-6}$ | $5{\cdot}5 \times 10^{-15}$ |
| 1·6 | $3{\cdot}5 \times 10^{-7}$ | $7{\cdot}1 \times 10^{-15}$ |
| 1·8 | $1{\cdot}3 \times 10^{-7}$ | $9{\cdot}1 \times 10^{-15}$ |
| 2·0 | $5{\cdot}6 \times 10^{-8}$ | $11{\cdot}7 \times 10^{-15}$ |

Beenakker, Taconis, Lynton, Dokoupil and van Soest (1952) have measured the average thermal conductivity for average concentrations in the range from 0·6 to $2{\cdot}8 \times 10^{-4}$. The thermal resistance was linearly proportional to the average concentration and so the computed diffusion coefficients were independent of concentration. The values shown in fig. 94 are taken from Beenakker and Taconis (1955) and include an experimental correction omitted in the 1952 paper.

Treating the $He^3$ as an ideal classical gas

$$D = \tfrac{1}{2}\bar{v}_t l_3 = \tfrac{1}{2} l_3 \sqrt{\frac{3kT}{\mu}}, \tag{9.62}$$

where $\mu$ is the effective mass of a $He^3$ atom, $\bar{v}_t$ is its mean thermal velocity and $l_3$ its mean free path. The values of $l_3$ derived from the experimental values of $D$ are shown in Table XI. The values of the effective mass $\mu$ used in these calculations were those deduced by King and Fairbank (1954) from their second sound measurements. Notice that the mean free path at 2·0° K. is very little larger than the average distance away of the nearest $He^4$ atom. If the mean

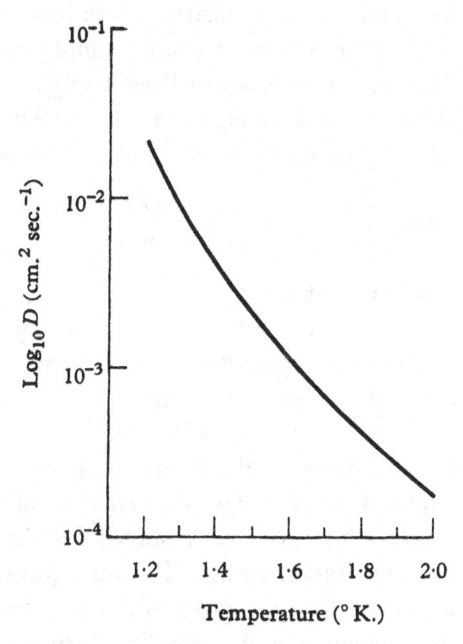


Fig. 94. The concentration diffusion coefficient $D$ (after Beenakker and Taconis, 1955).

free path is terminated by a collision with a roton, then the effective collision cross-section $\sigma_{3r}$ is

$$l_3 = \frac{1}{N_r \sigma_{3r}}, \tag{9.63}$$

where $N_r$ $(=\rho_{nr} 3kT/p_0^2)$ is the number of rotons per cm.$^3$. Values of $\sigma_{3r}$ are also shown in Table XI. Near 1° K. $\sigma_{3r}$ is similar in magnitude to the cross-section of a roton for collision with another roton (equation (4.25)). $\sigma_{3r}$ increases with temperature and so the concept of an ideal gas of hard spheres is a poor approximation. This is further emphasized by the fact that the effective diameter at 2° K. is larger than the mean free path.

A detailed theoretical discussion of diffusion and thermal conduction in mixtures will be found in a paper by Khalatnikov and Zharkov (1957).

## 9.5. Second sound in $He^3$-$He^4$ mixtures

The theory of second sound in mixtures was first treated by Pomeranchuk (1949) and has been cast into a more general form by Khalatnikov (1952*c*,*e*; 1956*b*,*c*) and Price (1953). If we look for periodic solutions of the thermohydrodynamical equations, as we did in § 5.1.1, we find that the velocity of second sound is

$$u_2^2 = \frac{\rho_s}{\rho_n}\left[\frac{T}{c_m}\left(s_{4,0}+\frac{kx_3}{m_4}\right)^2+\frac{kTx_3}{m_4}\right]. \tag{9.64}$$

Here $c_m$ is the specific heat per g. of the mixture, $s_{4,0}$ is the entropy per g. of pure liquid $He^4$, and $\rho_s$ and $\rho_n$ may be very different from the values these quantities assume in the pure liquid. In deriving this expression it is necessary to make certain assumptions in addition to the usual approximation of ignoring irreversible effects and second-order terms in the velocities. First, it has been assumed that the $He^3$ makes no contribution to the flow of the superfluid component. Secondly, it has been assumed that the solutions are dilute enough to obey Raoult's law, and equation (9.54) has been used. (Price (1953) has made the generalization to higher concentrations and has shown that the coupling between first and second sound may become quite large at high concentrations, so that a second sound wave may be accompanied by detectable pressure fluctuations.) Thirdly, the velocity of the $He^3$ has been assumed to be exactly the same as the velocity of the normal component of $He^4$ ($\mathbf{v}_3 = \mathbf{v}_{n,4}$). This means that the $He^3$ atoms, like the phonons and rotons, congregate in regions of high temperature, and the second sound wave is therefore accompanied by a 'concentration wave' in the dissolved $He^3$. Actually, the diffusion equation (9.52) implies a phase lag between $\mathbf{v}_3$ and $\mathbf{v}_{n,4}$, but this can be shown to be a small effect, producing a negligible change in the velocity, but giving rise to an extra attenuation which Khalatnikov (1952*c*, *e*) quotes as

$$\alpha_{2D} = \frac{\omega^2}{2\rho u_2^3}\frac{D\rho c_m}{s_m^2}\frac{kx_3}{m_4}\left(1+\frac{Ts_m}{c_m}\frac{D_T}{D}\right)^2. \tag{9.65}$$

Price (1953) has pointed out that the diffusion equation is very important near a boundary (as was obvious when we discussed heat conduction) and must therefore be taken into account in treatments of the generation, reception and reflexion of second sound.

The velocity of second sound has been measured from the λ-point down to 0·2° K. for concentrations up to 4·3 % (Lynton and Fairbank, 1950; King and Fairbank, 1954). Some of the results

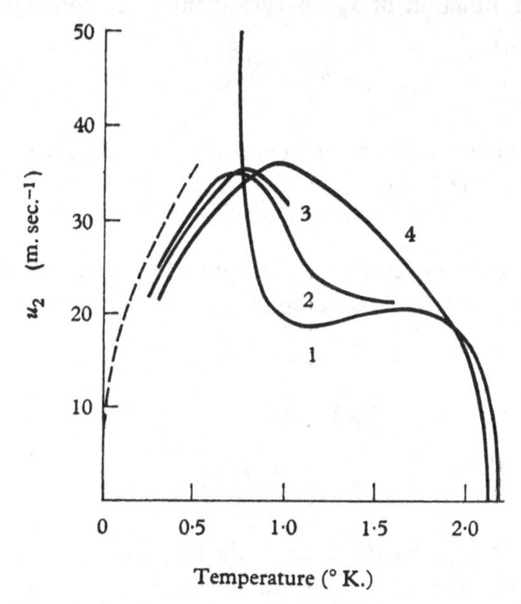


Fig. 95. The velocity of second sound in $He^3$-$He^4$ mixtures. $He^3$ concentration: 1, pure $He^4$; 2, 0·32 %; 3, 0·62 %; 4, 4·30 %. ----, $u_2^2 = kT/m_3$ (Mikura, 1954*a*). (After King and Fairbank, 1954.)

are shown in fig. 95 and can be readily understood in terms of equation (9.64). Above 1° K., $\rho_{n,4} \gg \rho_3$ and the dissolved $He^3$ produces very little change in $\rho_n$ and $\rho_s$. The main effect therefore comes from the expression inside the square bracket, which increases with $x_3$, so that $u_2$ also increases with $x_3$ (except in a narrow temperature range near the λ-point). At some temperature below 1° K., however, $\rho_n/\rho$ for pure liquid $He^4$ becomes very small, whereas $\rho_n/\rho$ for the mixture approaches a constant value because $\rho_n$ always includes the contribution from the $He^3$ atoms. The presence of the factor $\rho_s/\rho_n$ in front of the square bracket then results in a much smaller value of $u_2$ for the mixture than for the pure liquid. In fact, instead of $u_2$ rising rapidly towards $u_1/\sqrt{3}$ as it does in the pure liquid, it goes through a maximum and begins to decrease again.

All the quantities in equation (9.64) have been measured except $\rho_s/\rho_n$, and so we may use the experimental results to calculate $\rho_n$ as a function of concentration $x_3$. For concentrations up to 4%, $\rho_n$ is a linear function of $x_3$, in agreement with Pomeranchuk's equation (9.34)

$$\rho_n = \rho_{n,0} + \frac{\mu\rho}{m_4} x_3. \tag{9.66}$$

The effective mass $\mu$ is not independent of temperature but can be expressed in the form

$$\mu = m_3(T + 1{\cdot}8). \tag{9.67}$$

This is sufficient to exclude an alternative hypothesis of Landau and Pomeranchuk (1948) based on the equation

$$\epsilon_3 = E_0 + (p - p_0)^2/2\mu$$

for the $He^3$ atoms, since this leads to

$$\rho_n = \rho_{n,0} + \frac{\rho}{m_4} \frac{p_0^2}{3kT} x_3. \tag{9.68}$$

Equations (9.64) and (9.67) are able to explain the measured values of $u_2$ over the whole temperature range from 0·2 to 1·8° K. At temperatures well below 1° K., we may ignore $\rho_{n,0}$, $s_{4,0}$ and $c_{4,0}$ and take

$$\rho_n = \frac{\mu\rho}{m_4} x_3, \tag{9.69}$$

$$c_m = \frac{3k}{2m_4} x_3. \tag{9.70}$$

Equation (9.64) then reduces to

$$u_2^2 = \frac{5}{3} \frac{kT}{\mu} \quad (T \ll 1^\circ \text{K.}). \tag{9.71}$$

Below 0·6° K. the experimental results are in agreement with this inasmuch as $u_2$ is almost independent of concentration and varies approximately as $T^{\frac{1}{2}}$, small departures from this $T^{\frac{1}{2}}$ law being associated with the variation of $\mu$ with $T$. Since $u_2$ is independent of $x_3$, it is relevant to inquire what happens when $x_3 \to 0$. Working with a 0·017 % solution, King and Fairbank found that $u_2$ followed equation (9.71) approximately down to about 0·45° K. and then rose rapidly towards $u_1/\sqrt{3}$. They ascribed this rapid rise to a heat flush effect which swept the $He^3$ out of the second sound beam below 0·45° K. Another complication at low temperatures is that a second sound wave involves a periodic variation in concentration, and the required fluctuation in concentration for a finite temperature swing may become larger than the mean concentration. Equation (9.71) suggests that $u_2 \to 0$ as $T \to 0°$ K., but Pomeranchuk has pointed out that the $He^3$ probably becomes degenerate at a sufficiently low temperature, and $u_2$ then tends towards a small finite value:

$$u_2^2 = \frac{5}{9} \frac{\hbar^2}{\mu^2} \left( \frac{3\pi^2 \rho x_3}{m_4} \right)^{\frac{2}{3}}. \tag{9.72}$$

We must also remember that when a mixture of fixed finite concentration is cooled down it will eventually reach the phase separation curve (fig. 92) and then separate out into two phases of different concentrations (Fairbank and Elliott, 1957).

Koide and Usui (1951 *a*, *b*) have shown that the second sound measurements above 1° K. are consistent with the Taconis hypothesis only if one assumes that the $He^3$ remains at rest ($v_3 = 0$), which seems unreasonable. Mikura (1954*a*) has pointed out that such an assumption leads to $u_2 \to u_1/\sqrt{3}$ at 0° K. in strong disagreement with experiment. On the other hand, Mikura has shown that the Heer and Daunt theory, combined with the assumption $v_3 = v_{n,4}$, gives a very satisfactory result for $u_2$ near 0° K., namely

$$u_2^2 \to kT/m_3 \quad \text{as} \quad T \to 0° \text{ K.}, \tag{9.73}$$

which is very close to the experimental values.

Second sound pulses in pure liquid $He^4$ below 0·5° K. are considerably elongated and distorted. This is because the normal component consists entirely of phonons with very long mean free paths (§ 5.3.2). In a $He^3$-$He^4$ mixture, however, the normal component consists almost entirely of $He^3$ atoms, and, for the concentrations used in the second sound experiments, it is easily shown that the mean free path for collision of the $He^3$ atoms with one another is quite small. The second sound pulses propagated through the mixtures were therefore found to be well-defined and undistorted down to the lowest temperature reached ($\sim$0·2° K.).

The theory of shock waves of second sound in $He^3$-$He^4$ mixtures has been discussed by Khalatnikov (1952*d*).

Weinstock and Pellam (1953) have investigated the behaviour of a thermal Rayleigh disk (§ 5.3.5) in a 4 % mixture. They obtained values of the wave velocity in approximate agreement with the results of King and Fairbank (1953). At a constant heat current density, the addition of $He^3$ reduced the torque on the disk, which at 0·9° K. was only a few tenths per cent of its value in the pure liquid. The torque can readily be shown to be

$$\tau' = \frac{4}{3}\frac{a^3}{\rho}\frac{\rho_n}{\rho_s T^2}\frac{\overline{W}^2}{[(1-x_3)s_4+x_3s_3]} \qquad (9.74)$$

and the increase in the entropy term in the denominator outweighs the increase in $\rho_n$ in the numerator.

*First* sound has been investigated in an 11 % mixture at 14 Mc./s. (Flicker and Atkins, 1957) and in a 2·7 % mixture at 2 Mc./s. (Lim, Hallett and Guptill, 1957). The $He^3$ lowers the velocity and moves the $\lambda$-point anomaly (fig. 41) to the $\lambda$-temperature of the solution. Preliminary attenuation measurements indicate the absence of the rapid rise in attenuation near 1° K. found in pure $He^4$ (fig. 43). The $He^3$ atoms are probably responsible for some process which shortens the relaxation time for creation and annihilation of rotons and phonons (§ 5.2.3).

## 9.6. Normal density, $\rho_n$, and viscosity, $\eta_n$

Pellam (1955*a*) has measured $\rho_n/\rho$ directly in an Andronikashvili type experiment (see fig. 2 and § 1.3). The measurements extended from the $\lambda$-point down to 0·95° K. and the concentration used was 3·3 % by mass, corresponding to a mole fraction of 4·4 %. A correction was made for the finite depth of penetration of the viscous wave into the liquid. At 1·9° K. this correction was 4·3 %, which is comparable with the difference between $\rho_n$ and $\rho_{n,0}$, but below 1·5° K. the correction was negligible. Throughout the whole temperature range $\rho_n/\rho$ was greater than for the pure liquid and near 1° K. levelled off to a constant value of approximately 9·5 %, as compared with 0·7 % for the pure liquid. This demonstrates rather nicely that, at 1° K. and below, the density of the normal component arises principally from the constant density of $He^3$ atoms, with the phonons and rotons making a negligible contribution. A value of 9·5 % at 1° K. corresponds to an effective mass of $2{\cdot}7m_3$ which compares very favourably with $2{\cdot}8m_3$ from second sound measurements (King and Fairbank, 1954).

Extensive measurements of $\rho_n$ by the Andronikashvili method have been made by Berezniak and Eselson (1956) and by Dash and Taylor (1957*c*). Since $\rho_n$ always varies linearly with concentration for very dilute solutions, it is always possible to define a quantity $\mu$ by the Pomeranchuk equation

$$\rho_n = \rho_{n,0} + n_3\mu. \tag{9.75}$$

These oscillating disk experiments give $\mu/m_3 = 2{\cdot}7 \pm 0{\cdot}3$ between 1·3° K. and the $\lambda$-point. If $\mu$ increases at all with increasing temperature, the rate of increase is certainly less rapid than in equation (9.67), which is based on the second sound results. However, a more realistic definition of the effective mass of a $He^3$ atom in solution would be

$$\rho_n = \rho_{n,4} + n_3\mu^*. \tag{9.76}$$

Since $\rho_{n,4}$, the contribution of the $He^4$ to $\rho_n$, is undoubtedly different from the value $\rho_{n,0}$ for pure liquid $He^4$, the two effective masses may be very different, except at low temperatures where $\rho_{n,4} \ll n_3\mu^*$. Since $\rho_{n,4}$ is not known, $\mu^*$ cannot be deduced directly at higher temperatures.

Feynman (1954) has put forward an atomistic explanation of the large effective mass of the $He^3$ atom in solution. It is well known in classical hydrodynamics that a sphere of mass $m$ moving with velocity $v$ through a non-viscous liquid has an effective mass $m + \frac{1}{2}m'$, where $m'$ is the mass of fluid displaced by the sphere. This is because, in addition to the kinetic energy $\frac{1}{2}mv^2$ of the sphere, there is a kinetic energy $\frac{1}{4}m'v^2$ associated with the liquid flowing round the sphere. Feynman suggests that the $He^4$ atoms are similarly pushed out of the way by the $He^3$ atom and flow round it. He then uses the variational principle to find an approximate wave function to describe this flow and to calculate the energy associated with it. The effective mass of the $He^3$ atom is calculated to be approximately $1{\cdot}9m_3$ which agrees well with the extrapolation of the experimental results to 0° K. Dash and Taylor (1957*c*) have suggested that this argument should be applied to the motion of the $He^3$ atom through the superfluid component only, and that the effective mass should therefore be written

$$\mu^* = m_3 + b\rho_s. \tag{9.77}$$

This equation is consistent with their particular method of analysing their oscillating disk results.

From the damping of the oscillations of the disk system, Dash and Taylor deduced the viscosity of the mixture. The relevant theory has been given by Zharkov (1957) for temperatures below 1·6° K., where the phonons and rotons can be treated as an ideal gas. The treatment is analogous to that given in § 4.5.2 for the viscosity of pure liquid $He^4$. The viscosity of a mixture can be written as the sum of three parts associated with the rotons, the phonons and the $He^3$ atoms respectively:

$$\begin{aligned} \eta_n &= \eta_r + \eta_{ph} + \eta_3 \\ &= \tfrac{1}{10}\pi\rho_r\bar{v}_r l_r + \alpha\rho_{ph} c l_{ph} + \beta\rho_3\bar{v}_3 l_3. \end{aligned} \tag{9.78}$$

Rotons and $He^3$ atoms are 'heavy' particles of comparable character, whereas a phonon is a 'light' particle and is not very effective in terminating the free path of another particle. The roton mean free path, $l_r$, is now determined by collisions with both rotons and $He^3$ atoms, and is reduced below its value for pure $He^4$. Similarly, the $He^3$ mean free path, $l_3$, is determined by a collision with either another $He^3$ atom or a roton. The mean free path of a phonon, $l_{ph}$, is determined by a collision of the phonon with a roton or a $He^3$ atom, and at concentrations greater than $10^{-6}$ phonon-phonon collisions are always negligible. In pure $He^4$ there is a rapid increase in $\eta_{ph}$ near 1° K. where the rotons become scarce and $l_{ph}$ becomes large. In a mixture the fixed number of $He^3$ atoms prevents $l_{ph}$ from increasing so rapidly and the increase in $\eta_{ph}$ is less pronounced. The experimental results are complicated, but seem to be in general qualitative agreement with the above ideas. For small concentrations the effect of the $He^3$ atoms is to reduce the viscosity by reducing $l_r$ and $l_{ph}$. At higher concentrations the viscosity increases again as the term $\eta_3$ becomes important.

Wansink and Taconis (1957*b*) have deduced $\rho_n$ and $\eta_n$ from observations on the flow of mixtures through a slit about $3 \times 10^{-5}$ cm. wide. As we have explained in § 6.3.2, the effective viscosity in a narrow channel may be less than for the bulk liquid, because of mean free path effects. The viscosities measured by Wansink and Taconis do appear to be abnormally low at 1·2° K.

## 9.7. Methods of concentrating $He^3$-$He^4$ mixtures

Now that pure $He^3$ can be produced in large quantities in a nuclear reactor, less importance is attached to the enrichment of the $He^3$ content of naturally occurring helium. We shall therefore give only a brief summary of the various attempts at enrichment.

The starting concentrations are small: well helium contains about 1·4 parts of $He^3$ in $10^7$ parts of $He^4$, whereas atmospheric helium contains 12 parts in $10^7$, but is less readily available. Using the standard technique of the thermal diffusion column, the concentration can be raised to about 1 %, with a production rate of a few standard ml. of the mixture per day (Jones and Furry, 1946; McInteer, Aldrich and Nier, 1948; Rollin and Hatton, 1948; Andrew and Smythe, 1948; Schuette, Zucker and Watson, 1950). A type of fractional distillation may also be effected by vaporizing the liquid under the right conditions and collecting the vapour, which, because of the large distribution coefficient $C_v/C_l$, is rich in $He^3$ (Rollin and Hatton, 1948; Eselson and Lasarew, 1950*c*). This method has produced concentrations up to 1·5 %.

The most promising methods of separating the two isotopes depend upon the fact that the $He^4$ is superfluid and the $He^3$ is not. Successful use has been made of the heat flush effect (§ 9.4.1), in which a heat current is used to flush the $He^3$ towards the surface of the liquid, where it is removed by vaporization (Lane, Fairbank, Aldrich and Nier, 1948; Eselson and Lasarew, 1950*a*, *b*; Taconis, 1950). Soller, Fairbank and Crowell (1953) modified a helium liquefier for this purpose and were able to produce about 70 standard gaseous ml. per hour of a $\frac{1}{2}$ % mixture, starting with well helium. Reynolds, Fairbank, Lane, McInteer and Nier (1949) obtained a concentration of $10^{-4}$ by thermal diffusion and then raised this to 4% by the heat flush method, but their rate of processing was not high.

Starting with an initial concentration of a few per cent, much higher concentrations can be obtained by allowing the $He^4$ to drain away through a superleak (Abraham, Weinstock and Osborne, 1949). This method has been developed by Atkins and Lovejoy (1954) to raise the concentration from 2 to 95% in a single operation, with a production rate of about 2 standard gaseous ml. of the 95% mixture per hour. Actually, the flow through the superleak ceased when the liquid concentration reached about 24%, perhaps because the solution was then at its $\lambda$-temperature. However, the vapour in equilibrium with a 24% solution has a concentration of 95% and the process was therefore designed in such a way that, when the flow ceased, most of the helium at the entrance to the superleak was in the vapour phase.

The final step to concentrations near 100% is probably best achieved by fractional distillation. Fairbank, Ard, Dehmelt, Gordy and Williams (1953) have obtained an enrichment to better than 99% by distilling a mixture initially containing a few per cent of $He^3$. Peshkov (1956) has described a series of processes involving heat flush, superflow and fractional distillation plus a rectifying column. He was able to enrich well helium to a final concentration of 99·95% $He^3$. Kuznetsov (1957) has made an experimental and theoretical study of the operation of such a rectifying column.

## 9.8. Ions in liquid helium

Another type of impurity that can be studied in liquid helium is a charged ion. Careri, Reuss, Scaramuzzi and Thomson (1957) have shown that ions, like $He^3$ atoms, move with the normal component in a heat flush experiment, and have also made estimates of the ionic mobilities. A more direct measurement of the ionic mobilities has been made by Meyer and Reif (1958). Above the $\lambda$-point the mobility is approximately constant, but below the $\lambda$-point it increases rapidly with decreasing temperature, varying approximately as $1/\rho_n$. This strongly suggests that the mean free path of an ion is determined by collisions with rotons. The mobility of a positive ion is slightly larger than that of a negative ion, and the order of magnitude of both indicates an effective mass several times the mass of the $He^4$ atom. The nature of the ions is not known, but they may be electrons, $He^+$ or $He^-$, or even complexes in which several helium atoms are bound together. They obviously provide us with a very promising technique for investigating many of the properties of liquid helium.

# REFERENCES


Abraham, B. M., Osborne, D. W. and Weinstock, B. (1950). *Phys. Rev.* **80**, 366.

Abraham, B. M., Osborne, D. W. and Weinstock, B. (1953). *Science*, **117**, 121.

Abraham, B. M., Osborne, D. W. and Weinstock, B. (1955). *Phys. Rev.* **98**, 551.

Abraham, B. M., Weinstock, B. and Osborne, D. W. (1949). *Phys. Rev.* **76**, 864.

Abrikosov, A. A. and Khalatnikov, I. M. (1957). *J. exp. theor. Phys.* USSR [Russian], **32**, 1083 (translated in Soviet Physics, *JETP*, **5**, 887).

Allcock, G. R. and Kuper, C. G. (1955). *Proc. Roy. Soc.* A, **231**, 226.

Allen, J. F. and Jones, H. (1938). *Nature, Lond.*, **141**, 243.

Allen, J. F. and Misener, A. D. (1938*a*). *Nature, Lond.*, **141**, 75.

Allen, J. F. and Misener, A. D. (1938*b*). *Proc. Camb. Phil. Soc.* **34**, 299.

Allen, J. F. and Misener, A. D. (1939). *Proc. Roy. Soc.* A, **172**, 467.

Allen, J. F. and Reekie, J. (1939). *Proc. Camb. Phil. Soc.* **35**, 114.

Ambler, E. and Kurti, N. (1952). *Phil. Mag.* **43**, 260.

Andrew, A. and Smythe, W. R. (1948). *Phys. Rev.* **74**, 496.

Andronikashvili, E. L. (1946). *J. Phys., Moscow*, **10**, 201.

Andronikashvili, E. L. (1948*a*). *J. exp. theor. Phys.* USSR [Russian], **18**, 424.

Andronikashvili, E. L. (1948*b*). *J. exp. theor. Phys.* USSR [Russian], **18**, 429.

Andronikashvili, E. L. (1952). *J. exp. theor. Phys.* USSR [Russian], **22**, 62.

Andronikashvili, E. L. and Kaverkin, I. P. (1955). *J. exp. theor. Phys.* USSR [Russian], **28**, 126.

Andronikashvili, E. L. and Mirskaia, G. G. (1955). *J. exp. theor. Phys.* USSR [Russian], **29**, 490.

Arkhipov, R. G. (1954). *Dokl. Akad. Nauk, SSSR* [Russian], **98**, 447.

Atkins, K. R. (1948). *Nature, Lond.*, **161**, 925.

Atkins, K. R. (1950*a*). *Proc. Roy. Soc.* A, **203**, 119.

Atkins, K. R. (1950*b*). *Proc. Roy. Soc.* A, **203**, 240.

Atkins, K. R. (1951). *Proc. Phys. Soc.* A, **64**, 833.

Atkins, K. R. (1953*a*). *Phys. Rev.* **89**, 526.

Atkins, K. R. (1953*b*). *Canad. J. Phys.* **31**, 1165.

Atkins, K. R. (1954). *Canad. J. Phys.* **32**, 347.

Atkins, K. R. (1955*a*). *Phys. Rev.* **98**, 319.

Atkins, K. R. (1955*b*). *Conférence de Physique des Basses Temperatures* (Paris), p. 100.

Atkins, K. R. (1957). *Phys. Rev.* **108**, 911.

Atkins, K. R. and Chase, C. E. (1951*a*). *Proc. Phys. Soc.* A, **64**, 826.

Atkins, K. R. and Chase, C. E. (1951*b*). *Proceedings of the International Conference on Low Temperature Physics* (Oxford, England), p. 60.

Atkins, K. R. and Edwards, M. H. (1955). *Phys. Rev.* **97**, 1429.
Atkins, K. R. and Hart, K. H. (1954). *Canad. J. Phys.* **32**, 381.
Atkins, K. R. and Lovejoy, D. R. (1954). *Canad. J. Phys.* **32**, 702.
Atkins, K. R. and Osborne, D. V. (1950). *Phil. Mag.* **41**, 1078.
Atkins, K. R. and Seki, H. (1956). *Phys. Rev.* **102**, 582.
Atkins, K. R. and Seki, H. (1957). *Proceedings of the Fifth International Conference on Low Temperature Physics and Chemistry* (Madison, U.S.A.), p. 4.
Atkins, K. R. and Stasior, R. A. (1953). *Canad. J. Phys.* **31**, 1156.

Band, W. (1949). *Phys. Rev.* **76**, 441.
Beenakker, J. J. M. and Taconis, K. W. (1955). *Progress in Low Temperature Physics*, vol. I (edited by C. J. Gorter), chapter VI. Amsterdam: North Holland Publishing Co.
Beenakker, J. J. M., Taconis, K. W., Lynton, E. A., Dokoupil, Z. and van Soest, G. (1952). *Physica*, **18**, 433.
Benson, C. B. and Hallett, A. C. H. (1956). *Canad. J. Phys.* **34**, 668.
Berezniak, N. G. and Eselson, B. N. (1956). *J. exp. theor. Phys. USSR* [Russian], **31**, 902 (translated in Soviet Physics, *JETP*, **4**, 766).
van den Berg, G. J. and de Haas, W. J. (1949). *Rev. Mod. Phys.* **21**, 524.
van den Berg, G. J., van Itterbeck, A., van Aardenne, G. M. V. and Herfkens, J. H. J. (1955). *Physica*, **21**, 860.
Berman, R. and Poulter, J. (1952). *Phil. Mag.* **42**, 1047.
Bijl, A. (1940). *Physica*, **7**, 869.
Bijl, A., de Boer, A. and Michels, J. (1941). *Physica*, **8**, 655.
Blatt, J. M., Butler, S. T. and Schafroth, M. R. (1955). *Phys. Rev.* **100**, 476, 481, 495.
Bleaney, B. and Simon, F. (1939). *Trans. Faraday Soc.* **35**, 1205.
Bloembergen, N., Purcell, E. M. and Pound, R. V. (1948). *Phys. Rev.* **73**, 679.
de Boer, J. (1948). *Physica*, **14**, 139.
de Boer, J. and Cohen, E. G. D. (1955). *Physica*, **21**, 79.
de Boer, J. and Gorter, C. J. (1950). *Physica*, **16**, 225, 667.
de Boer, J. and Lunbeck, R. J. (1948). *Physica*, **14**, 318, 510, 520.
Bogoliubov, N. N. (1947). *J. Phys., Moscow*, **11**, 23.
Bogoliubov, N. N. and Zubarev, D. N. (1955). *J. exp. theor. Phys. USSR* [Russian], **28**, 129 (translated in Soviet Physics, *JETP*, **1**, 83).
Bots, G. J. C. and Gorter, C. J. (1956). *Physica*, **22**, 503.
Bowers, R. (1953*a*). *Phil. Mag.* **44**, 1309.
Bowers, R. (1953*b*). *Phil. Mag.* **44**, 465, 487.
Bowers, R., Brewer, D. F. and Mendelssohn, K. (1951). *Phil. Mag.* **42**, 1445.
Bowers, R. and Mendelssohn, K. (1949). *Proc. Phys. Soc.* A, **62**, 394.
Bowers, R. and Mendelssohn, K. (1950*a*). *Proc. Roy. Soc.* A, **204**, 366.
Bowers, R. and Mendelssohn, K. (1950*b*). *Proc. Phys. Soc.* A, **63**, 1318.
Bowers, R. and Mendelssohn, K. (1951). *Nature, Lond.*, **167**, 111.
Bowers, R. and Mendelssohn, K. (1952*a*). *Proc. Phys. Soc.* A, **65**, 551.

BOWERS, R. and MENDELSSOHN, K. (1952*b*). *Proc. Roy. Soc.* A, **213**, 158.
BOWERS, R. and WHITE, G. K. (1951). *Proc. Phys. Soc.* A, **64**, 558.
BREWER, D. F. and MENDELSSOHN, K. (1953). *Phil. Mag.* **44**, 340, 559.
BREWER, D. F., SREEDHAR, A. K., KRAMERS, H. C. and DAUNT, J. G. (1958). *Phys. Rev.* **110**, 282.
BROESE VAN GROENOU, A., POLL, J. D., DELSING, A. M. G. and GORTER, C. J. (1956). *Physica*, **22**, 905.
BRUECKNER, K. A. and GAMMEL, J. L. (1957). *Symposium on Liquid and Solid* $He^3$ (Ohio State University), p. 186.
BRUECKNER, K. A. and GAMMEL, J. L. (1958). *Phys. Rev.* **109**, 1040.
BRUECKNER, K. A. and SAWADA, K. (1957). *Phys. Rev.* **106**, 1117, 1128.
BUCKINGHAM, M. J. (1955). *Phys. Rev.* **98**, 1855.
BUCKINGHAM, M. J. (1957). *Symposium on Liquid and Solid* $He^3$ (Ohio State University), p. 50.
BUCKINGHAM, R. A. and TEMPERLEY, H. N. V. (1950). *Phys. Rev.* **78**, 482.
BUCKTHOUGHT, K. (1953*a*). Ph.D. Thesis, University of Toronto.
BUCKTHOUGHT, K. (1953*b*). *Canad. J. Phys.* **31**, 932.
BURGE, E. J. and JACKSON, L. C. (1951). *Proc. Roy. Soc.* A, **205**, 270.

CARERI, G., REUSS, J., SCARAMUZZI, F. and THOMSON, J. O. (1957). *Proceedings of the Fifth International Conference on Low Temperature Physics and Chemistry* (Madison, U.S.A.), p. 79.
CHALLIS, L. J. and WILKS, J. (1957). *Symposium on Liquid and Solid* $He^3$ (Ohio State University), p. 38.
CHAMPENEY, D. C. (1957). *Proceedings of the Fifth International Conference on Low Temperature Physics and Chemistry* (Madison, U.S.A.), p. 3.
CHANDRASEKHAR, B. S. (1952). *Phys. Rev.* **86**, 144.
CHANDRASEKHAR, S. and DONNELLY, R. J. (1957). *Proc. Roy. Soc.* A, **241**, 9.
CHANDRASEKHAR, B. S. and MENDELSSOHN, K. (1952). *Proc. Phys. Soc.* A, **65**, 226.
CHANDRASEKHAR, B. S. and MENDELSSOHN, K. (1953). *Proc. Roy. Soc.* A, **218**, 18.
CHANDRASEKHAR, B. S. and MENDELSSOHN, K. (1955). *Proc. Phys. Soc.* A, **68**, 857.
CHASE, C. E. (1953). *Proc. Roy. Soc.* A, **220**, 116.
CHASE, C. E. (1956). *Amer. J. Phys.* **24**, 136.
CHASE, C. E. (1957). *Proceedings of the Fifth International Conference on Low Temperature Physics and Chemistry* (Madison, U.S.A.), p. 92.
CHASE, C. E. and HERLIN, M. A. (1955). *Phys. Rev.* **97**, 1447.
CHESTER, G. V. (1954). *Phys. Rev.* **93**, 1412.
CHESTER, G. V. (1955). *Phys. Rev.* **100**, 455.
COHEN, M. and FEYNMAN, R. P. (1957). *Phys. Rev.* **107**, 13.

DANA, L. I. and KAMERLINGH ONNES, J. (1926). *Proc. Roy. Acad. Amsterdam*, **29**, 1061.
DASH, J. G. and BOORSE, H. A. (1951). *Phys. Rev.* **82**, 851.
DASH, J. G. and TAYLOR, R. D. (1955). *Phys. Rev.* **99**, 598.

DASH, J. G. and TAYLOR, R. D. (1957*a*). *Phys. Rev.* **105**, 7.
DASH, J. G. and TAYLOR, R. D. (1957*b*). *Phys. Rev.* **106**, 398.
DASH, J. G. and TAYLOR, R. D. (1957*c*). *Phys. Rev.* **107**, 1228.
DAUNT, J. G. (1952). *Phil. Mag.* Suppl. **1**, 209.
DAUNT, J. G. and HEER, C. V. (1950). *Phys. Rev.* **79**, 46.
DAUNT, J. G. and HEER, C. V. (1952). *Phys. Rev.* **86**, 205.
DAUNT, J. G. and MENDELSSOHN, K. (1939*a*). *Nature, Lond.*, **143**, 719.
DAUNT, J. G. and MENDELSSOHN, K. (1939*b*). *Proc. Roy. Soc.* A, **170**, 423, 439.
DAUNT, J. G. and MENDELSSOHN, K. (1946). *Nature, Lond.*, **157**, 829.
DAUNT, J. G., PROBST, R. E. and JOHNSTON, H. L. (1948). *Phys. Rev.* **73**, 638.
DAUNT, J. G., PROBST, R. E., JOHNSTON, H. L., ALDRICH, L. T. and NIER, A. O. (1947*a*). *Phys. Rev.* **72**, 502.
DAUNT, J. G., PROBST, R. E., JOHNSTON, H. L., ALDRICH, L. T. and NIER, A. O. (1947*b*). *J. Chem. Phys.* **15**, 759.
DAUNT, J. G., PROBST, R. E. and SMITH, S. R. (1948). *Phys. Rev.* **74**, 495.
DAUNT, J. G. and TSENG, T. P. (1955). *Conférence de Physique des Basses Temperatures* (Paris), p. 22.
DESSLER, A. J. and FAIRBANK, W. M. (1956). *Phys. Rev.* **104**, 6.
VAN DIJK, H. and DURIEUX, M. (1957). *Progress in Low Temperature Physics*, vol. II (edited by C. J. Gorter), chapter XIV. Amsterdam: North Holland Publishing Co.
VAN DIJK, H. and SHOENBERG, D. (1949). *Nature, Lond.*, **164**, 151.
DINGLE, R. B. (1948). *Proc. Phys. Soc.* A, **61**, 9.
DINGLE, R. B. (1950). *Proc. Phys. Soc.* A, **63**, 638.
DINGLE, R. B. (1952*a*). *Phil. Mag.* Suppl. **1**, 111.
DINGLE, R. B. (1952*b*). *Physica*, **18**, 841.
DOKOUPIL, Z., VAN SOEST, G., WANSINK, D. H. N. and KAPADNIS, D. G. (1954). *Physica*, **20**, 1181.
DONNELLY, R. J., CHESTER, G. V., WALMSLEY, R. H. and LANE, C. T. (1956). *Phys. Rev.* **102**, 3.
DONNELLY, R. J. and PENROSE, O. (1956). *Phys. Rev.* **103**, 1137.
DRANSFELD, K. and WILKS, J. (1957). *Proceedings of the Fifth International Conference on Low Temperature Physics and Chemistry* (Madison, U.S.A.), p. 31.
DUGDALE, J. S. and SIMON, F. E. (1953). *Proc. Roy. Soc.* A, **218**, 291.
DUYCKAERTS, G. (1943). *Mém. Soc. Sci. Liége* [1], **2**, 349.
DYBA, R. V., LANE, C. T. and BLAKEWOOD, C. H. (1954). *Phys. Rev.* **95**, 1365.

EDWARDS, M. H. (1956). *Canad. J. Phys.* **34**, 898.
EGELSTAFF, P. A. and LONDON, H. (1953). *Proceedings of the International Conference on Low Temperature Physics* (Houston), p. 35.
EGELSTAFF, P. A. and LONDON, H. (1957). *Proc. Roy. Soc.* A, **242**, 374.
EHRENFEST, P. (1933). *Proc. Acad. Sci. Amst.* **36**, 153.
EISELE, K. M. and HALLETT, A. C. H. (1957). *Canad. J. Phys.* **36**, 25.
ELLIOTT, S. D. and FAIRBANK, H. A. (1958). *Bulletin of the American Physical Society*, **3**, 225.

ESELSON, B. N. and BEREZNIAK, N. G. (1956). *J. exp. theor. Phys.* USSR [Russian], **30**, 628 (translated in Soviet Physics, *JETP*, **3**, 568).

ESELSON, B. N., BEREZNIAK, N. G. and KAGANOV, M. I. (1956). *Dokl. Akad. Nauk, SSSR* [Russian], **111**, 568 (translated in Soviet Physics, *Doklady*, **1**, 683).

ESELSON, B. N., KAGANOV, M. I. and LIFSHITZ, I. M. (1957). *J. exp. theor. Phys.* USSR [Russian], **33**, 938.

ESELSON, B. N. and LASAREW, B. G. (1950*a*). *J. exp. theor. Phys.* USSR [Russian], **20**, 748.

ESELSON, B. N. and LASAREW, B. G. (1950*b*). *Dokl. Akad. Nauk, SSSR* [Russian], **72**, 265.

ESELSON, B. N. and LASAREW, B. G. (1950*c*). *J. exp. theor. Phys.* USSR [Russian], **20**, 742.

ESELSON, B. N. and LASAREW, B. G. (1952). *J. exp. theor. Phys.* USSR [Russian], **23**, 552.

ESELSON, B. N., LASAREW, B. G. and ALEKSEEVSKI, N. E. (1950). *J. exp. theor. Phys.* USSR [Russian], **20**, 1055.

ESELSON, B. N., SVETZ, A. D. and BABLIDZE, R. A. (1958). *J. exp. theor. Phys.* USSR [Russian], **34**, 233.

FAIRBANK, W. M., ARD, W. B., DEHMELT, H. G., GORDY, W. and WILLIAMS, S. R. (1953). *Phys. Rev.* **92**, 208.

FAIRBANK, W. M., ARD, W. B. and WALTERS, G. K. (1954). *Phys. Rev.* **95**, 566.

FAIRBANK, H. A. and ELLIOTT, S. D. (1957). *Symposium on Liquid and Solid* $He^3$ (Ohio State University), p. 156.

FAIRBANK, H. A., LANE, C. T., ALDRICH, L. T. and NIER, A. O. (1947). *Phys. Rev.* **71**, 911.

FAIRBANK, H. A., LANE, C. T., ALDRICH, L. T. and NIER, A. O. (1948). *Phys. Rev.* **73**, 729.

FAIRBANK, H. A., LANE, C. T., ALDRICH, L. T. and NIER, A. O. (1949). *Phys. Rev.* **75**, 46.

FAIRBANK, H. A. and LEE, D. M. (1957). *Symposium on Liquid and Solid* $He^3$ (Ohio State University), pp. 26, 32.

FAIRBANK, H. A. and WILKS, J. (1955). *Proc. Roy. Soc.* A, **231**, 545.

FEYNMAN, R. P. (1948). *Rev. Mod. Phys.* **20**, 367.

FEYNMAN, R. P. (1953*a*). *Phys. Rev.* **91**, 1291.

FEYNMAN, R. P. (1953*b*). *Phys. Rev.* **91**, 1301.

FEYNMAN, R. P. (1954). *Phys. Rev.* **94**, 262.

FEYNMAN, R. P. (1955). *Progress in Low Temperature Physics*, vol. I (edited by C. J. Gorter), chapter II. Amsterdam: North Holland Publishing Co.

FEYNMAN, R. P. and COHEN, M. (1956). *Phys. Rev.* **102**, 1189.

FINDLAY, J. C., PITT, A., GRAYSON-SMITH, H. and WILHELM, J. O. (1938). *Phys. Rev.* **54**, 506.

FINDLAY, J. C., PITT, A., GRAYSON-SMITH, H. and WILHELM, J. O. (1939). *Phys. Rev.* **56**, 122.

FLICKER, H. and ATKINS, K. R. (1957). *Symposium on Liquid and Solid* $He^3$ (Ohio State University), pp. 11, 164.

FRANCHETTI, S. (1955). *Il Nuovo Cimento*, **10**, 1127.
FRASER, A. R. (1951). *Phil. Mag.* **42**, 156, 165.
FREDERIKSE, H. P. R. (1949). *Physica*, **15**, 860.
FREDERIKSE, H. P. R. and GORTER, C. J. (1950). *Physica*, **16**, 403.
FRENKEL, J. (1940*a*). *J. Phys., Moscow*, **2**, 345.
FRENKEL, J. (1940*b*). *J. Phys., Moscow*, **3**, 355.
FRENKEL, J. (1946). *The Kinetic Theory of Liquids*, p. 308. Oxford University Press.
FRENKEL, J. (1948). *J. exp. theor. Phys.* USSR [Russian], **18**, 659.
FRÖHLICH, H. (1949). *Theory of Dielectrics*, p. 169. Oxford University Press.

GARWIN, R. L. and REICH, H. A. (1958). *Bulletin of the American Physical Society*, **3**, 133.
GINSBURG, V. L. (1949). *Dokl. Akad. Nauk, SSSR* [Russian], **69**, 161.
GINSBURG, V. L. (1956). *J. exp. theor. Phys.* USSR [Russian], **29**, 244 (translated in Soviet Physics, *JETP*, **1**, 170).
GOLDSTEIN, L. (1954). *Phys. Rev.* **96**, 1455.
GOLDSTEIN, L. (1955). *Phys. Rev.* **100**, 981.
GOLDSTEIN, L. (1956). *Phys. Rev.* **102**, 1205.
GOLDSTEIN, L. and REEKIE, J. (1955). *Phys. Rev.* **98**, 857.
GOLDSTEIN, L., SWEENEY, D. and GOLDSTEIN, M. (1950). *Phys. Rev.* **77**, 319.
GORDON, W. L., SHAW, C. H. and DAUNT, J. G. (1954). *Phys. Rev.* **96**, 1444.
GORTER, C. J. (1949). *Physica*, **15**, 523.
GORTER, C. J. (1951). *Proceedings of the International Conference on Low Temperature Physics*, p. 97. Oxford, England.
GORTER, C. J. (1952). *Phys. Rev.* **88**, 681.
GORTER, C. J., KASTELEIJN, P. W. and MELLINK, J. H. (1950). *Physica*, **16**, 113.
GORTER, C. J. and MELLINK, J. H. (1949). *Physica*, **15**, 285.
GORTER, C. J., TACONIS, K. W. and BEENAKKER, J. J. M. (1951). *Physica*, **17**, 841.
GREBENKEMPER, C. J. and HAGEN, J. P. (1950). *Phys. Rev.* **80**, 89.
GRENIER, C. (1951). *Phys. Rev.* **83**, 598.
GRILLY, E. R., HAMMEL, E. F. and SYDORIAK, S. G. (1949). *Phys. Rev.* **75**, 1103.
DE GROOT, S. R. (1951). *Thermodynamics of Irreversible Processes*. Amsterdam: North Holland Publishing Co.
GUGGENHEIM, E. A. (1950). *Thermodynamics*. Amsterdam: North Holland Publishing Co.
GUTTMAN, L. and ARNOLD, J. R. (1953). *Phys. Rev.* **92**, 547.

TER HAAR, D. (1954). *Phys. Rev.* **95**, 895.
TER HAAR, D. and WERGELAND, H. (1949). *Phys. Rev.* **75**, 886.
HALL, H. E. (1955). *Conférence de Physique des Basses Temperatures*, (Paris), p. 63.
HALL, H. E. (1957). *Phil. Trans.* A, **250**, 359.

HALL, H. E. and VINEN, W. F. (1956). *Proc. Roy. Soc.* A, **238**, 204, 215.
HALLETT, A. C. H. (1950). *Proc. Phys. Soc.* A, **63**, 1367.
HALLETT, A. C. H. (1951). Ph.D. Thesis, University of Cambridge.
HALLETT, A. C. H. (1952). *Proc. Roy. Soc.* A, **210**, 404.
HALLETT, A. C. H. (1953). *Proc. Camb. Phil. Soc.* **49**, 717.
HALLETT, A. C. H. (1955). *Progress in Low Temperature Physics*, vol. I (edited by C. J. Gorter), chapter IV. Amsterdam: North Holland Publishing Co.
HAM, A. C. and JACKSON, L. C. (1953). *Phil. Mag.* **44**, 214.
HAM, A. C. and JACKSON, L. C. (1954). *Phil. Mag.* **45**, 1084.
HAM, A. C. and JACKSON, L. C. (1957). *Proc. Roy. Soc.* A, **240**, 243.
HAMMEL, E. F. (1955). *Progress in Low Temperature Physics*, vol. I (edited by C. J. Gorter), chapter V. Amsterdam: North Holland Publishing Co.
HAMMEL, E. F., LAQUER, H. L., SYDORIAK, S. G. and MCGEE, W. E. (1952). *Phys. Rev.* **86**, 432.
HAMMEL, E. F. and SCHUCH, A. F. (1952). *Phys. Rev.* **87**, 154.
HANSON, W. B. and PELLAM, J. R. (1954). *Phys. Rev.* **95**, 321.
HARASIMA, A. (1951). *J. Phys. Soc. Japan*, **6**, 271.
HEBERT, G. R., CHOPRA, K. L. and BROWN, J. B. (1957). *Phys. Rev.* **106**, 391.
HEER, C. V. and DAUNT, J. G. (1951). *Phys. Rev.* **81**, 447.
HEIKKILA, W. J. and HALLETT, A. C. H. (1955). *Canad. J. Phys.* **33**, 420.
HERCUS, G. R. and WILKS, J. (1954). *Phil. Mag.* **45**, 1163.
HERREY, E. M. J. (1955). *Conférence de Physique des Basses Temperatures*, (Paris), p. 87.
HILDEBRAND, J. H. and SCOTT, R. L. (1950). *Solubility in Non-Electrolytes.* New York: Reinhold Publishing Co.
HILL, T. L. (1949). *J. Chem. Phys.* **17**, 580, 668.
HILL, R. W. and LOUNASMAA, O. V. (1957). *Phil. Mag.* **2**, 143.
HOFFMAN, C. J., EDESKUTY, F. J. and HAMMEL, E. F. (1956). *J. Chem. Phys.* **24**, 124.
HOLLAND, F. A., HUGGILL, J. A. W. and JONES, G. O. (1951). *Proc. Roy. Soc.* A, **207**, 268.
HOUSTON, W. V. and RORSCHACH, H. E. (1955). *Phys. Rev.* **100**, 1003.
HUANG, K. and YANG, C. N. (1957). *Phys. Rev.* **105**, 767.
HUANG, K., YANG, C. N. and LUTTINGER, J. M. (1957). *Phys. Rev.* **105**, 776.
HULL, R. A., WILKINSON, K. R. and WILKS, J. (1951). *Proc. Phys. Soc.* A, **64**, 379.
HUNG, C. S., HUNT, B. and WINKEL, P. (1952). *Physica*, **18**, 629.
HURST, D. G. and HENSHAW, D. G. (1955*a*). *Phys. Rev.* **100**, 994.
HURST, D. G. and HENSHAW, D. G. (1955*b*). *Canad. J. Phys.* **33**, 797.

INGHRAM, M. C., LONG, E. and MEYER, L. (1955). *Phys. Rev.* **97**, 1453.
VAN ITTERBEEK, A., VAN DEN BERG, G. J. and LIMBURG, W. (1954). *Physica*, **20**, 307.
VAN ITTERBEEK, A. and FORREZ, G. (1954). *Physica*, **20**, 133.

JACKSON, L. C. and HENSHAW, D. G. (1953). *Phil. Mag.* **44**, 14.
JOHNS, H. E. and WILHELM, J. O. (1938). *Canad. J. Res.* A, **16**, 131.
JONES, R. C. and FURRY, W. H. (1946). *Rev. Mod. Phys.* **18**, 151.

KAMERLINGH ONNES, H. (1908). *Proc. Acad. Sci. Amst.* **11**, 168.
KAMERLINGH ONNES, H. and BOKS, J. D. A. (1924). Commun. Phys. Lab. Univ. Leiden, no. 170*b*.
KAPADNIS, D. G. and DOKOUPIL, Z. (1955). *Conférence de Physique des Basses Temperatures* (Paris), p. 19.
KAPITZA, P. L. (1938). *Nature, Lond.*, **141**, 74.
KAPITZA, P. L. (1941). *J. Phys., Moscow*, **4**, 181.
KAPITZA, P. L. (1942). *J. Phys., Moscow*, **5**, 59.
KEESOM, W. H. (1933). *Proc. Acad. Sci. Amst.* **36**, 147.
KEESOM, W. H. (1936). Commun. Phys. Lab. Univ. Leiden, Suppl. no. 80*b*.
KEESOM, W. H. (1942). *Helium*. Amsterdam: Elsevier.
KEESOM, W. H. and CLUSIUS, K. (1931). Commun. Phys. Lab. Univ. Leiden, no. 216*b*.
KEESOM, W. H. and CLUSIUS, K. (1932). *Proc. Acad. Sci. Amst.* **35**, 307.
KEESOM, W. H. and KEESOM, A. P. (1932). *Proc. Acad. Sci. Amst.* **35**, 736.
KEESOM, W. H. and KEESOM, A. P. (1933). Commun. Phys. Lab. Univ. Leiden, no. 224*d,e*.
KEESOM, W. H. and KEESOM, A. P. (1935). *Physica*, **2**, 557.
KEESOM, W. H. and KEESOM, A. P. (1936). *Physica*, **3**, 105.
KEESOM, W. H. and MACWOOD, G. E. (1938). *Physica*, **5**, 737.
KEESOM, W. H., SARIS, B. F. and MEYER, L. (1940). *Physica*, **7**, 817.
KEESOM, W. H. and SCHWEERS, J. (1941). *Physica*, **8**, 1020, 1032.
KEESOM, W. H. and TACONIS, K. W. (1938*a*). *Physica*, **5**, 161.
KEESOM, W. H. and TACONIS, K. W. (1938*b*). *Physica*, **5**, 270.
KEESOM, W. H. and WESTMIJZE, W. K. (1941). *Physica*, **8**, 1044.
KELLER, W. E. (1955). *Phys. Rev.* **98**, 1571.
KELLER, W. E. (1956). *Nature, Lond.*, **178**, 883.
KERR, E. C. (1954). *Phys. Rev.* **96**, 551.
KERR, E. C. (1957*a*). *J. Chem. Phys.* **26**, 511.
KERR, E. C. (1957*b*). *Proceedings of the Fifth International Conference on Low Temperature Physics and Chemistry* (Madison, U.S.A.), p. 80.
KHALATNIKOV, I. M. (1950). *J. exp. theor. Phys.* USSR [Russian], **20**, 243.
KHALATNIKOV, I. M. (1951). *Dokl. Akad. Nauk, SSSR* [Russian], **79**, 237.
KHALATNIKOV, I. M. (1952*a*). *J. exp. theor. Phys.* USSR [Russian], **22**, 687.
KHALATNIKOV, I. M. (1952*b*). *J. exp. theor. Phys.* USSR [Russian], **23**, 8, 21.
KHALATNIKOV, I. M. (1952*c*). *J. exp. theor. Phys.* USSR [Russian], **23**, 169.
KHALATNIKOV, I. M. (1952*d*). *J. exp. theor. Phys.* USSR [Russian], **23**, 253.

KHALATNIKOV, I. M. (1952*e*). *J. exp. theor. Phys.* USSR [Russian], **23**, 265.

KHALATNIKOV, I. M. (1956*a*). *J. exp. theor. Phys.* USSR [Russian], **30**, 617 (translated in Soviet Physics, *JETP*, **3**, 649).

KHALATNIKOV, I. M. (1956*b*). *Usp. Fiz. Nauk*, **59**, 673.

KHALATNIKOV, I. M. (1956*c*). *Usp. Fiz. Nauk*, **60**, 69.

KHALATNIKOV, I. M. and ABRIKOSOV, A. A. (1957). *J. exp. theor. Phys.* USSR [Russian], **33**, 110 (translated in Soviet Physics, *JETP*, **6**, 84).

KHALATNIKOV, I. M. and ZHARKOV, V. N. (1957). *J. exp. theor. Phys.* USSR [Russian], **32**, 1108 (translated in Soviet Physics, *JETP*, **5**, 905).

KIKUCHI, R. (1954). *Phys. Rev.* **96**, 563.

KING, J. C. and FAIRBANK, H. A. (1953). *Phys. Rev.* **91**, 489.

KING, J. C. and FAIRBANK, H. A. (1954). *Phys. Rev.* **93**, 21.

DE KLERK, D., HUDSON, R. P. and PELLAM, J. R. (1954). *Phys. Rev.* **93**, 28.

KOIDE, S. and USUI, T. (1951*a*). *Progr. theor. Phys., Osaka*, **6**, 506.

KOIDE, S. and USUI, T. (1951*b*). *Progr. theor. Phys., Osaka*, **6**, 622.

KONTOROVICH, V. M. (1956). *J. exp. theor. Phys.* USSR [Russian], **30**, 805 (translated in Soviet Physics, *JETP*, **3**, 770).

KRAMERS, H. C. (1955). Ph.D. Thesis, Leiden University.

KRAMERS, H. C., TINEKE VAN PESKI-TINBERGEN, WIEBES, J., VAN DEN BURG, F. A. W. and GORTER, C. J. (1954). *Physica*, **20**, 743.

KRAMERS, H. C., WASSCHER, J. D. and GORTER, C. J. (1952). *Physica*, **18**, 329.

KRONIG, R. (1951). *Proceedings of the International Conference on Low Temperature Physics*, p. 99. Oxford, England.

KRONIG, R. and THELLUNG, A. (1950). *Physica*, **16**, 678.

KRONIG, R. and THELLUNG, A. (1952). *Physica*, **18**, 749.

KURTI, N. and MCINTOSH, J. (1955). *Phil. Mag.* **46**, 104.

KUZNETSOV, V. M. (1957). *J. exp. theor. Phys.* USSR [Russian], **32**, 1001 (translated in Soviet Physics, *JETP*, **5**, 819).

LAMB, H. (1895). *Hydrodynamics*, p. 222. Cambridge University Press.

LANDAU, L. D. (1941). *J. Phys., Moscow*, **5**, 71.

LANDAU, L. D. (1947). *J. Phys., Moscow*, **11**, 91.

LANDAU, L. D. (1956). *J. exp. theor. Phys.* USSR [Russian], **30**, 1058 (translated in Soviet Physics, *JETP*, **3**, 920).

LANDAU, L. D. (1957). *J. exp. theor. Phys.* USSR [Russian], **32**, 59 (translated in Soviet Physics, *JETP*, **5**, 101).

LANDAU, L. D. and KHALATNIKOV, I. M. (1949). *J. exp. theor. Phys.* USSR [Russian], **19**, 637, 709.

LANDAU, L. D. and KHALATNIKOV, I. M. (1954). *Dokl. Akad. Nauk, SSSR*, **96**, 469.

LANDAU, L. D. and LIFSHITZ, E. (1938). *Statistical Physics*. Oxford University Press.

LANDAU, L. D. and LIFSHITZ, E. (1955). *Dokl. Akad. Nauk, SSSR* [Russian], **100**, 669.

LANDAU, L. D. and POMERANCHUK, I. (1948). *Dokl. Akad. Nauk, SSSR* [Russian], **59**, 669.

LANE, C. T., FAIRBANK, H. A., ALDRICH, L. T. and NIER, A. O. (1948). *Phys. Rev.* **73**, 256.
LANE, C. T., FAIRBANK, H. A. and FAIRBANK, W. M. (1947). *Phys. Rev.* **71**, 600.
LANE, C. T., FAIRBANK, H. A., SCHULTZ, H. and FAIRBANK, W. M. (1946). *Phys. Rev.* **70**, 431.
LANE, C. T., FAIRBANK, H. A., SCHULTZ, H. and FAIRBANK, W. M. (1947). *Phys. Rev.* **71**, 600.
LAQUER, H. L., SYDORIAK, S. G. and ROBERTS, T. R. (1957). *Symposium on Liquid and Solid* $He^3$ (Ohio State University), p. 15.
LAWSON, A. W. and MEYER, L. (1954). *Phys. Rev.* **93**, 259.
LEE, T. D., HUANG, K. and YANG, C. N. (1957). *Phys. Rev.* **106**, 1135.
LESENSKY, L. and BOORSE, H. A. (1952). *Phys. Rev.* **87**, 1135.
LIFSHITZ, I. M. and KAGANOV, M. I. (1956). *J. exp. theor. Phys.* USSR [Russian], **29**, 259 (translated in Soviet Physics, *JETP*, **1**, 172).
LIM, C. C., HALLETT, A. C. H. and GUPTILL, E. W. (1957). *Canad. J. Phys.* **35**, 1343.
LINHART, P. B. and PRICE, P. J. (1956). *Physica*, **22**, 57.
LONDON, F. (1936). *Proc. Roy. Soc.* A, **153**, 576.
LONDON, F. (1938*a*). *Nature, Lond.*, **141**, 643.
LONDON, F. (1938*b*). *Phys. Rev.* **54**, 947.
LONDON, F. (1939). *J. Phys. Chem.* **43**, 49.
LONDON, F. (1943). *J. Chem. Phys.* **11**, 203.
LONDON, F. (1954). *Superfluids*, vol. II. New York: John Wiley and Sons.
LONDON, H. (1939). *Proc. Roy. Soc.* A, **171**, 484.
LONG, E. and MEYER, L. (1949). *Phys. Rev.* **76**, 440.
LONG, E. and MEYER, L. (1950). *Phys. Rev.* **79**, 1031.
LONG, E. and MEYER, L. (1951). *Phys. Rev.* **83**, 860.
LONG, E. and MEYER, L. (1952). *Phys. Rev.* **85**, 1030.
LONG, E. and MEYER, L. (1953). *Phil. Mag.* Suppl. **2**, 1.
LONG, E. and MEYER, L. (1955). *Phys. Rev.* **98**, 1616.
LOVEJOY, D. R. (1955). *Canad. J. Phys.* **33**, 49.
LYNTON, E. A. and FAIRBANK, H. A. (1950). *Phys. Rev.* **80**, 1043.

MCCRUM, N. G. and EISENSTEIN, J. C. (1955). *Phys. Rev.* **99**, 1326.
MCCRUM, N. G. and MENDELSSOHN, K. (1954). *Phil. Mag.* **45**, 102.
MCINTEER, B. B., ALDRICH, L. T. and NIER, A. O. (1948). *Phys. Rev.* **74**, 946.
MARGENAU, H. (1939). *Phys. Rev.* **56**, 1000.
MARKHAM, A. H., PEARCE, D. C., NETZEL, R. G. and DILLINGER, J. R. (1957). *Proceedings of the Fifth International Conference on Low Temperature Physics and Chemistry* (Madison, U.S.A.), p. 118.
MASTRANGELO, S. V. R. and ASTON, J. G. (1951). *J. Chem. Phys.* **19**, 1370.
MATHIAS, E., CROMMELIN, C. A., KAMMERLINGH ONNES, H. and SWALLOW, J. C. (1925). *Proc. Acad. Sci. Amst.* **28**, 526.
MATSUBARA, T. (1951). *Progr. theor. Phys., Osaka*, **6**, 714.
MAURER, R. D. and HERLIN, M. A. (1949). *Phys. Rev.* **76**, 948.
MAURER, R. D. and HERLIN, M. A. (1951). *Phys. Rev.* **82**, 329.

MAYPER, V. and HERLIN, M. A. (1953). *Phys. Rev.* **89**, 523.
MAZO, R. M. and KIRKWOOD, J. G. (1955). *Phys. Rev.* **100**, 1787.
MAZUR, P. and PRIGOGINE, I. (1951). *Physica*, **17**, 680.
MELLINK, J. H. (1947). *Physica*, **13**, 180.
MENDELSSOHN, K. and WHITE, G. K. (1950). *Proc. Phys. Soc.* **63**, 1328.
MEYER, L. (1955). *Conférence de Physique des Basses Temperatures* (Paris), p. 95.
MEYER, L. (1956). *Phys. Rev.* **103**, 1593.
MEYER, L. and REIF, F. (1958). *Phys. Rev.* **110**, 279.
MIKURA, Z. (1954*a*). *Progr. theor. Phys., Osaka*, **11**, 25.
MIKURA, Z. (1954*b*). *Progr. theor. Phys., Osaka*, **11**, 504.
MIKURA, Z. (1955). *Progr. theor. Phys., Osaka*, **14**, 337.
MILLS, R. L. and GRILLY, E. R. (1955). *Phys. Rev.* **99**, 480.
MILLS, R. L. and GRILLY, E. R. (1957). *Symposium on Liquid and Solid* $He^3$ (Ohio State University), p. 100.
MORROW, J. C. (1951). *Phys. Rev.* **84**, 502.
MORROW, J. C. (1953*a*). *Phys. Rev.* **89**, 1034.
MORROW, J. C. (1953*b*). *Phys. Rev.* **92**, 1.
MOTT, N. F. (1949). *Phil. Mag.* **40**, 61.

NANDA, V. S. (1954). *Phys. Rev.* **94**, 241.
NANDA, V. S. (1955). *Phys. Rev.* **97**, 571.
NEWELL, J. A. (1955). *Conférence de Physique des Basses Temperatures* (Paris), p. 80.
NEWELL, J. A. and WILKS, J. (1956). *Phil. Mag.* **1**, 588.

ONSAGER, L. (1944). *Phys. Rev.* **64**, 114.
ONSAGER, L. (1949). *Il Nuovo Cimento*, **6**, Suppl. 2, 249.
OSBORNE, D. V. (1948). *Nature, Lond.*, **162**, 213.
OSBORNE, D. V. (1950). *Proc. Phys. Soc.* A, **63**, 909.
OSBORNE, D. V. (1951). *Proc. Phys. Soc.* A, **64**, 114.
OSBORNE, D. V. (1956). *Phil. Mag.* **1**, 301.
OSBORNE, D. W., ABRAHAM, B. W. and WEINSTOCK, B. (1951). *Phys. Rev.* **82**, 263.
OSBORNE, D. W., ABRAHAM, B. W. and WEINSTOCK, B. (1952). *Phys. Rev*, **85**, 158.
OSBORNE, D. W., WEINSTOCK, B. and ABRAHAM, B. M. (1949). *Phys. Rev.* **75**, 988.

PALEVSKY, H., OTNES, K., LARSSON, K. E., PAULI, R. and STEDMAN, R. (1957). *Phys. Rev.* **108**, 1346.
PEKERIS, C. L. (1950). *Phys. Rev.* **79**, 884.
PELLAM, J. R. (1949). *Phys. Rev.* **75**, 1183.
PELLAM, J. R. (1950). *Phys. Rev.* **78**, 818.
PELLAM, J. R. (1955*a*). *Phys. Rev.* **99**, 1327.
PELLAM, J. R. (1955*b*). *Progress in Low Temperature Physics*, vol. I (edited by C. J. Gorter), chapter III. Amsterdam: North Holland Publishing Co.
PELLAM, J. R. and HANSON, W. B. (1952). *Phys. Rev.* **85**, 216.

PELLAM, J. R. and MORSE, P. (1950). *Phys. Rev.* **78**, 474.
PELLAM, J. R. and SCOTT, R. B. (1949). *Phys. Rev.* **76**, 869.
PELLAM, J. R. and SQUIRE, C. F. (1947). *Phys. Rev.* **72**, 1245.
PENROSE, O. (1957). *Symposium on Liquid and Solid* $He^3$ (Ohio State University), p. 85.
PENROSE, O. and ONSAGER, L. (1956). *Phys. Rev.* **104**, 576.
PESHKOV, V. P. (1944). *J. Phys., Moscow*, **8**, 131, 381.
PESHKOV, V. P. (1946). *J. Phys., Moscow*, **10**, 389.
PESHKOV, V. P. (1948*a*). *J. exp. theor. Phys.* USSR [Russian], **18**, 867.
PESHKOV, V. P. (1948*b*). *J. exp. theor. Phys.* USSR [Russian], **18**, 857.
PESHKOV, V. P. (1948*c*). *J. exp. theor. Phys.* USSR [Russian], **18**, 951.
PESHKOV, V. P. (1949). *J. exp. theor. Phys.* USSR [Russian], **19**, 270.
PESHKOV, V. P. (1956). *Soviet Physics*, **3**, 706.
PESHKOV, V. P. and ZINOVEVA, K. N. (1948). *J. exp. theor. Phys.* USSR [Russian], **18**, 438.
PESHKOV, V. P. and ZINOVEVA, K. N. (1957). *J. exp. theor. Phys.* USSR [Russian], **32**, 1256 (translated in Soviet Physics, *JETP*, **5**, 1024).
PICUS, G. S. (1954). *Phys. Rev.* **94**, 1459.
PIPPARD, A. B. (1951). *Phil. Mag.* **42**, 1209.
PIPPARD, A. B. (1956). *Phil. Mag.* **1**, 473.
PITAEVSKII, L. P. (1956). *J. exp. theor. Phys.* USSR [Russian], **31**, 536 (translated in Soviet Physics, *JETP*, **4**, 439).
POMERANCHUK, I. (1949). *J. exp. theor. Phys.* USSR [Russian], **19**, 42.
POMERANCHUK, I. (1950). *J. exp. theor. Phys.* USSR [Russian], **20**, 1919.
PRICE, P. J. (1953). *Phys. Rev.* **89**, 1209.
PRICE, P. J. (1955). *Phys. Rev.* **97**, 259.
PRIGOGINE, I. (1954). *Phil. Mag.* Suppl. **3**, 131.
PRIGOGINE, I. and PHILIPPOT, J. (1952). *Physica*, **18**, 729.
PRIGOGINE, I. and PHILIPPOT, J. (1953). *Physica*, **19**, 227.
PTUKHA, T. P. (1958). *J. exp. theor. Phys.* USSR [Russian], **34**, 33.

RAYLEIGH, LORD (1905). *Phil. Mag.* **10**, 366.
REEKIE, J. (1940). *Proc. Camb. Phil. Soc.* **36**, 236.
REEKIE, J. (1947). *Proc. Camb. Phil. Soc.* **43**, 262.
REEKIE, J., HUTCHINSON, T. S. and BEAUMONT, C. F. A. (1953). *Proc. Phys. Soc.* A, **66**, 409.
REEKIE, J., HUTCHINSON, T. S. and BEAUMONT, C. F. A. (1955). *Proc. Roy. Soc.* A, **228**, 363.
REYNOLDS, C. A., FAIRBANK, H. A., LANE, C. T., MCINTEER, B. B. and NIER, A. O. (1949). *Phys. Rev.* **76**, 64.
RICE, O. K. (1950). *Phys. Rev.* **79**, 1024.
RICE, O. K. (1954). *Phys. Rev.* **96**, 1460, 1464.
RICE, O. K. (1955). *Phys. Rev.* **97**, 263, 558.
RICE, O. K. (1956). *Phys. Rev.* **102**, 1416.
ROBERTS, T. R. and SYDORIAK, S. G. (1955). *Phys. Rev.* **98**, 1672.
ROBERTS, T. R. and SYDORIAK, S. G. (1957*a*). *Phys. Rev.* **106**, 175.
ROBERTS, T. R. and SYDORIAK, S. G. (1957*b*). *Symposium on Liquid and Solid* $He^3$ (Ohio State University), p. 117.

ROLLIN, B. V. (1936). *Actes du 7e Congr. intern du Froid* (La Haye: Amsterdam), **1**, 187.
ROLLIN, B. V. and HATTON, J. (1948). *Phys. Rev.* **74**, 508.
RUSHBROOKE, G. S. (1949). *Introduction to Statistical Mechanics*, chapters XIV, XVIII. Oxford University Press.

SCHAEFFER, W. D., SMITH, W. R. and WENDELL, C. B. (1949). *J. Chem. Soc.* **71**, 863.
SCHIFF, L. I. (1941). *Phys. Rev.* **59**, 838.
SCHMIDT, G. and KEESOM, W. H. (1937). *Physica*, **4**, 963.
SCHUETTE, O. F., ZUCKER, A. and WATSON, W. W. (1950). *Rev. Sci. Instr.* **21**, 1016.
SHERMAN, R. H. and EDESKUTY, E. J. (1957). *Symposium on Liquid and Solid* $He^3$ (Ohio State University), p. 44.
SILIN, V. P. (1957). *J. exp. theor. Phys.* USSR [Russian], **33**, 1227.
SIMON, F. (1934). *Nature, Lond.*, **133**, 529.
SIMON, F., RUHEMANN, M. and EDWARDS, W. A. M. (1929*a*). *Z. phys. Chem.* [German], **2**, 340.
SIMON, F., RUHEMANN, M. and EDWARDS, W. A. M. (1929*b*). *Z. phys. Chem.* [German], **6**, 62.
SIMON, F., RUHEMANN, M. and EDWARDS, W. A. M. (1930). *Z. phys. Chem.* [German], **6**, 331.
SINGWI, K. S. and KOTHARI, L. S. (1949). *Phys. Rev.* **76**, 305.
SLATER, J. C. and KIRKWOOD, J. G. (1931). *Phys. Rev.* **37**, 682.
SMITH, B. and BOORSE, H. A. (1955). *Phys. Rev.* **99**, 328, 346, 358, 367.
SOLLER, T., FAIRBANK, W. M. and CROWELL, A. D. (1953). *Phys. Rev.* **91**, 1058.
SOMMERS, H. S. (1952). *Phys. Rev.* **88**, 113.
SOMMERS, H. S., DASH, J. G. and GOLDSTEIN, L. (1955). *Phys. Rev.* **97**, 855.
SOMMERS, H. S., KELLER, W. E. and DASH, J. G. (1953). *Phys. Rev.* **92**, 1345.
SREEDHAR, A. K. and DAUNT, J. G. (1957). *Symposium on Liquid and Solid* $He^3$ (Ohio State University), p. 141.
STRAUSS, A. J. (1952). Ph.D. Thesis, Chicago University.
STOUT, J. W. (1949). *Phys. Rev.* **76**, 864.
SWENSON, C. A. (1950). *Phys. Rev.* **79**, 626.
SWENSON, C. A. (1952). *Phys. Rev.* **86**, 870.
SWENSON, C. A. (1953). *Phys. Rev.* **89**, 538.
SWIM, R. T. and RORSCHACH, H. E. (1955). *Phys. Rev.* **97**, 25.
SYDORIAK, S. G., GRILLY, E. R. and HAMMEL, E. F. (1949). *Phys. Rev.* **75**, 303.

TACONIS, K. W. (1950). *Ned. Tijdschr. Natuurk.* **16**, 101.
TACONIS, K. W., BEENAKKER, J. J. M., ALDRICH, L. T. and NIER, A. O. (1949). *Physica*, **15**, 733.
TACONIS, K. W., BEENAKKER, J. J. M. and DOKOUPIL, Z. (1950). *Phys. Rev.* **78**, 171.
TAYLOR, G. I. (1923). *Phil. Trans.* A, **223**, 289.

TEMPERLEY, H. N. V. (1951). *Proc. Phys. Soc.* A, **64**, 105.
TEMPERLEY, H. N. V. (1955*a*). *Phys. Rev.* **97**, 835.
TEMPERLEY, H. N. V. (1955*b*). *Conférence de Physique des Basses Temperatures* (Paris), p. 115.
THELLUNG, A. (1953). *Physica*, **19**, 217.
TISZA, L. (1938). *Nature, Lond.*, **141**, 913.
TISZA, L. (1940). *J. Phys. Radium*, **1**, 165, 350.
TISZA, L. (1947). *Phys. Rev.* **72**, 838.
TISZA, L. (1951). *Phase Transformations in Solids* (edited by Smoluchowski, Mayer and Weyl). New York: John Wiley and Sons.
TJERKSTRA, H. H. (1952). *Physica*, **18**, 853.
TODA, M. (1951). *Progr. theor. Phys., Osaka*, **6**, 458.
TOMONAGA, S. (1938). *Z. Phys.* [German], **110**, 573.
DE TROYER, A., VAN ITTERBEEK, A. and VAN DEN BERG, G. J. (1951). *Physica*, **17**, 50.
TWEET, A. G. (1954). *Phys. Rev.* **93**, 15.

USUI, T. (1951). *Progr. theor. Phys., Osaka*, **6**, 244.

VINEN, W. F. (1957*a*). *Proc. Roy. Soc.* A, **240**, 114.
VINEN, W. F. (1957*b*). *Proc. Roy. Soc.* A, **240**, 128.
VINEN, W. F. (1957*c*). *Proc. Roy. Soc.* A, **242**, 493.
VINEN, W. F. (1958). *Proc. Roy. Soc.* A, **243**, 400.
DE VRIES, G. and DAUNT, J. G. (1953). *Phys. Rev.* **92**, 1572.
DE VRIES, G. and DAUNT, J. G. (1954). *Phys. Rev.* **93**, 631.

WALTERS, G. K. and FAIRBANK, W. M. (1956*a*). *Phys. Rev.* **103**, 263.
WALTERS, G. K. and FAIRBANK, W. M. (1956*b*). *Phys. Rev.* **103**, 262.
WALTERS, G. K. and FAIRBANK, W. M. (1957*a*). *Bulletin of the American Physical Society*, **2**, 183.
WALTERS, G. K. and FAIRBANK, W. M. (1957*b*). *Symposium on Liquid and Solid* $He^3$ (Ohio State University), p. 99.
WANSINK, D. H. N. (1957). *Physica*, **23**, 140.
WANSINK, D. H. N. and TACONIS, K. W. (1955). *Conférence de Physique des Basses Temperatures* (Paris), p. 33.
WANSINK, D. H. N. and TACONIS, K. W. (1957*a*). *Physics*, **23**, 125.
WANSINK, D. H. N. and TACONIS, K. W. (1957*b*). *Physica*, **23**, 273.
WANSINK, D. H. N., TACONIS, K. W. and STAAS, F. A. (1956). *Physica*, **22**, 449.
WANSINK, D. H. N., TACONIS, K. W., STAAS, F. A. and REUSS, J. (1955). *Physica*, **21**, 596.
WARD, J. C. and WILKS, J. (1951). *Phil. Mag.* **42**, 314.
WARD, J. C. and WILKS, J. (1952). *Phil. Mag.* **43**, 48.
WARING, R. K. (1955). *Phys. Rev.* **99**, 1704.
WEINSTOCK, B., OSBORNE, D. W. and ABRAHAM, B. M. (1950). *Phys. Rev.* **77**, 400.
WEINSTOCK, B. and PELLAM, J. R. (1953). *Phys. Rev.* **89**, 521.
WELBER, B. and QUIMBY, S. L. (1957). *Phys. Rev.* **107**, 645.

WHEELER, R. G., BLAKEWOOD, C. H. and LANE, C. T. (1955). *Phys. Rev.* **99**, 1667.
WHITE, D., GONZALES, O. D. and JOHNSTON, H. L. (1953). *Phys. Rev.* **89**, 593.
WHITNEY, W. M. (1957). *Phys. Rev.* **105**, 38.
WHITWORTH, R. W. (1955). *Conférence de Physique des Basses Températures* (Paris), p. 56.
WINKEL, P. (1955). *Physica*, **21**, 322.
WINKEL, P., BROESE VAN GROENOU, A. and GORTER, C. J. (1955). *Physica*, **21**, 345.
WINKEL, P., DELSING, A. M. G. and GORTER, C. J. (1955). *Physica*, **21**, 312.
WINKEL, P., DELSING, A. M. G. and POLL, J. D. (1955). *Physica*, **21**, 331.
WOLFKE, M. and KEESOM, W. H. (1928). *Proc. Acad. Sci. Amst.* **31**, 81, 800.

ZHARKOV, V. N. (1957). *J. exp. theor. Phys.* USSR [Russian], **33**, 929.
ZIMAN, J. M. (1953*a*). *Proc. Roy. Soc.* A, **219**, 257.
ZIMAN, J. M. (1953*b*). *Phil. Mag.* **45**, 100.
ZINOVEVA, K. N. (1954). *J. exp. theor. Phys.* USSR [Russian], **28**, 25.
ZINOVEVA, K. N. (1955). *J. exp. theor. Phys.* USSR [Russian], **29**, 899 (translated in Soviet Physics, *JETP*, **2**, 774).
ZINOVEVA, K. N. (1956). *J. exp. theor. Phys.* USSR [Russian], **31**, 31 (translated in Soviet Physics, *JETP*, **4**, 36).
ZINOVEVA, K. N. (1958). *J. exp. theor. Phys.* USSR [Russian], **34**, 609.
ZUBAREV, D. N. (1955). *J. exp. theor. Phys.* USSR [Russian], **29**, 881 (translated in Soviet Physics, *JETP*, **2**, 745).
ZWANNIKEN, G. C. J. (1951). *Physica*, **16**, 805.

# INDEX

Printed in the United States
By Bookmasters